Handbook of Drying for Dairy Products

Handbook of Drying for Dairy Products

Edited by C. Anandharamakrishnan

Indian Institute of Crop Processing Technology
Thanjavur, Tamil Nadu
India

WILEY Blackwell

Registered Offices
John Wiley & Sons Ltd, The Atrium, Southern Gate, Chichester, West Sussex, PO19 8SQ, UK

Editorial Office
9600 Garsington Road, Oxford, OX4 2DQ, UK

For details of our global editorial offices, customer services, and more information about Wiley products visit us at www.wiley.com.

Wiley also publishes its books in a variety of electronic formats and by print-on-demand. Some content that appears in standard print versions of this book may not be available in other formats.

Library of Congress Cataloging-in-Publication Data:

Names: Anandharamakrishnan, C., editor.
Title: Handbook of drying for dairy products / edited by
Dr. C. Anandharamakrishnan.
Description: Chichester, UK ; Hoboken, NJ : John Wiley & Sons, 2017. |
Includes bibliographical references and index.
Identifiers: LCCN 2016047616| ISBN 9781118930496 (cloth) |
Subjects: LCSH: Dairy products–Drying.
Classification: LCC SF250.5 .H36 2017 | DDC 637–dc23 LC record available at
https://lccn.loc.gov/2016047616
Cover image: Atropat/Gettyimages
Cover design: Wiley

Set in 10/12pt Warnock by SPi Global, Chennai, India

10 9 8 7 6 5 4 3 2 1

Contents

Contributors

Aadinath
Department of Food Engineering
CSIR – Central Food Technological Research Institute
Mysore, Karnataka, India

P.H. Amaladhas
Engineering Section
National Dairy Research Institute
Southern Campus
Bangalore, Karnataka, India

D. Anand Paul
Nestle Research and Development Centre
Manesar
Gurgaon, India

C. Anandharamakrishnan
Indian Institute of Crop Processing Technology
Ministry of Food Processing Industries
Government of India
Thanjavur, Tamil Nadu, India

A. Bhushani
Department of Food Engineering
CSIR – Central Food Technological Research Institute
Mysore, Karnataka, India

N. Chhanwal
Department of Food Engineering
CSIR – Central Food Technological Research Institute
Mysore, Karnataka, India

Triroopa Ghosh
Department of Food Engineering
CSIR – Central Food Technological Research Institute
Mysore, Karnataka, India

J. Gimbun
Centre of Excellence for Advanced Research in Fluid Flow
Universiti Malaysia Pahang
Gambang, Pahang, Malaysia

R. Gopirajah
Department of Food Engineering
CSIR – Central Food Technological Research Institute
Mysore, Karnataka, India

P. Karthik
Department of Food Engineering
CSIR – Central Food Technological Research Institute
Mysore, Karnataka, India

U.K. Kolli
ITC Limited
Agri-Business Division
Guntur, Andhra Pradesh, India

W.P. Law
Centre of Excellence for Advanced Research in Fluid Flow
Universiti Malaysia Pahang
Gambang, Pahang, Malaysia

F. Magdaline Eljeeva Emerald
Engineering Section
National Dairy Research Institute
Southern Campus
Bangalore, Karnataka, India

S. Padma Ishwarya
Department of Food Engineering
CSIR – Central Food Technological Research Institute
Mysore, Karnataka, India

S. Parthasarathi
Department of Food Engineering
CSIR – Central Food Technological Research Institute
Mysore, Karnataka, India

I. Roy
Department of Food Engineering
CSIR – Central Food Technological Research Institute
Mysore, Karnataka, India

A.G.F. Stapley
Department of Chemical Engineering
Loughborough University
Loughborough, Leicestershire, UK

M.W. Woo
Department of Chemical Engineering
Faculty of Engineering
Monash University
Clayton Campus, Victoria, Australia

About the editor

Dr C. Anandharamakrishnan is currently the Director of Indian Institute of Crop Processing Technology, Thanjavur, Tamil Nadu, India. He obtained his doctorate in chemical engineering from Loughborough University, UK, for his work on experimental and computational fluid dynamics studies on spray freeze drying and spray drying of whey proteins. Formerly he was a Principal Scientist in the Department of Food Engineering at the CSIR – Central Food Technological Research Institute, Mysore, India. He specialises in the fields of drying, encapsulation of bioactive food ingredients and computational modelling of food processes. He has been actively involved in research on employing drying as an encapsulation technique for the protection and delivery of food bioactives, probiotics and flavours. He also has expertise in handling drying technology classes at the graduate and post graduate levels. Hitherto, he has published three books, and has nine patents and many research articles in international journals to his credit. He is the elected Fellow of the Royal Society of Chemistry (FRSC) and the Institute of Engineers (FIE) and is a recipient of several awards, including the Professor Jiwan Singh Sidhu Award 2010 from the Association of Food Scientists and Technologists (India).

Preface

Drying, although it has its origin in prehistoric times, has paramount significance in the modern food processing industry. With respect to dairy processing, drying has certainly carved its own niche. Drying not only improves the shelf life and microbial quality of milk and milk products, but also makes them convenient for storage and transportation. Additionally, drying technology brings a variety of dried dairy products for the global consumer market alongside its vast applications in beverages, bakery and confectionery industries.

The subjects drying and dairy complement each other and are a developing field of research in both academia and industry. With enormous advances in drying techniques, any individual with expertise in the drying of dairy products can decipher new developments and improve their work. Hence, the *Handbook of Drying for Dairy Products* is an attempt to amalgamate the fundamental (theoretical) and technological (application) aspects of dairy-specific drying processes in a detailed fashion.

Drum drying and spray drying are the chief drying techniques used for dairy products. Apart from these, non-thermal processes such as freeze drying and the relatively modern spray freeze drying methods are applied for the preservation of starter cultures and the development of functional dairy ingredients. This book includes dedicated chapters for each of these techniques along with the characterization of dried powders and its packaging methods. Furthermore, industrial-scale drying of dairy products and its associated challenges are discussed in detail. Insights into the mathematical and computational tools adopted to optimize and predict the performance of drying process are also provided. Overall, this book showcases the significance of drying in the production of dairy powders and illustrates the further scope still existing in this vibrant technology.

Acknowledgments

First and foremost, I would like to thank Professor Ram Rajasekharan, Director, CSIR – Central Food Technological Research Institute, Mysore, for his encouragement and able guidance for my academic endeavours. I also register thanks to my colleagues at the Indian Institute of Crop Processing Technology, Thanjavur, for their motivation.

I express my gratitude to Professor Chris Rielly and Dr Andy Stapley, Loughborough University, UK, for their constant support.

I thank all the chapter contributors and those involved at John Wiley & Sons Inc. for their work in bringing this book to fruition. My thanks also to Ms Anu Bhushani and all students of Computational Modelling and Nanoscale Processing Lab at CSIR–CFTRI, Mysore.

1

Introduction to Drying

C. Anandharamakrishnan

Indian Institute of Crop Processing Technology, Ministry of Food Processing Industries, Government of India, Thanjavur, Tamil Nadu, India

1.1 Introduction

The history of drying dates back to 12000 BC, when people dried meat and fish under the sun. Since then, drying technology has evolved and it is presently an indispensable process in the industrial sector. Drying involves removal of relatively small proportions of volatile liquid from a product by thermal means via the vapour phase. With respect to food products, the volatile liquid is almost always water, present in bound or unbound form. Bound water is that which is physically and/or chemically entrapped within the microstructure of the food matrix; anything in excess of this is known as unbound water. In addition, the fraction of bound and unbound water that is removable at a given temperature is known as the free water content. The term *drying* is often used synonymously with *evaporation* and *dehydration*. However, it differs from evaporation in that the final product is a solid rather than a highly viscous liquid; it differs from dehydration in that the final moisture content of the dried food product is more than 2.5%, while that of dehydrated is less than 2.5% (Vega-Mercado *et al.* 2001).

Drying is one of the oldest methods of food preservation, resulting in shelf-stable products that are of utmost relevance in this era of convenience. The preservation effect is achieved by reducing the water content to a level that reduces the incidence of microbial growth and retards deteriorative chemical reactions such as enzymatic and non-enzymatic browning and rancidity due to lipid oxidation. Nevertheless, owing to the complex nature of foods, the effect of drying is not just limited to reduction in water content. The application of heat during drying causes structural modifications of macromolecular components in foods (carbohydrates, proteins and lipids), by which the final product acquires significant functional characteristics. Also, moisture diffusion from within the product is accompanied by various physical changes, including shrinkage, puffing, crystallization and glass transitions (Mujumdar 2007). Drying can also lead to the encapsulation of active components within a protective outer layer, owing to the difference in drying rate between the feed constituents. Spray drying of milk is a classical example of drying-mediated encapsulation wherein the fat component is encapsulated in the matrix of sugars and proteins. Apart from the above, drying enhances the transportation and packaging properties of food products, attributed to the reduction in weight and volume. Before proceeding to understand the drying process, it is important to become familiar with the important terminologies associated with drying (Box 1.1).

Box 1.1 Lexicon of drying process

- **Critical moisture content:** The moisture content at which the drying rate starts to drop under constant drying conditions.
- **Equilibrium moisture content:** At a given temperature and pressure, the moisture content of a moist solid is in equilibrium with the gas–vapour mixture (zero for non-hygroscopic solids).
- **Bound moisture:** The amount of moisture tightly bound to the food matrix with properties different from those of bulk water; this also represents the portion of water that is unfreezable.
- **Unbound moisture:** Moisture in excess of the equilibrium moisture content, corresponding to saturated humidity.
- **Free moisture:** Amount of moisture mechanically entrapped in the void spaces of the system, having nearly all properties similar to those of bulk water.
- **Relative humidity:** Ratio of the partial pressure of water vapour in a gas–vapour mixture to equilibrium vapour pressure at the same temperature.
- **Water activity:** Relative humidity divided by 100.
- **Moisture sorption isotherm:** A graphical representation of the relationship between moisture content and equilibrium humidity (or) water activity at a specified temperature.
- **Dry bulb temperature:** Temperature measured by a (dry) thermometer immersed in a vapour–gas mixture.
- **Wet bulb temperature (T_{wb}):** The liquid temperature attained when large amounts of air–vapour mixture is contacted with the surface. In purely convective drying, the drying surface reaches T_{wb} during the constant rate period.

1.2 Fundamental principles of drying: the concept of simultaneous heat and mass transfer

Drying is a simultaneous heat- and mass-transfer process. This is reasonable as the phase transition of any component is associated with the evolution of latent heat. In any process that involves a net transfer of mass from one phase to another, the heat-transfer rate is the limiting factor of the rate at which the mass is transferred (Foust *et al.* 2008).

1.2.1 Heat transfer during the drying process

Heat transfer occurs from the drying medium to the product surface and also from one point to another within the product. This is accomplished by one or a combination of the following mechanisms: conduction, convection, radiation and dielectric heating. These mechanisms vary with respect to the type of drying medium, mode of contact between the product and drying medium, and the scale of heat transfer within the product, that

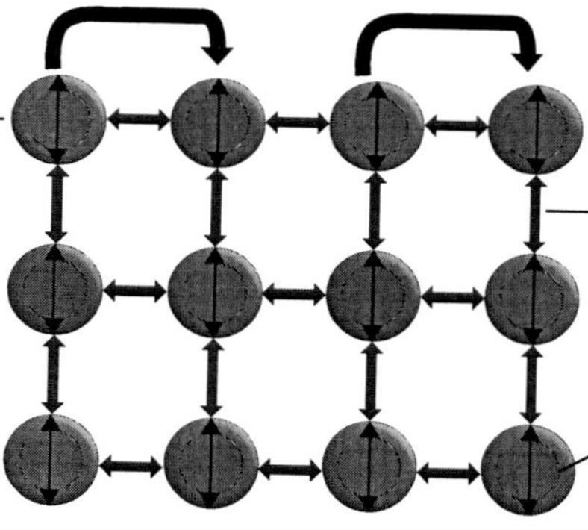

Figure 1.1 Principle of heating during conduction drying.

is, molecular or bulk transport, and the direction of heat transfer. However, irrespective of the above, the driving force for heat transfer is the temperature gradient that exists between the product and water surfaces at some location within the product (Singh & Heldman 2014). This can be appreciated from the equations governing heat transfer for each mechanism. The mode of heat transfer during drying also forms the basis of dryer classification, which is discussed in subsequent sections.

1.2.1.1 Conduction drying

Conduction drying, also known as indirect or contact drying, occurs when heat is transferred to the product through contact with a metal surface that separates the product and the heating medium. Conduction heating involves transport of energy in a solid medium through vibration and collision of molecules and free electrons. The molecules vibrate by virtue of the heat energy absorbed from the drying medium. As a result, the molecules at higher temperature vibrate faster and transfer part of their kinetic energy to those at lower temperature by means of collision. However, only energy is transferred between the molecules and there is no change in their position.

A typical conduction dryer is made of a metal-walled, heat-jacketed arrangement that is either stationary or rotating. The jacketed vessel is heated by the circulation of condensed steam, flue gases, hot water, combustion gas, electricity or thermal fluids (e.g. silicone oil), which in turn transfers the heat to the metal surface of the dryer. Removal of vaporized water is independent of the heating medium. These dryers can operate in both batch and continuous modes. A drum dryer is a classical example of continuous conduction dryer, where the feed slurry is spread as a thin sheet over the surface of a rotating drum heated by steam. Heat transfer by conduction occurs from the heated drum to the sheet of feed slurry. The typical pattern of heat transfer in a conduction dryer is depicted in Figure 1.1.

The governing equation for heat transfer by conduction is given by *Fourier's law* (Eq. 1.1).

$$Q_{\text{conduction}} = kA\frac{(T_{\text{o}} - T_{\text{m}})}{x} \tag{1.1}$$

where Q is the rate of heat transfer (W), k is the thermal conductivity (W/(m·K)), T_{m} is the surface temperature of the product, which is at or slightly above the boiling point of water (K), T_o is the temperature of the drying medium (K), A is the area of heat transfer and x is the characteristic dimension of the product. The above equation is important as the rate of heat transfer has a significant influence on the extent of drying. There exists a positive correlation between heat-transfer rate and the temperature of the metal surface, but this is limited by the *case-hardening* phenomenon. Case hardening is the formation of surface shell during the early stages of drying, which is influenced by the temperature and velocity of drying medium (Chen 2008).

The thermal efficiency of conduction dryers is high as they consume only as much energy as is required to heat the product to its drying temperature. The operational temperature range is quite wide, from below the freezing point of water to close to the temperature of steam. This facilitates handling of products with different levels of heat sensitivity. Conduction dryers are also capable of operating under sub-atmospheric pressure and inert atmosphere. This enables the drying of food products that are susceptible to volatile loss and oxidation. Apart from drum dryers, other examples of conduction dryers include steam tube rotary dryers and cylinder dryers (continuous operation), vacuum tray dryers, freeze dryers and agitated pan dryers (batch operation) (Chakraverty & Singh 2014).

1.2.1.2 Convection drying

In convection drying, also referred to as direct drying, heat transfer is accomplished by direct contact between the wet product and a stream of hot air (Das & Chakraverty 2003; Figure 1.2). On contrary to conduction, convection involves bulk heat transfer by movement of energy as eddies of currents inside a fluid. The molecules undergo displacement within the fluid as a function of their temperature and density. While the molecules at lower temperature are displaced downwards owing to their higher density, the opposite happens with those at higher temperature. In other words, molecules with lower kinetic energy replace the molecules with higher kinetic energy.

The rate of heat transfer by convection is given by:

$$Q_{\text{convection}} = \frac{hA}{\lambda}(T_o - T_m) \tag{1.2}$$

where h is the convective heat-transfer coefficient ($W/m^2/K$) and λ is the latent heat of vaporization (kJ/kg). The rate of heat transfer during convective drying is influenced by

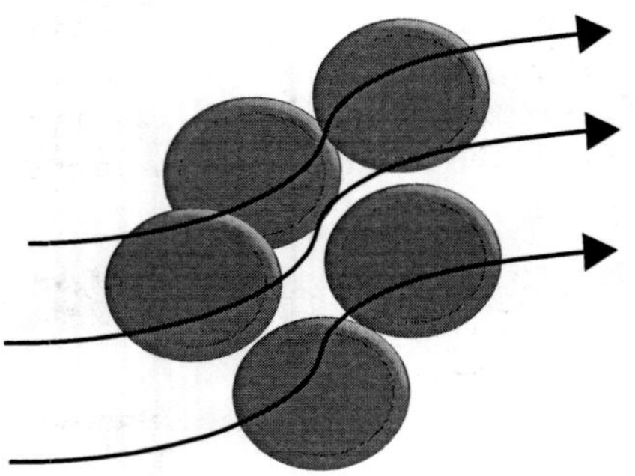
Flow of heating medium past the solid product

Figure 1.2 Principle of heating during convection drying.

the temperature, humidity, velocity and distribution pattern of the drying air and the product geometry and dimensions (Rahman & Perera 2007).

Convection drying is usually carried out as a continuous process in a closed chamber. Spray drying is a typical example of continuous convective drying, wherein the atomized feed droplets are dried by hot gas inside a spray chamber. Other examples of convective dryers include tray dryers, sheeting dryers, pneumatic convective dryers, rotary dryers, fluidized-bed dryers, tunnel dryers and compartment dryers. Convective drying is more energy intensive than conduction drying as the heat load is exerted by both the product and flow of the drying medium.

1.2.1.3 Radiation and dielectric drying

Radiation drying occurs when the product absorbs the radiant energy from a source and subsequently converts it into heat energy (Figure 1.3). There is no contact between the heated surface and the product, and this drying mode does not require a propagating medium. Solar drying is the most primitive form of radiation drying. The rate of heat transfer by radiation is given by the Stefan–Boltzmann law (Eq. 1.3):

$$Q_{\text{radiation}} = \varepsilon \sigma A({T_{\text{o}}}^4 - {T_{\text{m}}}^4) \tag{1.3}$$

where ε is the emissivity of the product and σ is the Stefan–Boltzmann constant (5.670373×10^{-8} W/m^2/K^4).

The concept of dielectric drying is a recent advancement with respect to the drying of food commodities, and comprises infra-red, microwave and radio-frequency drying methods. Here, the electromagnetic energy is selectively absorbed by the water present in a food product that undergoes volumetric heating. *Volumetric heating* refers to the generation and deposition of heat throughout the three-dimensional space within the product. This forms the primary difference to conventional drying, which is characterized by surface heating (outside-to-inside heating). The volumetric heating of dielectric drying overcomes the disadvantages of case hardening, which is associated with conventional drying (Figure 1.4). The surface heating caused by case hardening has a negative

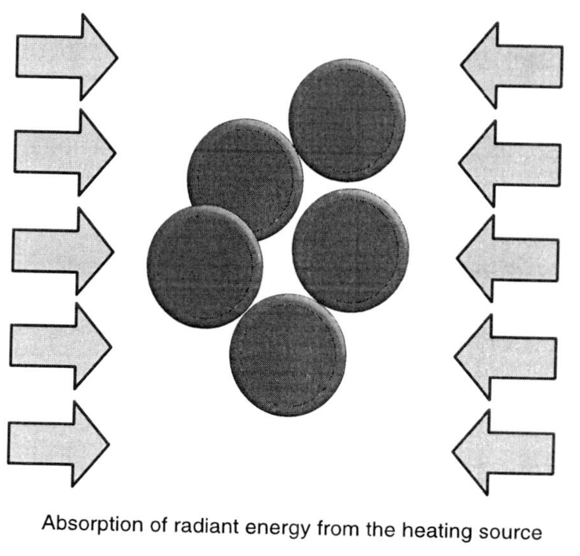

Figure 1.3 Principle of heating during radiation drying.

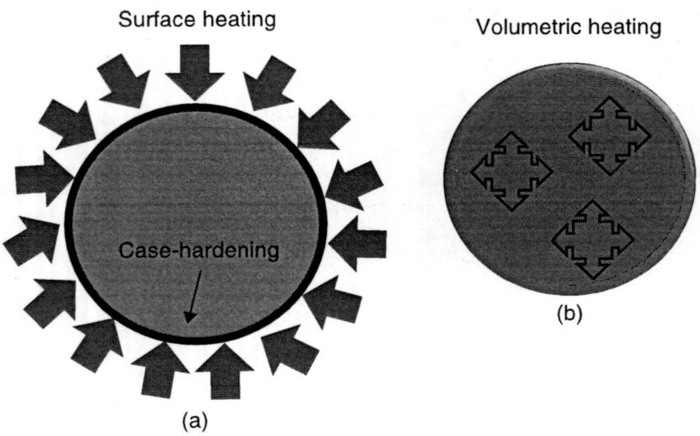

Figure 1.4 Pattern of heating in (a) conventional and (b) dielectric drying.

impact on both drying efficiency and product quality (Chakraverty & Singh 2014). Consequently, dielectric drying leads to higher thermal efficiency and shorter drying time. Furthermore, the selective and speedy energy absorption by water facilitates its rapid evaporation by the outward flux of water vapour. The outcomes are prevention of shrinkage and collapse in microstructure, leading to improved rehydration properties of the dried product. The dielectric drying methods are often employed at the end of the process lines to remove the residual moisture content of the final product.

1.2.2 Mass transfer during the drying process

The transfer of mass, in this case the moisture content, takes place from one phase to another, involving transitions among the solid, liquid and gaseous states. In most cases, drying is accomplished through evaporation-mediated phase transition of water from the liquid state to the vapour state by the addition of the latent heat of vaporization. Heat supplied by the drying medium is utilized to counterbalance the considerably large latent heat of vaporization for water (2200 kJ/kg). This increases the energy requirement of drying process. However, the withdrawal of latent heat from the drying medium maintains the product at relatively low temperature, more precisely at the wet-bulb temperature. The wet-bulb temperature (T_{wb}) is the temperature of air when it reaches 100% relative humidity, as water evaporates into it due to the supply of latent heat by the drying medium. T_{wb} is the lowest temperature that can be reached under ambient conditions by the evaporation of water. In certain exceptional cases, drying occurs by sublimation of the solid phase directly into the vapour phase, via the addition of latent heat of fusion (Mujumdar 2007). The removal of liquid water or water vapour from within the product to the surrounding medium follows a tortuous pathway (Figure 1.5) that is unique for each food product as a result of the structural changes that occur during the drying. The mechanisms that govern the transport of liquid water and water vapour are elaborated with respect to different drying stages in the subsequent sections.

From the above discussion, it is clear that the transport processes occur both within the product and also between the product and the drying medium. While the former is diffusion controlled due to the tortuous pathway that results during the later stages of drying, the latter is a function of the free moisture content at the surface. The limiting

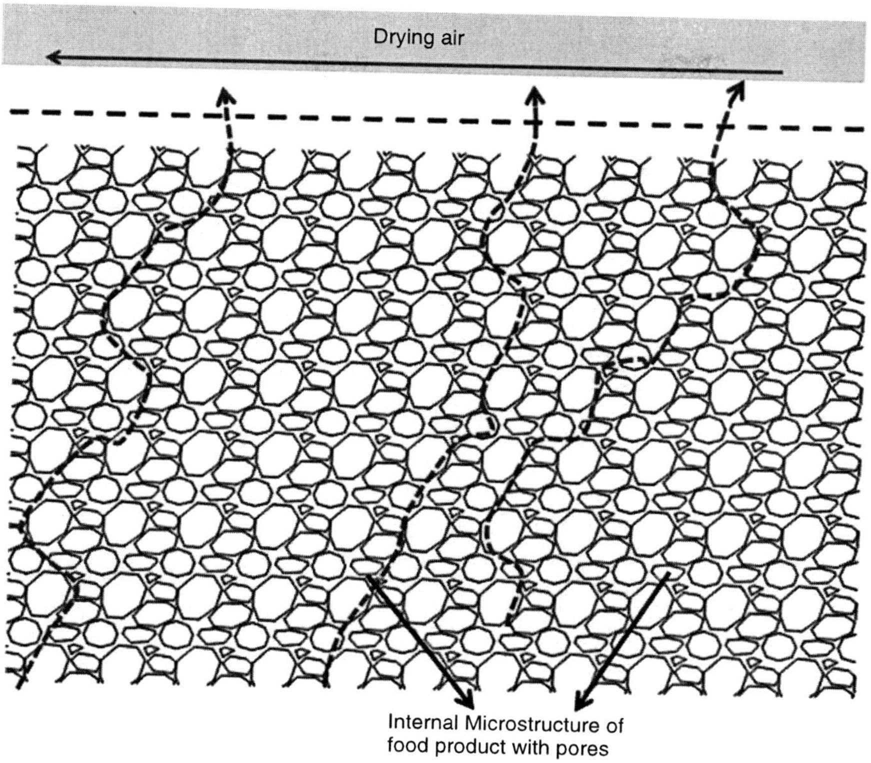

Figure 1.5 Tortuous pathway for moisture removal during drying.

parameter is the water activity (a_w), which presents the gradient for water movement and an equilibrium condition that marks the end point of the drying process. Water activity can be defined as the ratio of the vapour pressure of the solution to that of the solvent (water). In other words, a_w is equilibrium relative humidity divided by 100. This justifies the use of high temperature in most of drying methods, as higher temperatures result in lower equilibrium moisture content and hence present a larger concentration gradient for moisture removal (Singh & Heldman 2014).

The mass transport of water can occur in two ways: liquid water can be transported to the surface and then removed as vapour by evaporation or it can be evaporated within the product at the liquid–vapour interface and transported as vapour to the surface. Thus, different transport mechanisms have been elucidated for liquid and vapour transport. The two major mechanisms of mass transport are described below.

1.2.2.1 Diffusion mechanism

The driving force for mass transfer by diffusion is obtained by the difference in partial pressure or concentration that prevails between the product surface and the surrounding unsaturated gas phase. In addition to the pressure or concentration gradient, mass transfer is also dependent on the system properties, represented by the mass-transfer coefficient. This is analogous to the heat-transfer coefficient in the heat-transport process. The governing equation for mass transfer by diffusion is given by Fick's second law (Eq. 1.4):

$$\frac{dx}{dt} = D_{eff}\frac{\partial^2 x}{\partial^2 l} \tag{1.4}$$

where, x is the moisture content of the product (kg H_2O/kg product), t is time (s), l is the distance in the direction of mass transfer (m) and D_{eff} is the diffusion coefficient or moisture diffusivity (m^2/s).

Equation (1.4) is analogous to the Fourier equation of heat transfer (Eq. 1.1), but with temperature and thermal diffusivity replaced with concentration and moisture diffusivity, respectively. The diffusivity increases with an increase in the drying air temperature, defined by the Arrhenius relationship (Eq. 1.5):

$$D_{eff} = D_o \exp\left(\frac{-E_o}{RT}\right) \tag{1.5}$$

where D_o is the pre-exponent factor (m^2/s), E_o is the activation energy for moisture diffusion (kJ/mol), R is the ideal gas constant (kJ/(mol·K)) and T is the absolute temperature (K). In addition, diffusivity is also found to decrease with decreasing moisture content as the drying proceeds. The diffusion mechanism of mass transfer is observed during the removal of residual moisture from products such as flour and starch and also in other non-porous solids (Geankoplis 2006) during the latter stages of drying.

The diffusion can be that of liquid water or water vapour, with osmotic pressure and partial pressure as the driving forces, respectively. Liquid diffusion is limited by the attainment of equilibrium moisture content. The difference in total pressure produced by external pressure and temperature also aids in liquid or vapour transport. This is governed by Poiseuille's law, which explains the laminar flow of an incompressible and Newtonian fluid flowing through a long cylindrical pipe of constant cross-section, mediated by difference in pressure. In addition to the concentration and pressure gradient, the internal microstructure of product also plays a significant role in diffusion transport. According to the pore size, two types of diffusion mechanisms can be defined: Knudsen and mutual diffusion. The former mechanism occurs at low pressure in products with small pores of size less than the mean free path of the molecules and when the frequency of molecular collision with the walls is predominant over inter-molecular collision. Vapour transport by Knudsen diffusion is a function of vapour density, Knudsen diffusivity, size and number of pores, tortuosity, and characteristic dimension of the product. The latter type of diffusion occurs when the pore size is larger than the free path of diffusion of vapour molecules.

1.2.2.2 Capillary mechanism

Capillary forces are known to act in synergy with the diffusion mechanism to facilitate the drying of food products (Gorling 1956, 1958). In this mechanism, the moisture movement is arbitrated by surface tension rather than diffusion. The porous microstructure of the product plays a key role here. The interconnected pores communicate with the external surface by means of channels through which the moisture is removed. The water removal is accompanied by the formation of a meniscus across each pore and subsequent setting of interfacial tension at the solid–water boundary. This initiates the development of capillary forces at the interface in a direction perpendicular to the solid

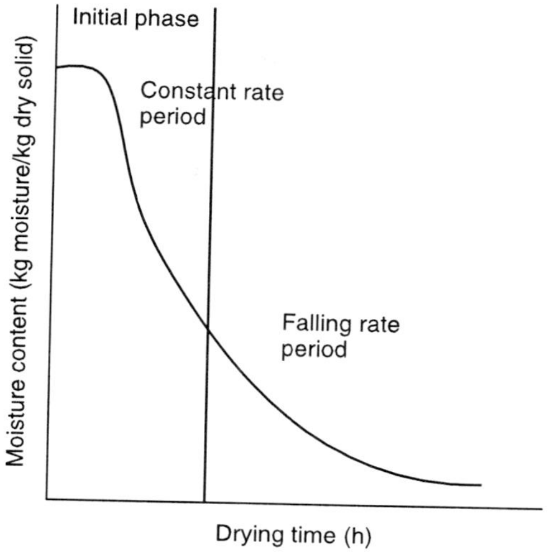

Figure 1.6 Typical drying curve.

surface. Thus, these capillary forces act as the driving force for moisture movement from within the porous structure of the product to the surface. However, the pore size plays an influential role, with the smaller pores leading to greater capillary forces than the larger ones (Geankoplis 2006). Capillary forces can be enhanced by the use of surfactants that reduce the surface tension of water (Rahman & Perera 2007).

1.3 The drying curve

Every product exhibits a typical curve that indicates its drying characteristics under specific conditions of the temperature, pressure, and velocity of the drying medium. Thus, it is evident that the dry-bulb and wet-bulb temperatures of the drying air play a vital role in determining the characteristic curve of a particular product. This is known as the drying curve, a plot of product moisture content versus drying time (Figure 1.6). The drying curve shows three distinct stages, which are described in the following section.

1.4 Stages of drying

Drying occurs in three stages: a short initial phase followed by the constant rate and falling rate drying periods. During the initial phase, the product heats up from the ambient temperature (T_a) to the evaporation temperature (T_e). In other words, the product temperature alters until it reaches a steady state. The solid temperature and the rate of drying may increase or decrease to reach the steady-state condition. When the product temperature is equal to the wet-bulb temperature of gas, the drying rate remains constant. This marks the beginning of the constant rate drying period (Foust *et al.* 2008). A comprehensible representation of the drying stages is given by the drying rate curve, which is a plot of free moisture content versus drying rate (Figure 1.7).

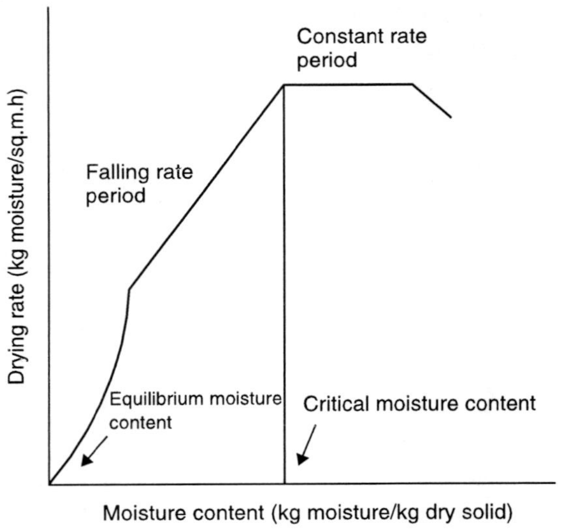

Figure 1.7 Drying rate curve.

1.4.1 Constant rate period

After the product attains its steady state, there is an initial linear reduction in the moisture content as a function of time, known as the constant drying rate period (Figures 1.6 and 1.7). During this period, as the name implies, the rate of water movement from within the food is equal to the rate at which the water evaporates from the surface. This stage is externally mass-transfer controlled while the surface is always wet. This is because there is a thin film of water at the product surface and no internal or external mass-transfer resistance (Rahman & Perera 2007). Thus, this stage predominantly involves the removal of free moisture that is available in the product at specific humidity and temperature. Because of the continuous replenishment of moisture, the surface essentially remains at the wet-bulb temperature of drying air. This can be ascribed to the phenomenon of *evaporative cooling*, by which the water film formed at the product surface as a result of adiabatic evaporation with the hot drying medium remains at the wet-bulb temperature of the drying gas.

The constant rate drying period proceeds as long as the surface remains wet and until the critical moisture content (M_c) is reached. During the constant rate period, the drying rate depends on the temperature, moisture content and mass flow of air. Overall, the drying rate during the constant rate period is considered to be heat-transfer limited.

Calculation of the drying rate during the constant rate period is based on the equilibrium that exists between rate of heat transfer and mass transfer (as moisture loss) between the food product and the surrounding drying medium. The rate of heat transfer is given by Eq. (1.2). Similarly, the mass-transfer rate is given by:

$$\frac{\partial M}{\partial t} = KA(H_s - H_a) \tag{1.6}$$

where $\frac{\partial M}{\partial t}$ is the rate of mass transfer or drying (kg/s), K is the mass-transfer coefficient (kg/(m^2·s)), A is the surface area for drying (m^2), and H_s and H_a (kg moisture/kg air) are the humidity of the product surface and drying medium, respectively. At equilibrium, the rate of moisture removal at constant product volume is given by:

$$\frac{\partial M}{\partial t} = \frac{hA}{\lambda}(T_o - T_m) \tag{1.7}$$

where h is the heat-transfer coefficient (W/m^2/k). Integration of Eq. (1.7) yields the expression for calculation of the duration of the constant rate drying period:

$$t = \frac{A\lambda}{h}\left(\frac{M_o - M_c}{T_o - T_m}\right) \tag{1.8}$$

where M_o and M_c are the initial moisture content and critical moisture content (kg water/kg dry product), respectively.

1.4.2 Falling rate period

Below M_c, the moisture content continues to decrease, but at a lower rate, since the mass transfer becomes internally mass-transfer controlled and diffusion governed. The surface of the product dries up and the temperature of the product rises close to the dry-bulb temperature of the air. At high moisture content, liquid flow due to capillary forces dominates. However, at decreasing moisture content, the amount of liquid in the pores decreases and a gas phase is built up, causing a decrease in liquid permeability. Gradually, the mass transfer is taken over by vapour diffusion in a porous structure (Rahman & Perera 2007). At saturation point, liquid is no longer available in the pores and mass transfer is taken over completely by vapour diffusion (Coumans & Kruf 1994). Thus, collectively, there is a decrease in the driving forces for heat and mass transfer. The transport processes cease to occur when the product temperature is nearly equal to the air temperature and the moisture content is equal to the equilibrium moisture content. Further drying of the product is not possible under ideal conditions. However, experimental determination of this end point or the state of equilibrium is tedious. The falling rate period takes a comparatively longer time than the constant rate period, although the amount of moisture removal is less. Thus, the product temperature and internal mass transfer are the rate-limiting factors during the falling rate period.

The critical moisture content at which the transition occurs from the constant to falling rate period varies with the food product. In general, it occurs at 58–65% of the equilibrium relative humidity of air, but most foods depict only a pseudo constant rate period with a steadily dropping drying rate. This indicates that food products predominantly undergo drying during the falling rate period.

The time for the falling rate period is given by:

$$t = \frac{D\pi^2}{4L^2} \ln\left(\frac{M_c - M_e}{M_f - M_e}\right) \tag{1.9}$$

where L is the characteristic dimension of the product, D is the diffusivity (m^2/h), and M_e and M_f are the equilibrium and final moisture contents, respectively. While the above equation is governed by the diffusion mechanism, that governed by the capillary flow of moisture is:

$$t = \frac{\rho L\lambda(M_c - M_e)}{h(T_o - T_m)} \ln\left(\frac{M_c - M_e}{M_f - M_e}\right) \tag{1.10}$$

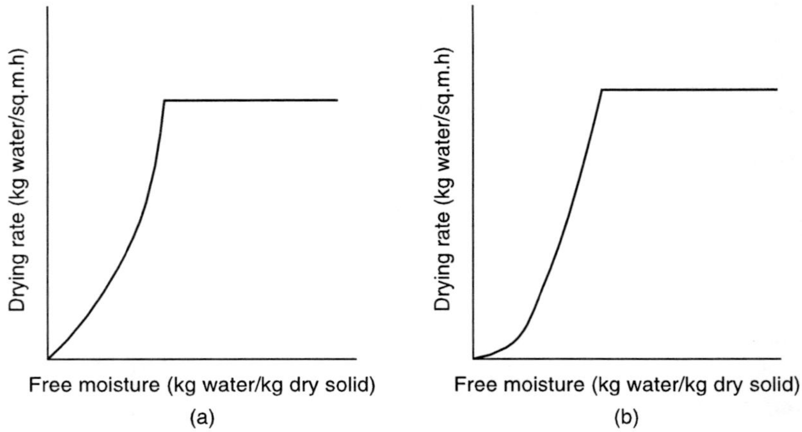

Figure 1.8 Drying rate curve with (a) diffusion mechanism and (b) capillary mechanism. Redrawn from Geankoplis (2006).

Two distinct falling rate periods may occur, depending on the product structure and drying conditions. A significant observation is that the falling rate period has a significant dependency on the mechanism of mass transfer. This is portrayed by the difference in drying rate curves resulting from drying processes governed by diffusion or capillary forces. In diffusion, the initial constant rate drying is predominant such that the first falling rate period comprising unsaturated surface evaporation does not occur (Figure 1.8a). With capillary force, the fine pores in the product cause the portion of the drying rate curve corresponding to the second falling rate period to turn upward (Figure 1.8b) (Geankoplis 2006).

1.5 Techniques for the drying of dairy products

Dried milk products are one of the major categories of tradable dairy commodities. The value addition of milk by means of drying is relevant owing to the perishable nature of liquid milk. Apart from enhanced shelf-life, as mentioned at the beginning of this chapter, drying of milk is also considered to be an effective marketing strategy as it reduces the bulkiness of the product and increases sensory appeal. This is the driving force behind a series of innovations in dairy drying techniques since they began in the 13th century. As with other food products, sun drying was initially used for milk. Industrial-scale drying of milk commenced only in the 19th century. The first dried milk was in the form of pills that were obtained after open-vessel evaporation, followed by air drying to remove the residual moisture (Olson 1950).

The full-fledged manufacture of dried milk powder began when spray-drying and drum-drying equipment was developed in 1872 and 1902, respectively. Since then the drying technology for milk products has undergone continuous evolution. It has included numerous innovations, such as the development of multiple-stage drying techniques for instantization and energy efficiency (Caric & Kalab 1987), and low-temperature drying techniques like freeze drying for enhanced quality of the final product. The range of dried dairy products includes but is not limited to skim-milk powder, partially skimmed milk powder, whole milk powder, whey powder, cheese powder and instant ice-cream mixes. The physicochemical and functional characteristics of dried dairy products depend on the drying technique used.

With this background, the choice of drying technique for dairy products depends on certain key aspects. Energy efficiency, drying time, product quality, process capability, and ease of scaling-up are often the major selection criteria. While a technique may excel in terms of continuous operation and industry-friendly operation, it may not be ideal with respect to quality characteristics and product functionality. On the other hand, the mode of heating, temperature of the drying medium and vapour withdrawal procedures employed increase the options for dairy drying techniques. The operation conditions, whether under atmospheric, sub-atmospheric or vacuum pressure, add to the classification of dairy drying techniques. Use of synergistic drying techniques is a well-established approach to overcome the limitations of conventional processes and to achieve novel product characteristics.

1.6 Conclusion

This handbook is intended to present readers with a mélange of the flavours of dairy drying. It will cover aspects of dairy drying operation over a wide scale, from the laboratory to the industrial level. The content of the following chapters aims to provide a blend of the science and engineering aspects of the dairy drying process.

References

Caric, M. & Kalab, M. (1987). Effects of drying techniques on milk powders quality and microstructure: A Review. *Food Structure*, **6**(**2**), 171–180.

Chakraverty, A. & Singh, R.P. (2014). Methods of Grain Drying. In: *Postharvest Technology and Food Process Engineering*, pp. 75–81, Taylor & Francis, Boca Raton, FL.

Chen, X.D. (2008). Food drying fundamentals. In: *Drying Technologies in Food Processing* (eds Chen, X.D. & Mujumdar, A.S.), pp. 1–54, Blackwell Publishing Ltd.

Coumans, W.J. & Kruf, W.M.A. (1994). Transport parameters and shrinkage in paper drying. *Drying '94*, Proceedings of the 9th International Drying Symposium, Gold Coast, Australia, pp. 1202–1205, Elsevier, Amsterdam.

Das, S.K. & Chakraverty, A. (2003). Grain-drying systems. In: *Handbook of Postharvest Technology: Cereals, Fruits, Vegetables, Tea, and Spices* (eds Chakraverty, A., Mujumdar, A.S. & Ramaswamy, H.S.), pp. 139–166, Marcel Dekker Inc., New York.

Foust, A.S., Wenzel, L.A., Clump, C.W., Maus, L. & Andersen, L.B. (2008). Simultaneous heat and mass transfer I: Humidification. In: *Principles of Unit Operations*, 2nd edn, pp. 289–317, John Wiley & Sons, Inc.

Geankoplis, C.J. (2006). Drying of process materials. In: *Transport Processes and Separation Process Principles*, 4th edn, pp. 559–624, Prentice Hall of India, New Delhi.

Gorling, P. (1956). Drying behavior of vegetable substances. *V.D.I. Forsch. Gebeite Ingenieurw*, **22**, 5.

Gorling, P. (1958). Physical phenomena during the drying of foodstuffs. *Fundamental Aspects of Dehydrated Foods*, pp. 42, Macmillan, London.

Mujumdar, A.S. (2007). Principles, classification and selection of dryers. In: *Handbook of Industrial Drying* (ed. Mujumdar, A.S.), 3rd edn, pp. 4–31, Taylor & Francis, Boca Raton, FL.

Olson, T.M. (1950). Miscellaneous dairy products. In: *Elements of Dairying*, pp. 526–536, Macmillan Co., New York.

Rahman, M.S. & Perera, C.O. (2007). Drying and food preservation. In: *Handbook of Food Preservation* (ed. Rahman, M.S.), pp. 404–427, Taylor & Francis, Boca Raton, FL.

Singh, R.P. & Heldman, D.R. (2014). Dehydration. In: *Introduction to Food Engineering*, 5th edn, pp. 675–708, Elsevier Inc.

Vega-Mercado, H., Góngora-Nieto, M.M. & Barbosa-Cánovas, G.V. (2001). Advances in dehydration of foods. *Journal of Food Engineering*, **49**(**4**), 271–289.

2

Dried Dairy Products and their Trends in the Global Market

Aadinath[1], *T. Ghosh*[1], *P.H. Amaladhas*[2] *and C. Anandharamakrishnan*[1,3]

[1] *Department of Food Engineering, CSIR – Central Food Technological Research Institute, Mysore, Karnataka, India*
[2] *Engineering Section, National Dairy Research Institute, Southern Campus, Bangalore, Karnataka, India*
[3] *Indian Institute of Crop Processing Technology, Ministry of Food Processing Industries, Government of India, Thanjavur, Tamil Nadu, India*

2.1 Introduction

Milk is known as nature's most complete food. It consists of 12–13% solids, mostly as proteins, fats, vitamins and minerals, and the rest is water. Cow's milk is composed of a complex mixture of proteins, fats, lactose, vitamins and trace minerals or salts. The composition may vary not only between breeds but also among the individual cows of the same breed. These major components of milk exist either in solution or as a colloidal suspension. *Proteins* are key constituents in milk. The proteins in milk consist of two broad types, namely, whey proteins and casein, in a ratio of 1:4. The whey proteins exist in solution while the caseins appear as a colloidal dispersion of aggregated casein molecules known as micelles. Casein is stabilized electrostatically and sterically by a hairy coat of κ-casein, therefore it is relatively heat stable, but it can precipitate out of solution at certain pH conditions (below pH 4.6). However, during processing, casein miscelles may be involved in interactions with other constituents of milk. The whey proteins, on the other hand, are susceptible to heat processing, which causes denaturation and protein–protein interactions involving disulphide interchange. The primary role of *lactose* in milk is in its colligative properties. Lactose also provides energy and can aid calcium absorption by the body. This reducing sugar turns into an amorphous product during drying. The *fat* is present in the form of an emulsion of fat globules, which are the largest (1–20 μm diameter) and lightest particles in milk. The *milk salts* include all the constituents that are present as ions or are in equilibrium with ions. Most of the *minerals* are in solution but a proportion of the calcium, magnesium and phosphate is associated with the casein. The functional properties of dried milk products are determined by the physical and chemical properties of its constituents, particularly moisture, proteins, fat and lactose, both individually and in combination (Oldfield & Singh 2005; Chessa *et al.* 2015).

2.2 Milk powders and dried milk products

Drying of milk has become an essential part of the long chain between the farmer and the consumer. Dried milk products that are popular are milk powders, whey powder, whey protein concentrates, infant formula, coffee whiteners, yoghurt powders, dry dairy-based beverages, casein, caseinates, lactose and various Indian products such as gulabjamun mix, basundi mix, etc. Milk powders are used in a wide range of products, such as reconstituted or recombined milk, ice-cream, processed cheese, confectionery such as chocolate, caramels, coatings and cakes, infant formula, health foods, etc. Whole milk powder (WMP) is also an important ingredient in the manufacture of products such as coffee whiteners, chocolates, desserts, confectionery and concentrated milks. Milk powders impart nutritive value, consistency, mouth feel and flavour in many of these products. For household use, milk powders are instantized by lecithination and agglomeration (if spray dried), so that they dissolve rapidly and uniformly in water at room temperature (Knipschildt & Andersen 1994; Oldfield *et al.* 2000). The marketing of these dairy products is dependent on key factors such as perishability, seasonal production variation, processing facilities, the balance between production and demand, competitive market, protectionism and value addition at farm level. With their benefits of easy transportation and higher shelf-life, dairy powders have several applications in the beverage, bakery, confectionary and allied industries.

These dairy products can be broadly classified into two types: primary and secondary products (Figure 2.1). Primary dairy powders are derived directly from milk and are further classified based on their content as fat (skimmed, whole and fat-filled milk powders with fat contents of 0%, 26% and more than 26%, respectively), protein (casein and caseinate powders, whey powders) or carbohydrate (lactose). Secondary dairy powders are derived from processed milk and include infant foods, fermented products and functional foods such as probiotic or dietetic powders.

2.2.1 Primary dairy powders

Whole milk powder consists of 26–42% fat and 34% protein, with less than 5% moisture content. Skimmed milk powder (SMP), obtained from raw milk after separation of milk fat, consists of 34% protein and less than 1.5% fat, with less than 5% moisture. Dried cream consists of 42% fat and 34% protein, with less than 5% moisture. In dry mixes where whipping or foaming is required, cream powder provides essential proteins to act as emulsifiers to stabilize and ensure proper whipping and foaming. Cream powder adds mouth feel, emulsification, colour and flavour to dairy-based dry mixes such as sauces, soups and salad dressings. Whey is a by-product of cheese and casein production, and finds diversified application in the food and pharmaceutical industries. In particular, the film-forming nature of whey contributed by lactoglobulin proteins makes it a suitable candidate as a wall material for the encapsulation process. Casein is a phosphoprotein and its proteinaceous part consists of α-, β- and κ-caseins. Similar to whey, casein finds diversified applications in various industries, including food.

2.2.2 Secondary dairy powders

Currently, the market for foods claiming gastrointestinal benefits is leading the functional food sector in terms of market size and activity. The dairy industry is in an excellent position to develop and exploit the functional food market. Probiotics are live

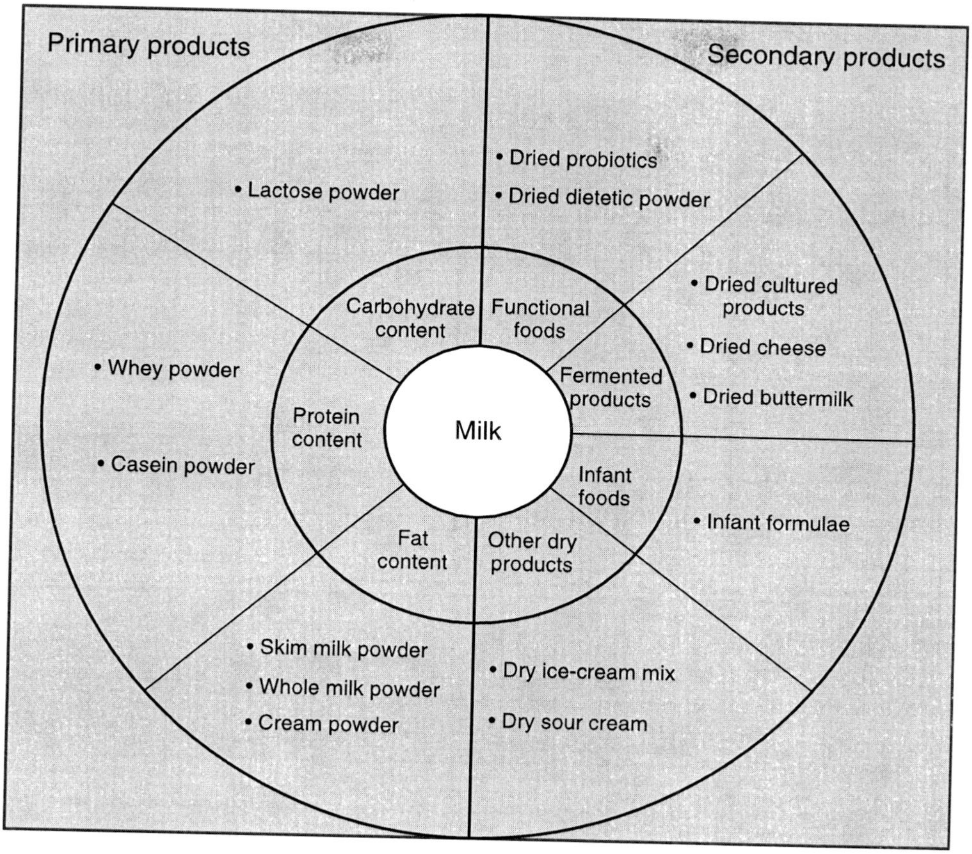

Figure 2.1 Classification of dried dairy powders.

microbial food ingredients that are beneficial to health. The gut flora play an important role in the maintenance of health by stimulating the immune system, protecting the host from invading bacteria and viruses, reducting lactose intolerance, alleviating gut mucosal inflammation, normalizing gut mucosal dysfunction, down-regulating hypersensitivity reactions and aiding digestion. However, probiotic stability during the storage period is inversely related to storage temperature (Gardiner *et al.* 2000). Although not published in the literature, certain companies producing starter cultures have the technology to produce freeze-dried lactic acid bacteria, including probiotics that are 'stabilized' and thus retain a high level of viability during drying and storage. The incorporation of such dried cultures into powdered milk may be the method of choice for preparing powdered milk products containing probiotics. However, research is needed, including storage testing, to confirm the feasibility of such processes.

The functional ingredients of milk also act as carriers for bioactive compounds. Dietetic products are known for the balanced ratio of all nutritive ingredients. For example, bioactive compounds like omega-3 fatty acids, vitamin A, vitamin D, curcumin and minerals are micro- and nanoencapsulated into whey protein powder, casein powder and lactose for functional food applications. The film-forming nature, good conductivity and high glass transition temperature of whey protein have been exploited as wall material in drying technologies like spray drying and freeze drying. Whey protein is one among the many wall materials used in spray drying because of its favorable properties that enable formation of microencapsulates with high encapsulation efficiency. Studies have also shown that addition of lactose to whey protein enhances crust formation by improving its drying properties.

Milk-based infant formulas are substitutes for breast milk as they include all the macrocomponents of milk, such as demineralized whey, lactose and casein. Milk fat is replaced by vegetable oil blends such as palm, coconut, oleo oil and safflower. The differences between human and cow's milk are their protein, lactose and fat contents. The casein to whey protein ratio in cow's milk is 3.7, whereas it is 0.54 in human milk. Whey protein concentration is increased by addition of denatured whey protein powder to provide a balanced nutrition for the infant. The source of whey protein is verified free from nitrate or sodium salts before addition. The fat component is tailored according to the desired physicochemical composition and calculated to have similar characteristics to those of breast milk. More saturated fatty acids and medium chain triglycerides are found in cow's milk than in human milk.

Distinct properties of dried dairy ingredients, such as emulsification, foaming, gelation and water-binding capacities, widen their applications. Dairy powders can be used as carriers for micronutrients such as iron. For instance, in Chile, a complementary feeding programme provides ascorbic acid (700 mg/kg) and iron (100 mg as ferrous sulfate/kg) fortified dried milk powders for infants (WHO 2006).

2.3 World market dynamics

The market trend of dried dairy products is based on statistics from major milk-producing and -consuming countries. Current milk production in India is approximately 146 million tonnes, as against 17 million tonnes during 1950–1951. According to the Food and Agricultural Organization (FAO), the quality and quantity of livestock have to be improved to increase the production up to 191 million tonnes of milk per year by 2020. The per capita consumption of milk and milk products in India is among the highest in Asia, but it is still below the world average of 285 g per day and the minimum nutritional requirement of 280 g per day recommended by the Indian Council of Medical Research (ICMR). This level of consumption results in a lower contribution to the world market due to the huge domestic demand, as evident from the SMP export ban in 2009. Further, a 6-year national dairy plan was approved in 2012 with a budget of 20 billion rupees to meet the increasing domestic demand for dairy products. Another reason for poor trading statistics in India is higher microbial spoilage. However, New Zealand, another major milk-producing country, exports 95% of its domestic production to other countries, mainly to China, giving it an important role in world trade (Southard 2004). The regulatory policies of various countries that contribute to the dairy market are provided in Table 2.1.

2.3.1 Production

2.3.1.1 Oceania

Australia is one of the major milk-producing nations in the world, with production of 9.9 billion tonnes for 2014, a 2% increase on the previous year. The Tasmanian branch

Table 2.1 Regulatory policies of various countries that contribute to the global dairy market

Country	Product	Regulatory policies implemented	Policy details
USA	Dairy products	State market regulation	Announced enrolment for the new dairy market protection programme which provides farmer support when profitability falls below predetermined levels
European Union	Dairy products	State market regulation	Announced an extension of private storage aid to include butter, SMP and certain types of cheese to provide assistance to companies affected by Russia's trade sanctions
Australia	Dairy products	Free trade agreement	Signed an economic partnership agreement with Japan, providing an additional 20,000 tonnes of duty-free import quota for processed cheese
Japan	Butter	State market regulation	Announced the emergency importation of 7000 tonnes of butter for industrial use, in addition its commitment of 3000 tonnes for the 2014 fiscal year
Bulgaria	Dairy products	State market intervention	Approved an allocation of Leva 114 million (US$ 80.6 million) to be paid to the dairy sector in 2014 Additional small top-ups will be paid to farmers with pasture land and selected farms will have access to minimal funds of US$ 5.3 million
Brazil	Dairy products	State market regulation	Announced a comprehensive dairy sector development strategy, "More milk", with short-, medium- and long-term goals to be implemented by a national research body
Russia	Dairy products	Import ban lifted	Lifted the ban on dairy imports from Lithuania, imposed in October 2013

of the Aurtralian dairy industry research and development organization has set a new goal of achieving 40% more production by 2017, which will boost the market scenario. Exports of processed dairy products are expected to rise in 2014 due to higher world prices and a fall in the Australian dollar. The New Zealand milk production forecast for 2014 is 20.6 million tonnes, which is a 4.5% increase on the previous year. Milk production in New Zealand cannot be predicted on a calendar year basis because it is highly monsoon dependent. Milk production in autumn and summer is highly dependent on rainfall, which adversely affects it. New Zealand is renowned for its spray-drying technology in producing milk powders, which make up 40% of the total world production of WMP. SMP is another major product expected to show an increase of 7.7% compared to previous year (USDA 2015).

2.3.1.2 India

India's economy is highly dependent on agriculture. It is the world's largest milk-producing country, contributing 17% of the world's total production. Milk production in India in 2015 was 146 million tonnes, which is 4.5% more than in 2014 production. This increase was driven by strong farm gate prices and rising domestic demand for value-added dairy products. Because of the withdrawal of the non-fat dry milk (NFDM) export ban in June 2012 and strong prices, production of NFDM for 2014 is expected to increase to 489,000 tonnes from 470,000 tonnes in 2013 (Mathews 2014).

2.3.1.3 European Union

The 27 countries of the European Union (EU-27) make up one of the highest milk-producing and exporting areas in the world. However, the percentage increase in production for the year 2013 is likely to be marginal due to increased production costs and the delayed spring. As the cheese market remains more profitable, the production of NFDM and butter remained stagnant in 2013. Production in Germany, France and Italy contributes more than 55% of the total output. Manufacturing of whole dry milk and NFDM is also expected to decrease due to international competition in the world market and a limited supply of raw milk (USDA 2015).

2.3.1.4 Argentina

Milk production for the calendar year 2014 is expected to show a 1.5% increase from the previous year and as domestic demand remains stagnant, the increased production will flow through to increased production of commodities and export. This modest growth in production is possibly due to standard weather conditions and stable international prices for WMP. Whole milk powder production for 2014 is expected to show an increase of 1.5% and the current industry's production capacity is 15 million tonnes per day. The second major dairy product of Argentina is cheese. Soft cheese holds 50% of the market, semi-soft varieties contributes 38% and hard cheese accounts for 18% (USDA 2015).

2.3.2 Consumption

2.3.2.1 Algeria

Algeria is the largest importer of milk powder in the world, averaging about 290 million tonnes for the last 5 years. Dairy products contribute 13% of the total food imports for the calendar year 2013 and the USA holds 9% of this market. Efforts have been made to increase milk production by implementing artificial insemination, embryo transfer, and importing pregnant heifers and dairy cattle. These measures led to the production goal of 3.04 million tonnes for the year 2013 being exceeded. Concerns such as proper animal husbandry, feed and nutrition management, improved genetics and modern milk collection systems need to be looked into for effective milk production as the domestic demand is 5 million tonnes annually. The dairy industry of Algeria comprises 116 dairies and relies on imported milk powder for production of reconstituted milk. The dairies are owned by both government and the private sector, and milk is sold at the domestic level in three forms: pasteurized reconstituted milk in small bags, fresh milk and whole milk powder in tetra-pack boxes. As world prices for milk powder at the end of 2013 and beginning of 2014 decreased, this was the best time for the government to renew its stock and fulfil the domestic demand (USDA 2015).

2.3.2.2 Indonesia

The market share of the most-consumed milk products, namely, fluid ultra-high-temperature (UHT) milk, sweetened condensed milk and powdered milk, are 26%, 35% and 39%, respectively. According to the Indonesian Ministry of Industry, current milk production only meets 20% of the total domestic demand. A sharp increase of 17% in the domestic demand for fluid milk over the past 7 years and a lack of production to meet this demand has made Indonesia a major milk market. The annual growth of 4% for sweetened condensed milk demonstrates the huge market for concentrated dairy products. The increased domestic demand for dairy products is driven by the expansion of the middle-class population, increased awareness of the health benefits of milk and diverse marketing campaigns (USDA 2015).

2.3.2.3 China

China is the world's biggest market for dairy products and its domestic demand for fluid milk reached 14.5 million tonnes in 2013, which is 200,000 tonnes higher than the previous year. Such intense domestic demand has resulted various fluid milk products being produced in China and imports from the USA increased from 43 tonnes in 2010 to 2750 tonnes in 2012. Non-fat dry milk is mostly imported from Germany, but the Chinese government has decided to reduce imports by nearly 20% because of the reduced price of WMP (USDA 2015).

2.3.2.4 Mexico

Mexico is a milk-deficit nation due to huge domestic demand. NFDM demand for the year 2013 showed an increase up to 290,000 million tonnes due to strong industrial demand for use in value-added products. Whole milk powder production and consumption has declined according to the 2013 forecast due to consumer health interests and high consumer demand for other value-added products (USDA 2015).

References

Chessa, S., Nicolazzi, E. L., Nicoloso, L., Negrini, R., Marino, R., Vicario, D., Marsan, P.A., Valentini, A. & Stefanon, B. (2015). Analysis of candidate SNPs affecting milk and functional traits in the dual-purpose Italian Simmental cattle. *Livestock Science*, **173**, 1–8.

Gardiner, G.E., O'Sullivan, E., Kelly, J., Auty, M.A.E., Fitzgerald, G.F., Collins, J.K. & Stanton, C. (2000). Comparative survival rates of human-derived probioticlactobacillus paracasei and l. salivariusstrains during heat treatment and spray drying. *Applied and Environmental Microbiology*, **66**(**6**), 2605–2612.

Knipschildt, M.E. & Andersen, G.G. (1994). Drying of milk and milk products. In: *Modern Dairy Technology: Advances in milk processing* (ed. Robinson, R.K.), pp. 159–254, Springer.

Mathews, K. (2014). *Livestock, Dairy, and Poultry Outlook*, Economic Research Service, US Department of Agriculture.

Oldfield, D. & Singh, H. (2005). Functional properties of milk powders. In: *Encapsulated and Powdered Foods* (ed. Onwulata, C.), pp. 365–386, Taylor & Francis.

Oldfield, D.J., Teehan, C.M. & Kelly, P.M. (2000). The effect of preheat treatment and other process parameters on the coffee stability of instant whole milk powder. *International Dairy Journal*, **10**(**9**), 659–667.

Southard, L. (2004). *Livestock, Dairy, and Poultry Outlook*, Electronic Report 1(8), Economic Research Service, US Department of Agriculture.

USDA (2015). *Dairy: World Markets and trade*, http://usda.mannlib.cornell.edu/usda/current/dairy-market/dairy-market-12-17-2015.pdf.

WHO (2006). Part III – Fortificants – physical characteristics, selection and use with specific food vehicles. In: *Guidelines on Food Fortification with Micronutrients* (eds Allen, L., de Benoist, B., Dary, O. & Hurrell, R.), World Health Organization, Geneva.

3

Techniques for the Preconcentration of Milk

I. Roy[1], *A. Bhushani*[1] *and C. Anandharamakrishnan*[1,2]

[1] *Department of Food Engineering, CSIR – Central Food Technological Research Institute, Mysore, Karnataka, India*
[2] *Indian Institute of Crop Processing Technology, Ministry of Food Processing Industries, Government of India, Thanjavur, Tamil Nadu, India*

3.1 Introduction

Milk is concentrated by reducing its water content up to its flow limits, that is, 40–50% total solids. This is generally done by evaporating the water in evaporators where the mass and volume are reduced and the viscosity and density are increased. Concentration of milk dates back to the 19th century, when milk was concentrated by evaporating it in open pans in batches to improve the keeping quality during long journeys and to provide dietary support to soldiers. Gradually, the simple evaporated milk was fortified with vitamins, treated with ultraviolet radiation and had sugars added to improve shelf life and increase variety.

Preconcentration of milk has become an integral part of the production process for whole milk powder, whey powder, skimmed milk powder and other dried dairy powders. There are a few factors that need to be considered for any preconcentration method of milk to be viably scaled up to an industrial level, as listed by Spreer (1998):

a) The process should not lead to drastic reduction in the nutritional values or organoleptic properties of the product.
b) The steam and water requirements of the equipment/ process should be as low as possible.
c) Control and monitoring of the process should be easy and hassle-free.

Apart from traditional methods like evaporation and reverse osmosis, processes like ultrafiltration, reverse osmosis (hyperfiltration) and demineralization (electrodialysis, ion exchange) are now used industrially to produce dried dairy products.

3.2 Need for preconcentration

The rationale for employing concentration techniques prior to production of dairy powders is to reduce the amount of water (solvent) and in turn increase the amount of available dry matter in the feed for the dryer. By including this unit operation, there is an estimated three-fold reduction in energy requirement for drying as against the direct drying process (i.e. without preconcentration; Spreer 1998). Furthermore, excluding the preconcentration step would increase cost of the process because the drying time would be prolonged, the equipment is not fully utilized and energy consumption is markedly increased. As reported by Caric (1994), steam consumption per 1 kg of evaporated water is about three times higher during roller drying and about three to six times higher during spray drying, in comparison to the corresponding consumption of a double-effect vacuum evaporator.

Apart from the energy efficiency aspect, preconcentration imparts desirable physical properties to the dried powders. For instance, milk powder produced from milk preconcentrated by evaporation has a longer shelf life and larger powder particles with smaller amounts of trapped air. The larger size of the dried particles is caused by the increased viscosity of the preconcentrated milk. If the preconcentration step is omitted, low-quality milk powder with poor rehydration properties results. When milk is dried in a spray dryer, the huge amount of water (>85% total milk volume) is evaporated inside the drying chamber, causing the formation of large air pockets inside each powder particle. When mixed with water these particles float to the surface and are not dissolved easily (Hunziker 1920). Another advantage of preconcentrating milk is the crystallization of lactose, which aids faster powder production. Furthermore, the bulk of the raw material to be stored and transported is reduced, with increased shelf life.

3.2.1 Skim milk

For production of skim milk powder, as reported by Písecký (1981), concentration of skimmed milk is done by reverse osmosis, which brings up the total concentration to 25% total solids. This is then further concentrated by subsequent vacuum evaporation. This combination method ensures that the energy required is much lower than that needed by a multi-effect evaporator, but it requires the same investment costs. Modified methods of producing skim milk powder by foam spray drying make it possible to dry skim milk that has been evaporated (to 50% total solids) to a greater extent than in a conventional procedure (Early 1998). The resulting powder has desirable dispersibility and reconstitution properties. If feed concentration is further increased by preconcentration, the drying efficiency is improved, thus increasing the capacity of the chamber. However, foam-dried skim milk with feed concentration higher than 45% has poor solubility. Hence, the optimum solid content of feed for specific drying processes has to be evaluated to achieve high-quality dried powders.

3.2.2 Whey powders and infant formula

During the production of whey powder, whey is initially concentrated to 42–60% total solids depending on the procedure used, with each step in the evaporation system increasing investment costs by 15% and decreasing steam consumption by 25%. Whey can also be preconcentrated by the reverse osmosis technique (Morr & Ha 1993).

For infant formula produced from milk products, evaporation is necessary if the formula contains 40–48% dry matter after blending (Fox 2001). It is during this stage of processing that most of the necessary components are added to skim milk: vegetable oils, demineralized whey, fat-soluble vitamins, emulsifiers and stabilizers.

3.3 Concentration methods

Various methods exist for the concentration of milk, depending on the type and quality of the desired end product and the economy of the process. Water can be removed from milk in the following three forms (Varnam & Sutherland 2001):

i) as vapour (by evaporation)
ii) as liquid (by membrane separation)
iii) as solid ice crystals (by freeze concentration).

Depending on the use of heat energy to concentrate, the methods for concentration of milk can be classified into thermal and non-thermal processes.

3.4 Thermal methods

3.4.1 Evaporation

Evaporators can be classified in several ways (Westergaard 2004):

a) by tube position: horizontal or vertical
b) by tube length: short or long
c) by the direction of movement of film: rising or falling
d) by the number of evaporators or effects.

Evaporators can either work on a batch or a continuous basis. They can be designed to facilitate heat recovery (in single-stage plants by preheating installations or multistage plants by using the vapour from one stage to heat the following stage). For this purpose, the vapour is either compressed with new fresh steam or compressed mechanically. Compressed vapour (with higher enthalpy and higher temperature) is fed to the evaporator, where residual heat is used to preheat milk (Jeurnink & Brinkman 1994). Batchwise operating plants are used in cases where only a small amount of water is to be removed or other ingredients are to be added during the concentration, as they might have a defined dry matter level in the finished product. Large plants are generally installed with continuous evaporators (Bienvenue *et al.* 2003).

Evaporator installations can be either single-stage or multistage. A *single-stage evaporator* is a closed unit with a defined pressure and corresponding temperature. In *multistage plants* (up to seven stages), product from the first stage is transferred into the second stage and is heated with vapour from the first stage to the boiling point. This process makes it possible to reduce the steam consumption to a value of 0.08 kg steam/kg evaporated water (Dodeja *et al.* 1990).

Some of the evaporators that are most commonly used in dairy plants for the concentration of milk are described below.

3.4.1.1 Single-effect recirculation evaporator

Most of these are *climbing film evaporators* in which the product is pumped into a tubular heating section of the evaporator from the bottom. Heat is added to the steam–vapour mix via a pipe and then product rises in the tubes to the top, reaches the boiling point and goes via a connection into a centrifugal separator, where the liquid

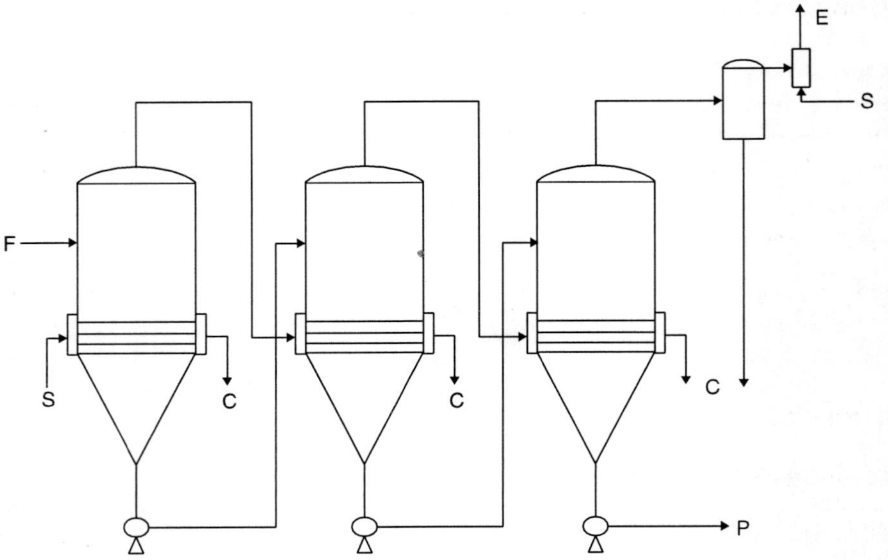

Figure 3.1 Schematic representation of a triple-effect evaporator with forward feed. F, feed; S, steam; C, condensate; P, product; E, steam jet ejector (Toledo 2007).

and the vapour are separated. Concentrated product leaves the separator via a pipe. If the desired concentration has not been reached, it is re-fed to the bottom through a pipe, mixed with fresh product and can be re-circulated (Hewitt *et al.* 1994). Part of the vapour is transferred by means of a pipe to the steam injection compressor and mixed with steam (injected via a port) in a mixing nozzle, where the enthalpy is raised, and transferred to the evaporator heating section. The balance of the vapour is transferred by a pipe to a condenser (Varnam & Sutherland 2001).

The product can be circulated several times in the same evaporator stage (especially in multistage evaporators) or it can leave after one passage. The resulting residence time is long and there also may be different residence times and thermal loads for different parts as some of the product can remain in the evaporator much longer and may deteriorate the milk quality (Robinson 2012).

3.4.1.2 Multiple-effect evaporator

Multiple-effect evaporators are constituted by two or more evaporators connected in series. The most common multiple-effect evaporator has three effects or units, as shown in Figure 3.1. Here, the evaporators are consecutively connected to move the heat in the vapour from the first effect to heat the second effect, and in turn the vapour from the second effect is used to heat the third effect. This flow of vapour is maintained by proper control of pressure and temperature because the pressure is maintained in a decreasing manner in the direction of the flow (Carl & Jenssen 1957).

According to McCabe *et al.* (1993), there are several methods of feeding the vapour and the product in the evaporator:

(i) *Parallel feed*: Milk is fed at the same time to all effects while steam moves from first to last effect. This type of feed is primarily used when the feed is saturated and mainly slurry remains in the product.

(ii) *Forward feed:* Milk and vapour move from first effect parallel to each other, through all the effects. This type of feed is generally used when the product may deteriorate if exposed to high temperature.
(iii) *Backward feed:* Milk enters the last effect and flows to the first one, counter-current to the steam flow. When the products are viscous and increased rate of heat transfer is required, this type of feed is employed.

Temperatures below 45 °C are not recommended because they do not prevent microbial growth. The following combinations of temperatures suggested by Caric (1994) are possible: (i) $t_1 = 75$ °C, $t_2 = 60$ °C, $t_3 = 45$ °C or (ii) $t_1 = 80$ °C, $t_2 = 65$ °C, $t_3 = 50$ °C, where t_1, t_2 and t_3 are the set temperatures of the first, second and third effects, respectively. Another recent study by Karimi *et al.* (2007) shows that a combination of $t_1 = 72$ °C, $t_2 = 58$ °C and $t_3 = 45$ °C also works well for this kind of evaporator. The main advantage of a multiple-effect system is that less steam is required for heating compared to a conventional high-capacity evaporator.

3.4.1.3 Falling-film evaporator

This is the most commonly used evaporator in the dairy industry. The product residence time in falling-film evaporators is always the same. Product is pumped into the top of the evaporation stages and passed through tubes in a thin layer (on the walls). The advantages of using falling-film evaporators for dairy products are short residence or passage time of the product on the heated tubes, good heat conduction due to a thin product layer and no difference in thermal load (Saravacos & Kostaropoulos 2002). Depending on the design, the tubes are 4–10 m long and have diameters ranging from 25 to 80 m. These plants can be operated with direct steam injection without vapour compression, with thermal vapour compression (steam injection) or with mechanical vapour compression (turbo compressors, high-capacity fans) (Kessler 1981). A schematic representation of a three-effect falling-film evaporator is given in Figure 3.2.

Milk that is to be concentrated is transferred by a pump from the buffer tank to a tubular preheater. Installation of the preheaters in the upper sections of the evaporators and condensers results in optimal energy utilization. Product is heated to 75–80 °C and transferred by pump to a tubular preheater, where preheating takes place at 110–130 °C (as required for evaporated milk) (Ranken *et al.* 1997). This preheating is eliminated when the concentrate is used subsequently for drying. As described by Spreer (1998), the steam pressure of the hot milk is reduced in the flash vessel, and the milk is cooled to evaporation temperature then evenly distributed into the tubes by a distribution device installed in the upper part of the evaporator above the calandria. The product flows downward in the tubes in a very thin film, is heated by the vapour (on the other side of the tube wall) and a part of the water evaporates. If the film of product on the heater surfaces is too thin, or if it does not cover the whole surface, then baked-on deposits will form, causing a drop in efficiency (Paramalingam *et al.* 2000). This can be avoided by a perforated distribution plate or a double-cone valve. As the liquid flows under gravity, the amount of vapour increases due to evaporation and influences the flow of the film. The vapour–liquid mix reaches the centrifugal separator, from which the product is transferred by pump into the second evaporator stage where the first stage is repeated. Vapour is extracted and compressed (Knipschildt 1986).

The third stage of the evaporator further concentrates the milk. Part of the concentrate can be recirculated by using the pump to make a more uniform product load and control

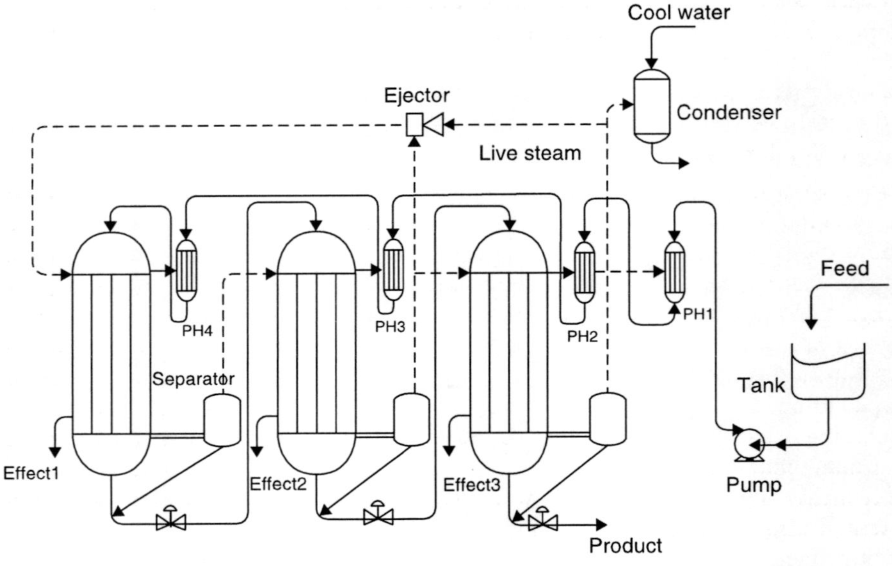

Figure 3.2 Schematic representation of a three-effect falling-film evaporator. PH, preheater (Karimi *et al.* 2007).

or further decrease water content. Concentrate is transferred from one stage to the next by pump. The plant is put under vacuum by a steam jet-operated vacuum nozzle. Two steam jet compressors compact the vapour and transfer it into the evaporation stages, then residual vapour is condensed in a surface condenser (Chen & Jebson 1997).

3.4.1.4 Plate evaporator

A plate evaporator uses a plate heat exchanger to transfer heat to the liquid for vapourization. The liquid is at the bottom of the plates, with steam between alternate plates. The vapour moves vertically through the plates and is removed in the vacuum path. The liquid rolls vertically along the plate and is heated quickly. Plate evaporators typically have low installation height and flexible capacity, achieved by increasing or reducing plates and having a short contact time between product and plates (Hoffman 2004).

After heating the milk, the concentrate/vapour mix reaches a drum-shaped centrifugal separator, which removes the concentrate or transfers it to another stage. Vapour can also be compressed and reused for heating the evaporator (Ribeiro & Andrade 2002).

Compression of vapour When vapour needs to be reused, it can be compressed by either of the following two methods:

(i) *Recompression*: The vapour is compressed a second time by a pump and returned to the evaporator for heating (Knipschildt 1986). Following compression, the temperature is increased so that the vapour removed will be at a higher temperature and therefore will transfer heat to the product.

(ii) *High-pressure steam method*: Steam, usually at 6.89 bar, is mixed with the low-pressure vapour that comes out of the evaporator at the temperature of vaporization, thus increasing the temperature. The mixture is then returned to the previous effect (Winchester & Marsh 1999).

Condensation of residual vapour Condensation of residual vapour is done in condensers, which are of mainly two types: surface condensers and mixing condensers. In a *surface condenser* there is no mixing of cooling water and the vapour; the vapour condenses on the surface of the heat exchanger and is removed from the system. Since the condensate remains free from adulteration, this type of condenser is useful in areas with brackish or contaminated water. In a *mixing condenser* cold water is economically used to condense the vapour, in which case there is a mixing of the cold water and the vapour which condenses (Kakac 1991). For manufacturing concentrate for drying, installations with 10–20,000 kg water evaporation per hour are used. For whey concentration the evaporation rate increases to 30,000 kg/h. A very important parameter is the steam consumption for evaporating 1 kg of water, which includes the entire steam consumption, including preheating and vacuum generation. Another parameter is cooling water consumption, which indicates the economy of the plant. Although a major part of the vapour enthalpy is recovered, part of the vapour (residual vapour) must be condensed with water (Mersmann *et al.* 2011).

Control of evaporation and concentrate As reported by Winchester & Marsh (1999), evaporation is assumed to be completed when the concentrate has reached the desired level of dry matter content. As the dry matter content is proportional to the specific density, this parameter is used for control of the concentrate, measured by either a lactodensimeter or a Beaumé spindle. A lactodensimeter is a graduated air-filled glass column with a heavy base, allowed to float freely in the container carrying the milk, with the level of submergence indicating the specific gravity (Doremus 1895). A Beaumé spindle consists of two graduated spindles, one for liquids heavier than water and one for liquids lighter than water, consecutively suspended in water. The solution is tested to compare specific gravity readings (Ernst 1914). Another indication of product quality is the degree of concentration, which is the factor by which the initial dry matter content has been raised. The evaporation process must be controlled such that the desired dry matter content or concentration degree is reached. For this purpose, the following process parameters must be considered, as listed by Mersmann *et al.* (2011):

a) pressure and temperature of the first evaporator effect
b) milk inlet temperature into the first evaporator effect
c) milk flow rate
d) temperature in the last effect
e) vacuum and stability in all evaporator effects
f) dry matter content or specific gravity.

All parameters must be kept as constant as possible to achieve a uniform product quality. At the end of the concentrate manufacture, a material balance must be established, which requires a comparison of the theoretical amount and the actual amount of concentrate using the following formula (Spreer 1998):

$$m_{\mathrm{c}} = m_{\mathrm{m}} \cdot \frac{DM_{\mathrm{m}}}{DM_{\mathrm{c}}}$$

where m_c is the mass of the concentrate (kg), m_m is the mass of the milk (kg), DM_m is the dry matter content of the milk (%) and DM_c is the dry matter content of the concentrate (%).

3.4.1.5 Horizontal tube evaporator

If there is a limitation in the vertical space for installation, a tubular heat exchanger can be used as a horizontal evaporator. To ensure that both the top and bottom tubes are uniformly covered with liquid, the evaporator must be operated in a flooded condition, that is, the feed inlet must be above the uppermost tube (Delyannis & Delyannis 1980). As illustrated by Wattelet (1994), the ratio of liquid to vapour at the discharge end is greater in this evaporator than in a falling-film evaporator. The film is moved by the vapour, making it necessary to use the greater pressure and temperature differences.

3.4.1.6 Mechanical film evaporator

When highly viscous products are to be evaporated, it may be that gravity or the propelling power of the liquid, which normally moves the liquid along, may be inadequate. This problem is overcome by mechanically creating a thin film. A shaft fitted with wiper blades, scrapers, vanes or other devices rotates within a vertical tube of relatively large diameter surrounded by a heating jacket (Janosfia & Buchi 1966). The purpose of the blades is to produce a thorough mixing of the film, to distribute it evenly and to transport the product through the evaporator (Komori *et al.* 1988). The specific rate of evaporation is high in spite of high viscosities. This is generally more economical if the product has been preconcentrated by some other equipment.

3.4.1.7 Low-temperature evaporator

This reduces the heat effect on the product during processing. This evaporation system is a compression refrigeration system, utilizing the heat from the condenser, the heat from the product, the cooling effect of the evaporator to condense the vapour and a vacuum pump to maintain a vacuum on the system (Varnam & Sutherland 2001). The unit can be operated at temperatures as low as 15–32 °C (Saravacos & Kostaropoulos 2002). Milk with concentration of up to 43% solid content can be produced by this method (Aceto *et al.* 1962). Because of the increased production cost, these evaporators are only used in the production of infant foods or other special powders requiring a low-temperature process.

3.5 Non-thermal methods

3.5.1 Freeze concentration

Freeze concentration is a relatively new method for removal of water from aqueous solutions like milk. It offers several advantages in product quality, aroma retention and product yields. The basic principle is to cool the solution to a temperature below its freezing point to produce ice crystals, which can then be removed, thus reducing the moisture content of the milk. Studies show that concentrations of total solids up to 36%, 38% and 49% could be achieved without greatly affecting the product quality or aroma retention and with minimum product losses (Van Mil & Bouman 1990).

At present freeze concentration plants have not been developed beyond the pilot scale. As described by Hartel & Espinel (1993), a stainless steel vessel with a cooling water jacket can be used as a batch crystallizer to produce freeze-concentrated milk products. Refrigerated ethylene glycol solution from a constant temperature bath is circulated through the crystallizer jacket to maintain constant temperature conditions. A stirrer comprising propeller blades maintains agitation in the crystallizer.

Ice crystal nuclei, used for the seeds in batch crystal growth, are formed in 10% lactose by spontaneous nucleation in a jacketed vessel. Glycol, which is used as a refrigerant, is cooled to −40 °C and pumped through the nucleator jacket containing lactose solution. Once nuclei are formed, they are allowed to grow for 10 min under these conditions. Part of this slurry is filtered to separate ice crystal seeds from the lactose solution. These nuclei are added to the milk, which is held at the appropriate temperature in the growth crystallizer to initiate the batch freeze concentration process. Crystallization is stopped when the liquid phase reaches the desired total solids, and the liquid is separated from the ice crystals by filtration (Sánchez *et al.* 2011).

Figure 3.3 shows a unidirectional progressive freeze concentration process for skim milk preconcentration. Here, a vertical freezing double-jacket vessel (90 mm diameter and 160 mm height) is connected to a recirculation cooling system containing polyethylene glycol with temperature control ranging from −8 to −20 °C. To achieve and maintain progressive freeze concentration of skim milk, about 450 ml of sample is introduced to the vessel at an initial temperature of 4 °C and cooled down to a fixed freezing temperature. The sample is collected once the ice front reaches a predetermined thickness (Chabarov & Aider 2014).

A continuous multilayer freezer has also been designed, constructed and operated under different conditions by Zhang & Hartel (1996). It uses a series of chambers, mounted as layers in a column, with individual coolant jackets for controlling ice-crystal growth and heat removal. Mixing effects and specific gravity differences between ice and liquid concentrates generate a natural counter-current flow, providing suitable conditions for ice-crystal growth and ice–liquid separation. This system is also currently limited to pilot-plant scale and industrial-level equipment is yet to be developed.

As reported by Chang & Hartel (1997), the texture and flow properties of freeze-concentrated milk depend on the optimization of the ice-crystal growth rate, with preferably rapid rates of ice crystallization. High rates of heat removal from the ice-crystal growth tanks results in rapid ice-crystal growth (Shi *et al.* 1990). If the appropriate operating conditions are varied, high rates of heat transfer can be maintained, resulting in rapid rates of ice-crystal growth (Hartel & Espinel 1993).

Various investigations have been carried out on the concentration of milk by freezing, with a notable one conducted by the Commonwealth of Independent States to freeze milk with Freon 12 in a vertical cylindrical crystallizer. The maximum dry matter concentration was obtained in a five-effect freezing process. The water fraction of the last degree contained 0.94% dry matter, mainly fat. In some attempts with skim milk, a concentrate with 18.6% dry matter was obtained in a triple-effect procedure, and the losses caused by the absence of fat were markedly reduced.

In other investigations of milk concentration by freezing, a mathematical correlation of solids concentrate, total solids in initial raw material and freezing temperature was found. The authors stated that this concentration method can be used for concentrating up to 26.9% total solids in a triple-effect freezing procedure, resulting in concentrates

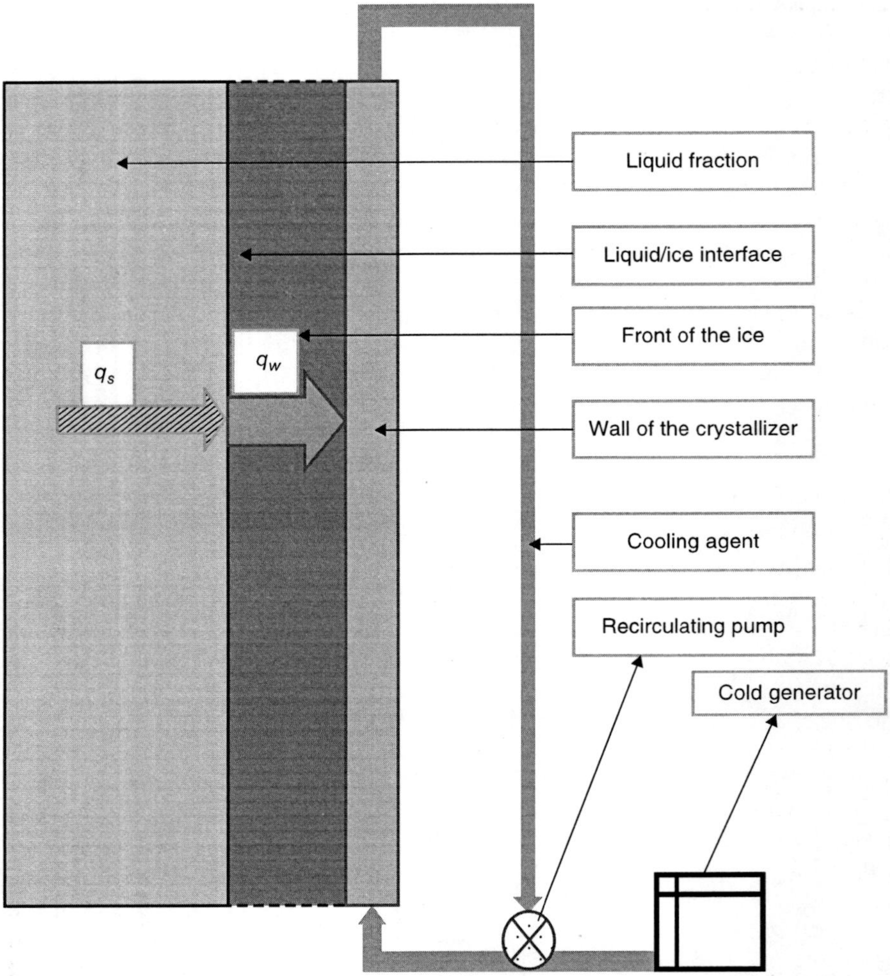

Figure 3.3 Pictorial representation of a unidirectional progressive freeze concentration process of skim milk. q_w, heat flux through the wall of the crystallizer; q_s, heat flux at the expense of the convective heat transfer from the boundary layer in W/m^2 (Chabarov & Aider 2014).

with better characteristics than those produced by vacuum evaporation (Shirai *et al.* 1987). However, this concentration method had no wide industrial application. According to Schwartzberg (1990), freeze concentration has the disadvantage of being an expensive process, current costs being three to four times greater than those for evaporation or reverse osmosis. The potential for microbial contamination is also high and the equipment requires careful cleaning.

3.5.2 Membrane separation techniques

Microfiltration, ultrafiltration, nanofiltration and reverse osmosis are membrane separation techniques that operate on the principle of hydrostatic pressure. All these

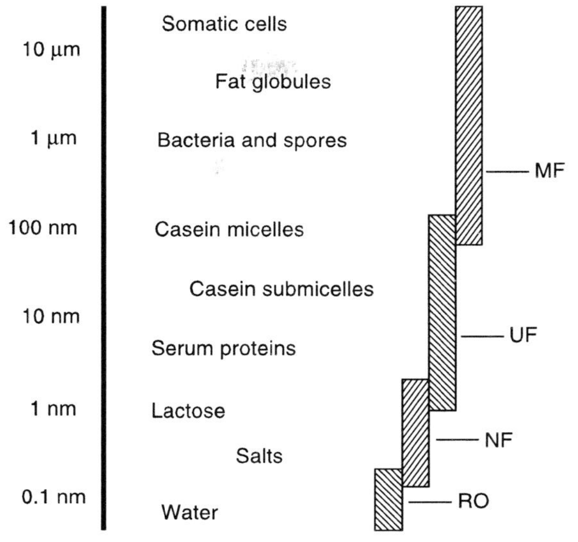

Figure 3.4 Various membrane separation techniques based on the size of major milk components. MF, microfiltration; UF, ultrafiltration; NF, nanofiltration; RO, reverse osmosis (Brans *et al.* 2004).

processes involve separation of fluid mixtures through selectively permeable membranes under a pressure gradient. Particles are separated based on a sieving effect and according to their dimensions (Rautenbach & Albrecht 1989). This concept is shown in Figure 3.4 using the dimensions of major dairy components (Brans *et al.* 2004). The physiological phenomenon responsible for this of type concentration is *osmosis*, which occurs in almost all living cells. In this process, liquids of two different concentrations (e.g. a solvent and a solution) are separated by a semi-permeable membrane, which is permeable to the solvent molecules. The solvent molecules pass from the less concentrated side to the side with higher concentration until the concentrations of dissolved substances are equal on both sides of the membrane. Application of pressure to the more concentrated side decreases the flow rate of the solvent molecules as the pressure increases, and when the applied pressure equals the osmotic pressure, the flow stops (Geankoplis 2003). Similarly, a vibrating membrane module was shown to improve the process efficiency (Al-Akoum *et al.* 2002). To prevent the formation of a stagnant gel layer on the retentate side, membranes are generally operated under a cross flow where the feed stream flows tangentially over the surface of the membrane to remove the rejected solutes from the membrane (Sirkar 1997).

As defined by Strathmann (1981), the *concentration factor* is the mode of expression of membrane filtration data, which is defined as the volume or weight reduction after concentration and is mathematically expressed as:

$$F_c = \frac{Q_m}{Q_r}$$

where F_c is the concentration factor, Q_m is the quantity of initial feed (kg) and Q_r is the quantity of retentate (kg).

A good membrane must have the following characteristics (Meares 1976):

a) Strict limit of separation: quantitatively separates molecules that are to be retained completely and allows others to pass through completely.

b) High flux with lowest possible hydrostatic pressure: leads to reduction in operation costs.
c) Good persistency: chemically, mechanically, thermally and bacteriologically stable.

One of the most important features determining the efficiency of membrane separation is the rate of passage of feed across the membrane, also known as the flux. *Flux* is a function of the rheological and physicochemical properties of the solutions being concentrated.

Membrane filtration units can also have a variety of modules in which the frames are arranged:

a) *Tubular modules:* A large-pore plastic or stainless steel pipe with diameter 20–25 mm holding an asymmetrical membrane is used, through which the process liquid is pumped at pressures of 3–10 bar. The tube holds another tube of 12.5 mm diameter membrane fitted to the internal surface of a synthetic fibre tube. The permeate leaves through the sides. The limitation of this module is its low packing density (Baker 2000).
b) *Plate module:* The flow through this is similar to that in a plate heat exchanger. It is built in a round or elliptical form and the membranes are inter-locked with carrier plates, which function as the inlet and outlet of the pipe. Easy replaceability of the membranes and the low height of the installations are the advantages of these systems (Brans *et al.* 2004).
c) *Spiral module:* A central perforated stainless-steel permeate-collecting pipe is present around which the membrane, the permeate-carrying layer, the separating layer (maintains the distance), generally a plastic mesh 1 mm thick, and the layer carrying the process liquid are stacked over each other and coiled up (Caric 1994). The process liquid maintains a laminar flow in an axial direction through the module, and permeate in the separating layer flows in a spiral fashion to the concentrically installed collecting pipe (Noble & Stern 1995).
d) *Capillary tube module:* With the advantage of having low hydrostatic pressure during operation this module has membranes arranged in a module with inner diameter 1.1 mm (Nath 2008). Standard dairy pumps can be used and cleaning is easily done by simple backflow or recirculation of cleaning liquid (Baker 2000).

3.5.2.1 Microfiltration

Microfiltration, like all other membrane separation processes, is a technique that allows differential concentration in the liquid retained by the membrane, called the microfiltration retentate, of components that are larger than the average pore diameter of the membrane (Saboyainsta & Maubois 2000). The hydrostatic pressure differences used are in the range 0.1–2 bar (0.01–0.2 MPa) (Spreer 1998). Particles like colloids, bacteria and suspended particles are allowed to pass though the membrane (Caric 1994). The pore diameter of microfiltration membranes ranges from 10 μm to 0.1 μm.

A basic microfiltration system comprises a feed tank, a feed pump and two recirculation loops, one for the microfiltration retentate, including one or a multiple of two modules, and a heat exchanger, and one for the microfiltration permeate (Figure 3.5a). Both loops have outlet valves regulated by sensors for continuous production of the two separated fluids. Microfiltration equipment must be started and used with care to avoid dramatic and fast fouling (Sirkar 1997). The microfiltration equipment is first filled with warm water (52 °C) with the air exhaust valve open for removing air bubbles. The hydraulic recirculation parameters which will be used on the milk are then adjusted on the recirculated water. Finally, milk for which the physicochemical equilibrium has been re-established by preheating at 50 °C for 20 min is progressively introduced in to the microfiltration retentate loop. The hydraulic parameters are readjusted during the first 10 min before running in stationary conditions (Goudédranche *et al.* 2000). This produces microfiltered milk, which can be further concentrated by evaporation and dried. Depending on the pore size of the membrane, whey concentrates can be obtained, which can be further fractionated to produce whey isolates and dried into powder (Pafylias 1996). One disadvantage of microfiltration is fouling of the membranes, which can be overcome by development of uniform and low transmembrane pressure (UTP). This breakthrough was brought about by the development of ceramic membranes with a multichannel geometry and a highly permeable support (Vadi & Rizvi 2001).

Cross-flow microfiltration (Figure 3.5b) is prominently used in the dairy industry for three main purposes: (i) removal of bacteria (Malmberg & Holm 1998; Trouvé *et al.* 1991), (ii) whey defatting (Pierre *et al.* 1994) and (iii) micellar casein enrichment in cheese making (Fauquant *et al.* 1988). Other applications, such as the removal of microorganisms from skim milk in order to produce a debacterized raw material, which is later transformed into fluid milks, cheeses or long storage dairy products such as powder or protein derivatives, are also possible (Grappin & Beuvier 1997).

3.5.2.2 Ultrafiltration

This is a separation process in which a pressure drop is created across a membrane with small pores, through which water and dissolved molecules are diffused, leaving colloidal and larger molecules in the concentrated product. Cellulose acetate membranes are used for simple separation processes which are cost-effective but have limited mechanical, thermal, chemical and functional properties with pH range 3–8 and temperature range 40–55 °C. Other membranes from synthetic materials such as polyamide, polysulfone and polyacrylinitryl/polyvinylchloride are more expensive but they have better characteristics, for example pH range 1–13, temperature range 85–100 °C and a wider utilization spectrum (Cheryan 1986). According to the way in which the layers are arranged, the membranes are classified into two categories: *symmetrical*, in which the filter layer goes through the entire cross-section and the products to be separated are retained only in the core of the membrane (also known as *depth filtration*), and *asymmetrical*, which has a porous frame structure and a flow rate inversely proportional to the membrane thickness, thus a higher flux (Maubois & Mocquot 1975).

Ultrafiltration membranes retain all the fat and almost all the protein in milk. Interestingly, protein plays a crucial role in controlling the rate of ultrafiltration and the retention coefficients of the non-protein nitrogen compounds are 20–40%, which are higher for high-concentration factors (Grandison & Glover 1994). The minerals and other ions retained during ultrafiltration of milk are those that are attached to the proteins, like calcium, magnesium, phosphate and citrate, while others pass into permeate. In a similar way, fat-soluble and protein-bound vitamins are completely retained. During 3× (the most commonly used concentration factor in the dairy industry) ultrafiltration of milk, 66–67% of the lactose passes to the permeate and 33–34% is retained in the retentate (Fox 2009). By combining ultrafiltration with a process known as diafiltration, where the

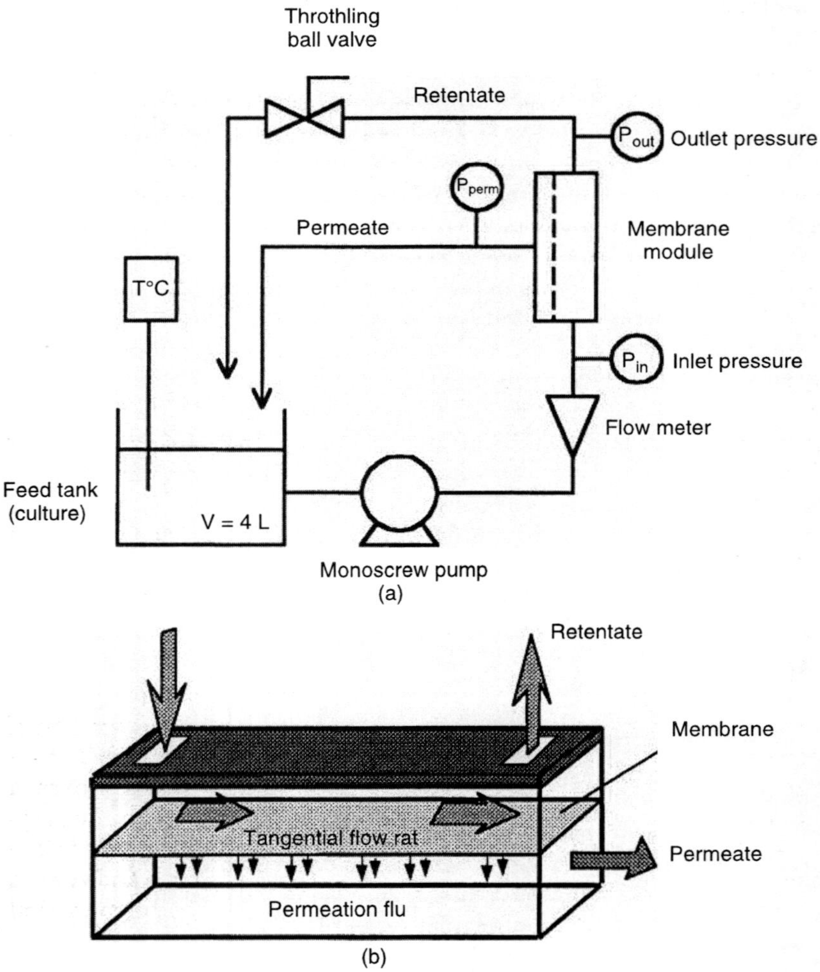

Figure 3.5 Illustration of (a) a pilot-scale microfiltration system and (b) cross-flow filtration in a flat membrane module (Rossignol *et al.* 1999).

milk or the ultrafiltration concentrate is diluted with water to allow washing out of components that are able to pass through the membranes, a concentrated variety of milk rich in protein (75% more than cow's milk) and 100% lactose-free with sweetness similar to milk has been produced (Jelen & Tossavainen 2003; Ur-Rehman *et al.* 2006; Dunker *et al.* 2007). Apart from producing lactose-free milk powder for the lactose-intolerant population, ultrafiltration can also be used to produce milk powders with very low sandiness due to the low level of lactose. (Sandiness is a condition observed in condensed milk when large, hard lactose particles are present. These particles reduce the smoothness and homogeneity of condensed milk and make them sandy, grainy or gritty in texture, which is unpleasant to the palate.) The colour of the milk powder also improves because reduction in lactose leads to less browning (Mistry & Hassan 1991).

3.5.2.3 Reverse osmosis

Also known as hyperfiltration, reverse osmosis works on the same principle as ultrafiltration. However, the membranes used here are semi-permeable, asymmetric and have smaller pores. Dissolved components like salt and lactose are retained at a level of 80–90% and only water molecules are allowed to pass as permeate (Spreer 1998). Reverse osmosis is widely used in the dairy sector because of its low cost compared to evaporation because of its lower energy requirements. Furthermore, there is no heat damage to the product and less fresh water is required. The waste produced requires less treatment due to low volume and hence is less expensive. Reverse osmosis requires relatively little floor space and capital investment. Considering all the advantages, reverse osmosis potentially increases the profit margins for new products (Glover 1971). However, the disadvantages are that it might be expensive and time-consuming to document product safety and to obtain the approval of new membranes. Fouling is also a very common problem with reverse osmosis of milk (Goosen *et al.* 2005).

The performance of reverse osmosis in the concentration of milk is limited by osmotic pressure, which is 30–40 bar in most commercial systems. This limits the concentration of milk to a factor of three to four. For production of skimmed or whole milk powder, reverse osmosis is used to concentrate milk to 45–50% total solids before spray drying. It is usually used as a preconcentration method before evaporation to reduce operational costs or to increase the capacity of existing plant (Sutherland & Chase 2011).

3.6 Conclusion

The preconcentration process has a crucial role in reducing the energy consumption of the drying process and also helps to impart desirable characteristics to the dried product, notably bulk density. With technological advancements, the dairy industry has seen the advent of numerous methodologies for concentrating milk prior to drying by evaporation, with an array of evaporators to choose from, freeze concentration or sophisticated membrane separation techniques like microfiltration, ultrafiltration and reverse osmosis. Careful selection of the preconcentration method can yield both economical and technological advantages.

References

Aceto, N.C., Sinnamon, H.I., Schoppet, E.F. & Eskew, R.K. (1962). Continuous vacuum drying of whole milk foam. *Journal of Dairy Science*, **45(4)**, 501–507.

Al-Akoum, O., Ding, L.H. & Jaffrin, M.Y. (2002). Microfiltration and ultrafiltration of UHT skim milk with a vibrating membrane module. *Separation and Purification Technology*, **28(3)**, 219–234.

Baker, R.W. (2000). *Membrane Technology*, John Wiley & Sons, Inc.

Bienvenue, A., Jimenez-Flores, R. & Singh, H. (2003). Rheological properties of concentrated skim milk: importance of soluble minerals in the changes in viscosity during storage. *Journal of Dairy Science*, **86(12)**, 3813–3821.

Brans, G.B.P.W., Schroën, C.G.P.H., Van der Sman, R.G.M. & Boom, R.M. (2004). Membrane fractionation of milk: state of the art and challenges. *Journal of Membrane Science*, **243(1)**, 263–272.

Caric, M. (1994). *Concentrated and Dried Dairy Products*, VCH.

Carl, J.L. & Jenssen, S. (1957). US Patent No. 2,796,120, US Patent and Trademark Office, Washington, DC.

Chabarov, A. & Aider, M. (2014). Mathematical modeling and experimental validation of the mass transfer during unidirectional progressive cryoconcentration of skim milk. *Innovative Food Science & Emerging Technologies*, **21**, 151–159.

Chang, Y.H. & Hartel, R.W. (1997). Flow properties of freeze-concentrated skim milk. *Journal of Food Engineering*, **31**(**3**), 375–386.

Chen, H. & Jebson, R.S. (1997). Factors affecting heat transfer in falling film evaporators. *Food and Bioproducts Processing*, **75**(**2**), 111–116.

Cheryan, M. (1986). *Ultrafiltration Handbook*, Technomic Publishing Co. Inc.

Delyannis, A. & Delyannis, E.E. (1980). Horizontal tube evaporators. In: *Seawater and Desalting*, pp. 52–53. Springer, Berlin/Heidelberg/New York.

Dodeja, A.K., Sarma, S.C. & Abichandani, H. (1990). Heat transfer during evaporation of milk to high solids in thin film scraped surface heat exchanger. *Journal of Food Process Engineering*, **12**(**3**), 211–225.

Doremus, R.O. (1895). The adulteration of milk, *Medico-Legal Journal*, **13**, 140.

Dunker, J.M., McCloskey, M.J. & Gomez, T.J. (2007). US Patent No. 7,169,428, US Patent and Trademark Office, Washington, DC.

Early, R. (1998). Milk concentrates and milk powders. In: *The Technology of Dairy Products*, 2nd edn (ed. Early, R.), pp. 228–300, Blackie Academic and Professional, Glasgow.

Ernst, W. (1914). *Text Book of Milk Hygiene*, A. Eger.

Fauquant, J., Maubois, J.L. & Pierre, A. (1988). Microfiltration of milk using a mineral membrane. *Technologie Lait Marketing*, **1028**, 21–23.

Fox, P.F. (2001). Milk proteins as food ingredients. *International Journal of Dairy Technology*, **54**(**2**), 41–55.

Fox, P.F. (2009). Lactose: Chemistry and properties. In: *Advanced Dairy Chemistry, Vol. 3: Lactose, Water, Salts and Minor Constituents*, 3rd edn (eds McSweeney, P. & Fox, P.F), pp. 1–15, Springer, New York.

Geankoplis, C.J. (2003). *Transport Processes and Separation Process Principles*, Prentice Hall Professional Technical Reference.

Glover, F.A. (1971). Concentration of milk by ultrafiltration and reverse osmosis. *Journal of Dairy Research*, **38**(**03**), 373–379.

Goosen, M.F.A., Sablani, S.S., Al-Hinai, H., Al-Obeidani, S., Al-Belushi, R. & Jackson, D. (2005). Fouling of reverse osmosis and ultrafiltration membranes: a critical review. *Separation Science and Technology*, **39**(**10**), 2261–2297.

Goudédranche, H., Fauquant, J. & Maubois, J.L. (2000). Fractionation of globular milk fat by membrane microfiltration. *Le Lait*, **80**(**1**), 93–98.

Grandison, A.S. & Glover, F.A. (1994). Membrane processing of milk. In: *Robinson: Modern Dairy Technology, Vol. 1: Advances in Milk Processing*, 2nd edn (ed. Robinson, R.K.), pp. 273–311, Springer.

Grappin, R. & Beuvier, E. (1997). Possible implications of milk pasteurization on the manufacture and sensory quality of ripened cheese. *International Dairy Journal*, **7**(**12**), 751–761.

Hartel, R.W. & Espinel, L.A. (1993). Freeze concentration of skim milk. *Journal of Food Engineering*, **20**(**2**), 101–120.

Hewitt, G.F., Shires, G.L. & Bott, T.R. (1994). *Process Heat Transfer, Vol. 113*, CRC Press, FL.

Hoffman, P. (2004). Plate evaporators in the food industry: theory and practice. *Journal of Food Engineering*, **61(4)**, 515–520.
Hunziker, O.F. (1920). *Condensed Milk and Milk Powder*. La Grange, IL.
Janosfia, P. & Buchi, W. (1966). US Patent No. 3,292,683, US Patent and Trademark Office, Washington, DC.
Jelen, P. & Tossavainen, O. (2003). Low lactose and lactose-free milk and dairy products-prospects, technologies and applications. *Australian Journal of Dairy Technology*, **58(2)**, 161.
Jeurnink, T.J. & Brinkman, D.W. (1994). The cleaning of heat exchangers and evaporators after processing milk or whey. *International Dairy Journal*, **4(4)**, 347–368.
Kakac, S. (1991). *Boilers, Evaporators, and Condensers*. John Wiley & Sons.
Karimi, M., Jahanmiri, A. & Azarmi, M. (2007). Inferential cascade control of multi-effect falling-film evaporator. *Food Control*, **18(9)**, 1036–1042.
Kessler, H.G. (1981). Food engineering and dairy technology. *International Journal of Dairy Technology*, **34**, 171.
Knipschildt, M.E. (1986). Drying of milk and milk products. *Modern Dairy Technology*, **1**, 131–234.
Komori, S., Takata, K. & Murakami, Y. (1988). Flow structure and mixing mechanism in an agitated thin-film evaporator. *Journal of Chemical Engineering of Japan*, **21(6)**, 639–644.
Malmberg, R. & Holm, S. (1988). Low bacteria skim milk by microfiltration. *North European Food and Dairy Journal*, **54**, 75–78.
Maubois, J.L. & Mocquot, G. (1975). Application of membrane ultrafiltration to preparation of various types of cheese. *Journal of Dairy Science*, **58**(7), 1001–1007.
McCabe, W.L., Smith, J.C. & Harriott, P. (1993). *Unit Operations of Chemical Engineering, Vol. 5*, p. 154, McGraw-Hill, New York.
Meares, P. (1976). *Membrane Separation Processes*, Elsevier Scientific Publishing Co.
Mersmann, A., Kind, M. & Stichlmair, J. (2011). *Thermal Separation Technology, Vol. 12*, Springer, Heidelberg.
Mistry, V.V. & Hassan, H.N. (1991). Delactosed, high milk protein powder. 1. Manufacture and composition. *Journal of Dairy Science*, **74(4)**, 1163–1169.
Morr, C.V. & Ha, E.Y.W. (1993). Whey protein concentrates and isolates: processing and functional properties. *Critical Reviews in Food Science & Nutrition*, **33(6)**, 431–476.
Nath, K. (2008). *Membrane Separation Processes*, PHI Learning Pvt. Ltd., New Delhi.
Noble, R.D. & Stern, S.A. (eds) (1995). *Membrane Separations Technology: Principles and Applications, Vol. 2*, Elsevier, Amsterdam.
Pafylias, I., Cheryan, M., Mehaia, M.A. & Saglam, N. (1996). Microfiltration of milk with ceramic membranes. *Food Research International*, **29(2)**, 141–146.
Paramalingam, S., Winchester, J. & Marsh, C. (2000). On the fouling of falling film evaporators due to film break-up. *Food and Bioproducts Processing*, **78(2)**, 79–84.
Pierre, A., Le Graet, Y., Daufin, G., Michel, F. & Gésan, G. (1994). Whey microfiltration performance: influence of protein concentration by ultrafiltration and of physicochemical pretreatment. *Le Lait*, **74(1)**, 65–77.
Písecký, J. (1981). Technology of skimmed milk drying. *International Journal of Dairy Technology*, **34(2)**, 57–62.
Ranken, M., Kill, R. & Baker, C. (1997) *Food Industries Manual. Great Britain*, Chapman & Hall.

Rautenbach, R. & Albrecht, R. (1989). *Membrane Separation Processes,* John Wiley and Sons Inc., New York.

Ribeiro, C.P. & Andrade, M.C. (2002). A heat transfer model for the steady-state simulation of climbing-falling-film plate evaporators. *Journal of Food Engineering,* **54(4)**, 309–320.

Robinson, R.K. (ed.) (2012). *Modern Dairy Technology, Vol. 2: Advances in Milk Products,* Springer Science & Business Media.

Rossignol, N., Vandanjon, L., Jaouen, P. & Quemeneur, F. (1999). Membrane technology for the continuous separation microalgae/culture medium: compared performances of cross-flow microfiltration and ultrafiltration. *Aquacultural Engineering,* **20(3)**, 191–208.

Saboyainsta, L.V. & Maubois, J.L. (2000). Current developments of microfiltration technology in the dairy industry. *Le Lait,* **80(6)**, 541–553.

Sánchez, J., Hernández, E., Auleda, J.M. & Raventós, M. (2011). Review: freeze concentration technology applied to dairy products. *Food Science and Technology International,* **17(1)**, 5–13.

Saravacos, G.D. & Kostaropoulos, A.E. (2002). *Handbook of Food Processing Equipment,* Springer Science & Business Media.

Schwartzberg, H.G. (1990). Food freeze concentration. In: *Biotechnology and Food Process Engineering* (eds Schwartzberg, H.G. & Rao, M.A.), pp. 127–202, Marcel Dekker Inc., New York.

Shi, Y., Liang, B. & Hartel, R.W. (1990). Crystallization of ice from aqueous solutions in suspension crystallizers. In: *Crystallization as a Separations Process* (eds Myerson, A.S. & Toyokura, K.), pp. 316–328, American Chemical Society, Washington, DC.

Shirai, Y., Sugimoto, T., Hashimoto, M., Nakanishi, K. & Matsuno, R. (1987). Mechanism of ice growth in a batch crystallizer with an external cooler for freeze concentration. *Agricultural and Biological Chemistry,* **51(9)**, 2359–2366.

Sirkar, K.K. (1997). Membrane separation technologies: current developments. *Chemical Engineering Communications,* **157(1)**, 145–184.

Spreer, E. (1998). *Milk and Dairy Product Technology, Vol. 83,* Marcel Dekker, Inc., New York.

Strathmann, H. (1981). Membrane separation processes. *Journal of Membrane Science,* **9(1)**, 121–189.

Sutherland, K.S. & Chase, G. (2011). *Filters and Filtration Handbook,* Elsevier.

Toledo, R.T. (2007). Evaporation. In: *Fundamentals of Food Process Engineering,* Springer Science & Business Media.

Trouvé, E., Maubois, J.L., Piot, M., Madec, M.N., Fauquant, J., Rouault, A. & Brinkman, G. (1991). Rétention de différentes espèces microbiennes lors de l'épuration du lait par microfiltration en flux tangentiel. *Le Lait,* **71(1)**, 1–13.

Ur-Rehman, S., Dunker, J., McCloskey, M., Gomez, T. & Seguin, R. (2006). Dairy compositions and method of making, US Patent Application No. 11/641,468, US Patent and Trademark Office, Washington, DC.

Vadi, P.K. & Rizvi, S.S.H. (2001). Experimental evaluation of a uniform transmembrane pressure crossflow microfiltration unit for the concentration of micellar casein from skim milk. *Journal of Membrane Science,* **189(1)**, 69–82.

Van Mil, P.J.J.M. & Bouman, S. (1990). Freeze concentration of dairy products. *Netherlands Milk and Dairy Journal,* **44(1)**, 21–31.

Varnam, A. & Sutherland, J.P. (2001). *Milk and Milk Products: Technology, Chemistry and Microbiology, Vol. 1,* Springer Science & Business Media.

Wattelet, J.P. (1994). *Heat Transfer Flow Regimes of Refrigerants in a Horizontal-Tube Evaporator*, Air Conditioning and Refrigeration Center, College of Engineering, University of Illinois at Urbana-Champaign.
Westergaard, V. (2004). *Milk Powder Technology: Evaporation and Spray Drying*, Niro A/S.
Winchester, J. & Marsh, C. (1999). Dynamics and control of falling film evaporators with mechanical vapour recompression. In: *Proceedings of the 1999 American Control Conference, Vol. 5*, pp. 3600–3604, IEEE.
Zhang, Z. & Hartel, R.W. (1996). A multilayer freezer for freeze concentration of liquid milk. *Journal of Food Engineering*, **29**(**1**), 23–38.

4

Drum Drying

P. Karthik[1], *N. Chhanwal*[1] *and C. Anandharamakrishnan*[1,2]

[1] *Department of Food Engineering, CSIR – Central Food Technological Research Institute, Mysore, Karnataka, India*
[2] *Indian Institute of Crop Processing Technology, Ministry of Food Processing Industries, Government of India, Thanjavur, Tamil Nadu, India*

4.1 Introduction

Drying is one of the food preservation techniques applied in the preservation of fruits, vegetables, grains, milk products and nutritive health-based products. Drying restricts the growth of microorganisms and prevents physiochemical changes during storage. Numerous types of dryer are available based on the raw material and final product applications. It is vital to understand the physiochemical properties of food products before selecting a dryer. The design of dryers differs with the nature of the product. Different types of dryer, such as tray dryers, spray dryers, freeze dryers, drum or roller dryers, microwave dryers, etc., are widely used in the food industry. Among these, drum dryers and spray dryers are largely used for milk and milk-based food products, weaning food, instant soup and health mixes, etc. The drum dryer is the oldest and best-established energy-efficient dryer for continuous processing operations. Compared to spray drying, drum drying needs less space and is more economical for producing less product. Although the freeze-drying process results in a higher yield of dried milk powder, it is commercially impractical owing to its high operating cost (Winder & Kielsmeier 1959).

Drum dryers, also known as roller dryers, were developed well before spray dryers in the early 1900s for the drying of liquid food materials (Tang *et al.* 2003). It should be noted that drum and rotary drum dryers are entirely different and the terminologies should not be used interchangeably as these dryers are entirely different in configuration, operation and applications. Rotary drum dryers are used for solid feed such as wet cloth and minerals whereas drum dryers are employed for viscous liquids. Rotary drum dryers use hot air whereas drum dryers use steam for the drying operation (Bhandari *et al.* 2013). Drum dryers use several features to dry a wide range of heavy pastes and thick liquids in the food, agriculture and chemical industries, for example precooked breakfast cereals, dry soup mixtures, weaning foods, milk products, yeast, fruit purees, polyacrylamides, sodium benzoate, acetates and many other chemical and food-based products (Moore 1995). Furthermore, drum drying has other advantages, such as low cost, compatibility with gelatinized food materials and efficient drying of highly viscous liquid foods. Drum drying is a continuous process in which the drums are heated by steam and simultaneously rotated at a constant speed to dry the product. Conduction heating governs the heat transfer during drum drying, as the product is in contact with the drum dryer for 5–30 s depending on the drum rotation speed. The feed material in liquid or paste form is applied to the rotating drum, either single or double, and the dried product is scrapped off by knife blades in the form of a powder or film (Tang *et al.* 2003; Daud 2014). Drum drying imparts unique characteristics in pregelatinized instant foods because the starch present in feed slurry gelatinizes on contact with the hot drum before the entire sheet dries. Moreover, due to the presence of a high heat gradient between the food material and the hot drum (150–180 °C), 'boiling-like' drying occurs which imparts a porous structure to the dried product due to the *bubble-formation* phenomenon within the sheet (Daud 2014). Drum drying is preferable over other drying methods for cold-water-soluble products such as baby foods, instant foods, starches and breakfast cereals.

Several patents have been filed on the drum drying of milk and its related products (Campbell 1917; Glen 1929; Peterson 1930; Robert & Lester 1968; Buter & Wiepking 2010). These patents depict the series of improvements that have taken place pertaining to the drum-drying process since its inception (Glen 1929; Balch 1975; David 1968; Duncan 1933). In this chapter, the application of drum drying to the drying of milk products is discussed. The chapter mainly focuses on drum drying operation, types of drum dryers and applications of drum drying in the production of whole milk and non-fat milk powders, and dairy-based infant food ingredients.

4.2 Drum-drying process

Drying is a unit operation where simultaneous heat and mass transfer occur to evaporate the moisture content in the food product. The heat of vaporization is supplied to the food product to vaporize the bound moisture. The vapour pressure gradient of air between the liquid water surface and the product surface acts as the driving force for moisture vapour diffusion. The thermal conductivity of the product limits the internal heat transfer, whereas molecular diffusion influences the mass transfer during drying.

Operation of the drum dryer involves applying a thin layer of liquid feed onto the hot rotating drum. The feed material is applied/supplied to the drum by dipping, splashing or spraying. During drum drying, heat is provided in the form of steam to the drum, which generates the heat of vaporization required to evaporate the water molecules from the product. The steam offers uniform temperature distribution throughout the drum and yields a consistently high-quality product. However, the condensed steam inside the drum has to be continuously removed to provide more surface area for heat transfer through the steam. During the drum-drying process the feed or product does not make contact with the steam or condensate due to the closed steam system. Based on the final product characteristics, the product is removed as a film or powder from the rotating drum surface with the doctor blade or knife that is fixed tightly to the rotating drum. The desired product after the drum drying could be powder, granules, sheets, flakes or film.

Drum dryers consist of either single or double drums depending on the concentration of feed material. For example, a higher degree of preconcentrated milk can be dried by single-drum dryer than by a double-drum dryer. After the drying process, the product has to be milled to break down the film. The microstructure of milled product varies from an irregular solid to a flaky structure (Knipschildt & Andersen 1994). The energy efficiency of a drum dryer can vary from 70% to 90% depending on the steam consumption (1.2–1.5 kg per kg of water evaporation) (Tang *et al.* 2003; Tang & Yang 2004). Factors influencing the drum-drying process are feed concentration, feed level in the tank, initial product temperature, solubility, moisture content, steam pressure and the rotation speed of the drum (Combs & Hubbard 1932). Drum drying is mainly applied to cooked starch or gelatinized starch slurry products such as potato, rice, maize, banana, soybean-banana, wheat, corn, etc. These precooked or pregelatinized dried products are mainly used in instant foods. In addition, non-sugary and non-starchy products like coffee, milk, tomato puree and beef broth are also subjected to drum drying.

Heat transfer during drum drying occurs mainly in three stages: an initial heating period, a constant product temperature period and finally a rising product temperature period. Because of the large temperature gradient present between the hot drum surface and the applied feed material, rapid heat transfer occurs, resulting in the initial rapid heating of wet product. The product temperature reaches the boiling point of free water. This marks the end of first heating stage and the beginning of the second. Once the product temperature reaches the boiling point of free water, free water evaporates and this keeps the wet product/feed temperature constant. Consequently, the drum surface temperature falls due to the evaporative cooling during the second stage, although the energy supplied by the steam remains same. On complete evaporation of free water from the product, due to the continuous heat supply by steam the drum surface temperature increases and the product temperature rises, which constitutes the third and final stage of the process. At this stage, bound water controls the heat transfer and drying rate of the product. Once the product is removed with the doctor's blade, the drum surface temperature increases to transfer the heat to the wet product.

The evaporation rate during the drum-drying process is estimated using Eq. (4.1) (Tang *et al.* 2003)

$$\frac{dM}{dt} = 30.94V^{0.8}\Delta P \tag{4.1}$$

where dM/dt is the rate of moisture removal per unit of drum surface (kg H_2O/h m^2), V is the velocity of ambient air (m/s) and ΔP is the difference between vapour pressure at the product surface and that in the ambient air.

The moisture removal rate can also be estimated based on the energy balance by Eq. (4.2)

$$\frac{dM}{dt} = 3.6\frac{h(T_D - T_P)}{L} \tag{4.2}$$

where T_D is the drum surface temperature (°C), T_P is the evaporating product temperature (°C), h is the overall heat-transfer coefficient (which ranges between 200 and 2000 W/m^2.°C) and L is the latent heat (kJ/kg H_2O).

4.2.1 Effect of operating parameters on product quality and the capacity of the drum dryer

When the influence of feed material (milk) temperature on drying rate was evaluated on cold skim milk (10–15.6 °C) and preheated skim milk (51.7–85 °C), the latter (85 °C) showed a decrease in processing time and an 18% increase in production rate compared to the cold milk. Steam pressure and the rotational speed of the drums also influence the product characteristics and capacity of the drum-drying operation. Drum-drying at low pressure and high drum rotation speed yields a wet product at a low production rate. Preconcentration of milk influences the operational capacity of the drum dryer. It is found that with preconcentrated milk, a drum dryer can be operated at higher speed. The production capacity of dried milk powder can be increased with higher drum speed and steam pressure (Combs & Hubbard 1932). The drying capacity of the drum-drying operation is influenced by the quantity or level of skim milk in the pool used for the drum. A lower level of skim milk in the pool results in low production and *vice versa*. Interestingly, a difference of almost £40/hour was observed between the lower and higher levels of the pool (Combs & Hubbard 1932).

4.3 Types of drum dryers

Drum dryers are classified based on the number and arrangement of drums (single, double, twin) and pressure (atmospheric or vacuum). Based on the product characteristics and drying requirement, the type of dryer is selected. The length of drum dryers varies from 1 to 3 m, thickness of the drum wall ranges from 0.02 to 0.04 m and the diameter in the range 0.4–1.5 m.

4.3.1 Single-drum dryers

Figure 4.1 shows a drum dryer containing a single drum that is used in the food industry. It consists of a rotating drum, an applicator roller to spread liquid feed evenly, a doctor blade or knife to remove the dried product and a conveyor to transfer dried product from the dryer to the filling section. Applicator rolls have a kneading effect, which prevents the formation of lumps in sticky products. A knife adjuster attached to the knife controls thickness of the spread film.

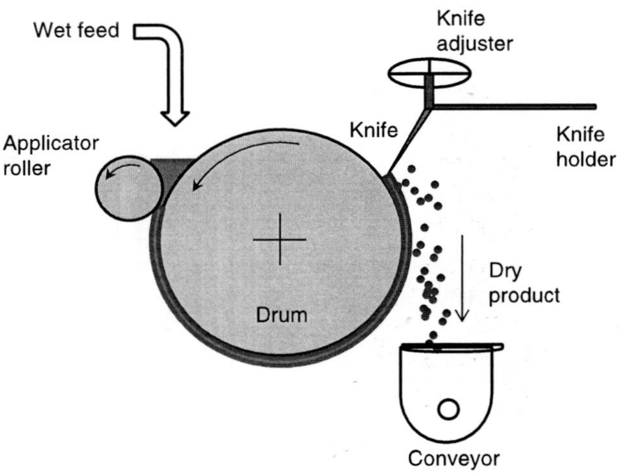

Figure 4.1 Single-drum dryer.

4.3.2 Double-drum dryers

The double-drum dryer contains two horizontal rotating cylinders with equal diameter and length rotating in opposite directions (Figure 4.2). The purpose of these hollow cylinders is to transport the saturated steam for heat generation (Gavrielidou *et al.* 2002).

The distance between two drums can be tuned appropriately with respect to film layer thickness. The knives are fixed individually on the two drums to collect the dried product, and move through conveyor on both drums. A double-drum dryer is often used for feed with lower viscosities. The liquid feed is fed into the sump between the two drums.

4.3.3 Twin-drum dryers

A twin-drum dryer looks similar to a double-drum dryer, but it differs in the distance between the two drums. The drums rotate away from each other (Figure 4.3). The wet

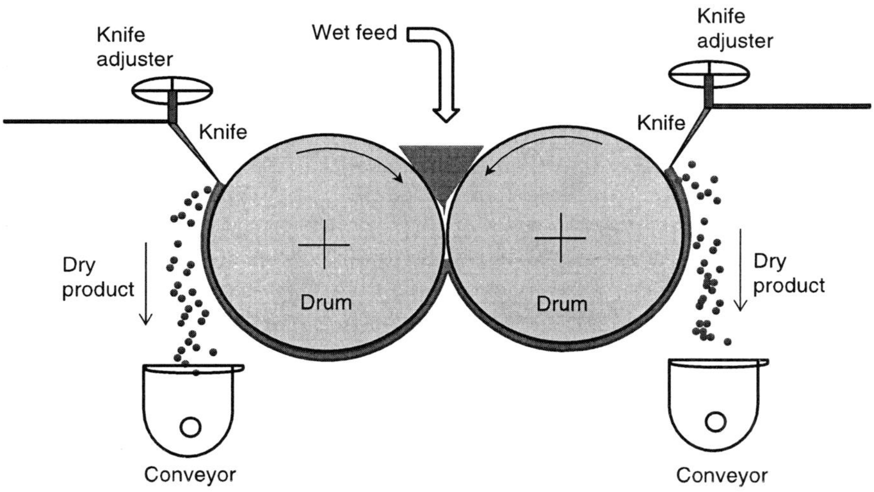

Figure 4.2 Double-drum dryer.

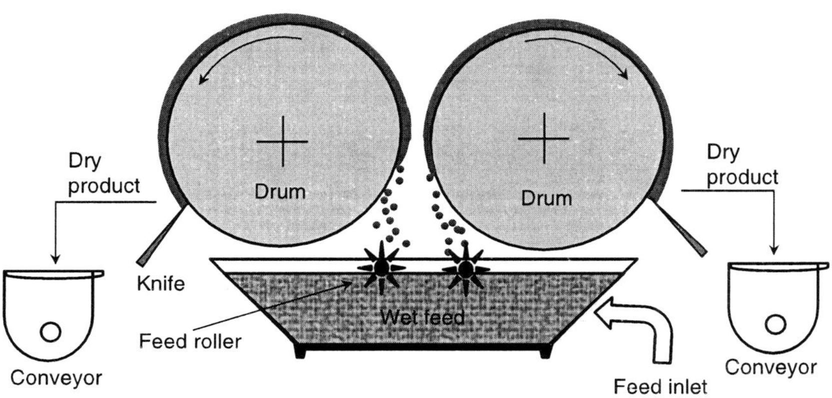

Figure 4.3 Twin-drum dryer.

feed is supplied from the bottom of the drum through splash using the feed roller. After drying, the product is removed by knives attached nearer to the rotating drum. A twin-drum dryer is used for product that turns dusty post drying, such as salt solutions. A twin-drum dryer can be used as auxiliary dryer to carry out the predrying process for other dryers.

4.3.4 Vacuum-drum dryers

Vacuum drying is a low-temperature drying process that operates at lower pressure and temperature compared to standard atmospheric drum drying (Figure 4.4). The dryer is surrounded by an airtight enclosure under vacuum and contains a condenser, a scrubber and a vacuum pump (Okos *et al.* 1992; Moore 1995).

Vacuum drum dryers are mainly used for heat-sensitive food products such as milk products, protein powder, enzymes, vitamins and highly hygroscopic products (Brown *et al.* 1973). Moreover, applying a vacuum during drum drying means the structural and functional properties of dairy ingredients can be retained significantly compared to regular drum drying (Laurent *et al.* 1999). The vacuum drum dryer provides better sustainability and energy-efficient drying, which helps in reducing the operating cost.

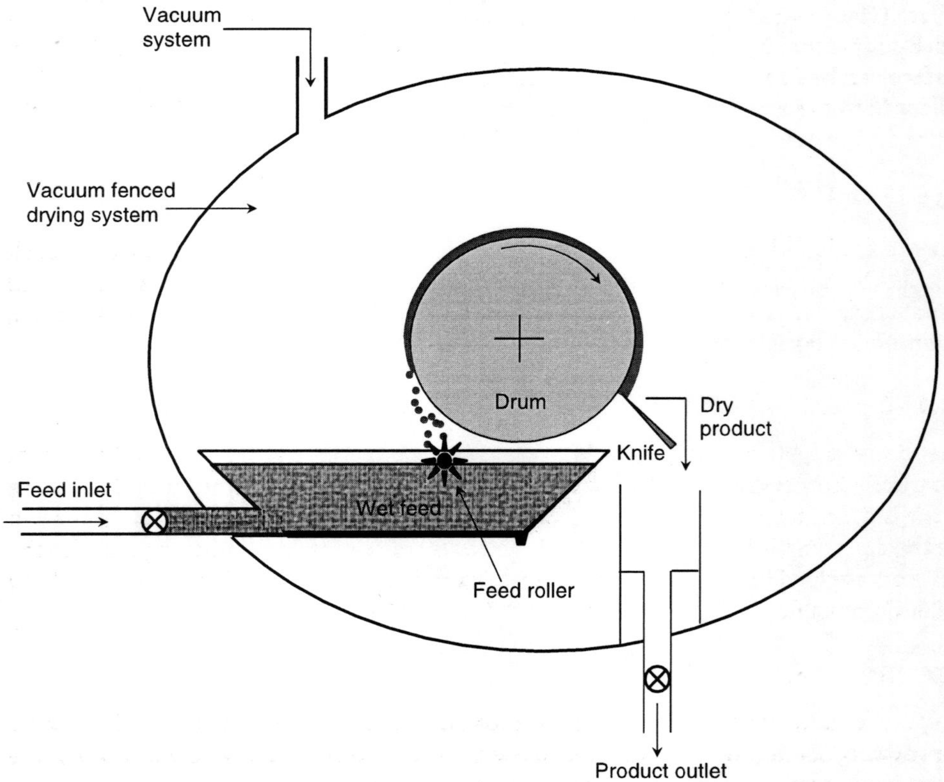

Figure 4.4 Vacuum-drum dryer.

4.3.5 Enclosed-drum dryers

As the name suggests, enclosed-drum dryers are placed in a vapour-tight enclosure. Enclosed-drum dryers are mainly used for drying operations that involve evaporation of solvents other than water or that result in dust particles. Solvents are recovered or removed using an appropriate condenser and dust particles are removed by a wet scrubber (Moore 1995).

4.4 Classification of the feeding method

Based on the physical properties of feed materials such as viscosity, solid content and wettability, the method of feeding onto the drum surface will vary. Feeding methods widely used in drum dryers are single-roll feeding, multiple-roll feeding, nip feeding, dip feeding, spray feeding and splash feeding. These feeding methods are shown in Figure 4.5.

4.4.1 Single- and multiple-roll feed

In these feeding methods the wet feed is supplied from the nip between the single applicator roller and drum surface (Figure 4.5a). The multiple-roll feeder uses more applicator rollers (Figure 4.5b). Both these systems can be used in single- and double-drum dryers. By adjusting the rollers individually, the distance between each roller and the drum surface can be fixed. A thin sheet can be obtained from a single applicator and multiple rollers increase the film thickness. This applicator feed method is mainly used in the drum drying of viscous and glutinous products (Tang *et al.* 2003).

4.4.2 Nip feed

The nip feed method is used in double-drum dryers for drying low-viscous fluids such as milk (Figure 4.5c). In this feed application method the film thickness is maintained by adjusting the gap between the two drums. The wet feed can be supplied from the top, and the pool level is maintained between the cylindrical drums.

4.4.3 Dip feed

The dip feed method can be used in single- or twin-drum dryers. The rotating drum is partially immersed in the feed material pool to apply the wet material to the surface of the hot drum (Figure 4.5d). The factors that influence the feed coating on the drum are the material viscosity, the density, the surface tension, the adhesive properties of the wet feed, angle of immersion and the revolving speed of the drum (Campanella & Cerro 1984; Campanella *et al.* 1986; Weinstein & Ruschak 2004).

4.4.4 Spray feed

In spray feeding the feed solution is atomized through a nozzle onto the surface of the hot rotating drum (Figure 4.5e). The nozzle is fixed in the feed tray at the bottom of the drum in the perpendicular direction. The coating thickness of the feed material can be varied by varying the spray level using a control system. The spray feeding method is not

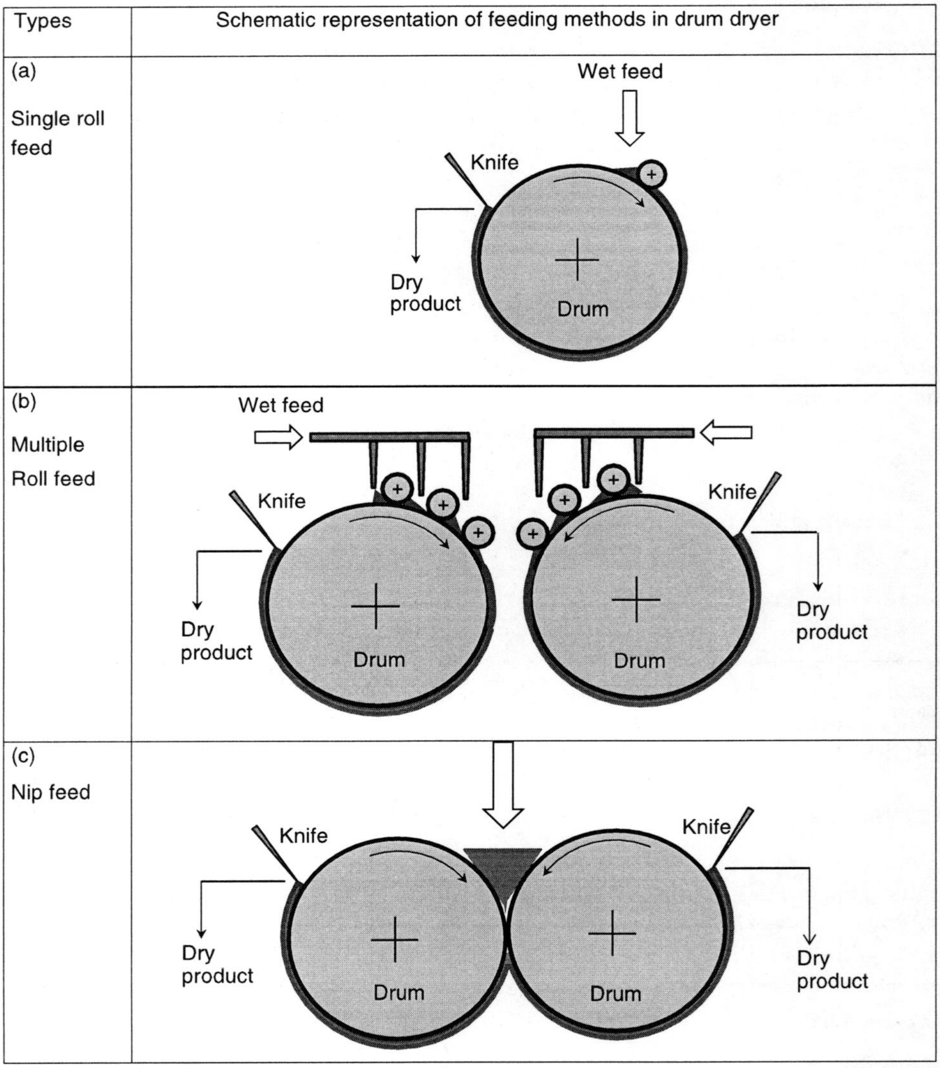

Figure 4.5 Different feeding methods in a drum dryer: (a) single-roll feed, (b) multiple-roll feed, (c) nip feed, (d) dip feed, (e) spray feed and (f) splash feed.

economical and is unproductive as most of the spray will bounce back into the wet feed tray. To improve the effectiveness of spraying and increase the product yield, the feed material should possess an adhesive property (Daud 2014).

4.4.5 Splash feed

The splash feeding method is similar to spray feeding. In this method, the feed material is splashed onto the rotating drum surface (Figure 4.5f). This method is more appropriate for products containing a high level of sediment content.

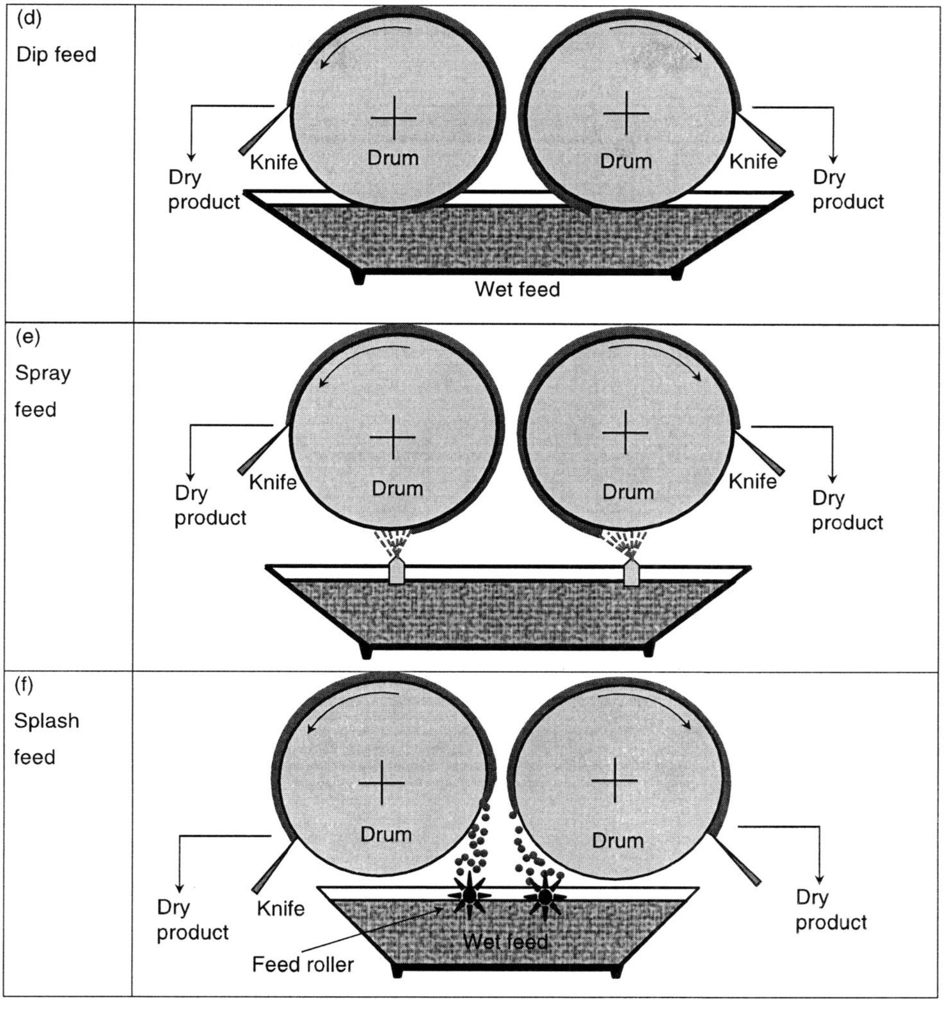

Figure 4.5 (*Continued*)

4.5 Operating parameters

Although the operating parameters for different drum dryers vary, some parameters are commonly used in all types of drum dryers, for example steam pressure, nip width, speed of drum rotation, type of feeding device, feed concentration, physical characteristics and the temperature of feed when it reaches the drum surface. The capacity of thin film drying in the drum dryer also depends on the rate of drying and degree of adherence to the drum surface (Abchir *et al.* 1988). To increase the drying rate or reduce the drying load, feed is preconcentrated or preheated. However, preconcentration may affect sheet formation during drum drying.

4.5.1 Important operational conditions in the drum drying of milk

Drum dryer types and operating conditions usually differ based on the native form of milk and the purpose of drying. All the major operational condition of the different drum dryers are listed in Table 4.1. Durand & Grandadam (1985) successfully used a double-drum dryer to dry partially and fully skimmed milk. The obtained milk powder is cooled and crushed by a hammer and can be utilized in cocoa and chocolate-based products. The addition of drum-dried milk powder improves (10–15%) the rheological properties of chocolate. Buter & Wiepking (2010) patented the frame arrangement mechanism of a double-drum dryer to produce dried milk, baby food, yeast and other food products. Generally, the gap between the drums is approximately 0.1–1 mm and they rotate in opposite directions. The quality and reproducibility of products are influenced by the gap between the drums. The gap may vary from product to product, and can be adjusted mechanically by a screw spindle. However, the major problem is the accuracy of the gap adjustment and reproducibility of the gap width is difficult during the drying process. To avoid contact between the two drums, the gap between them is increased, which may not be suitable for all products. Also, the arrangement of the drive is a very complicated process and relatively difficult to control. Hence, frame arrangement mechanisms have provided a new means of adjusting the gap between the drums in an accurate manner that is reproducible and within set tolerances.

Campbell (1917) proposed a useful improvement for drying milk, which produced milk powder with desirable characteristics such as high porosity, light and fluffy structure, and high solubility. The steps involved in the process are concentration of milk, convertion of the milk concentrate in to an aerated, adhesive mass, spreading of the mass on an air-permeable support and drying. Glen (1929) also transformed the drum-drying operation to make it more suitable for drying milk. He used a drum with a perfect surface and the cylinder was constructed in such a manner that the heat from the steam could be evenly distributed across the entire cylindrical surface. Furthermore, there was an improved head connection for the drum. Duncan (1933) developed a vacuum drum dryer for milk-based products. In this vacuum drum drying operation, feed material was fed to the exterior of an encased heated drum, scraped off in a dried condition and deposited on a conveyor which conveyed it to the outlet without breaking the vacuum around the drum. Whole milk powder is manufactured by different drying methods, such as drum drying and spray drying. In drum drying, preheated whole milk is applied to the surface of a heated drum, and the whole milk powder is scraped off by knives on drying. Whole milk powder manufactured by this process tends to have a cooked flavour due to the increased caramelization of the milk.

John's invention is related to the condensation and evaporation mechanism of liquids for milk, coffee solution and other beverages (John 1951). It was shown in this work that the continuous production of a condensate or powder was possible by quick drying. Infra-red rays with air and temperature control were used for evaporation of liquid by preventing it from reaching boiling point. In addition, the drum-drying process was carried out using steam-heated cylinders, which gave a dried product that was equivalent to that obtained from the more expensive spray drying method. David (1968) designed a drum dryer with either with single- or double-drum configuration for

Table 4.1 Important operational conditions for the drum drying of milk

Drum dryer type	Product	Operating conditions	Advantages of the process	Reference
Double-drum dryer	Partially or fully skimmed milk powder (55.5%)	Powder feed: 50 kg/h Water feed: 5 kg/h Temperature: 55–65 °C.	Rapid drying process for film formation Rheological properties are improved after addition to chocolate	Durand & Grandadam (1985)
Double-drum dryer	Milk, cereal-based baby food, potato flakes, cereals, etc.	Gap approx. 0.1–1 mm	Gap between the drums can be precisely adjusted with frame arrangement mechanisms	Buter & Wiepking (2010)
Vacuum drum dryer	Milk	Subatmospheric pressures	Drying takes place in clean and sanitary conditions	Duncan (1933)
Single- or double-drum dryer	Milk, tomatoes, lemons, etc.	–	An economical process with uniform moisture and film thickness	David (1968)
Single-drum dryer	Milk	Thin mass has been used to achieve more porous material	Produces dry milk that is highly porous, less dense, light, fluffy and with high solubility	Campbell (1917)
Drum dryer	Milk	–	Steam heat will be evenly distributed to the entire cylindrical surface	Glen (1929)
Rotary drum dryer	Milk, coffee solution, beverages	–	Continuous production of powder by quick drying Dried product is equivalent to spray-dried product	John (1951)
Atmospheric or vacuum drum dryer	Dairy products, chemicals and other materials	–	To dry materials effectively under either atmospheric or vacuum conditions Can be readily changed over from one type of operation to another	Overton (1959)
Double-drum dryer	Non-fat milk	Drum rotates for two to three revolutions per minute	Enhance the flavor and nutritional value of the product	Hing (1970)
Drum dryer	Milk	–	High quality in colour, favour and solubility, and also economical	Peterson (1930)
Vacuum drum dryer	Milk	Vacuum	All the controls are provided outside the vacuum chamber and dried powder is free from boiled or cooked flavor	Overton (1952)
Atmospheric double-drum dryer	Milk	60 pounds of steam pressure used on the drums (24 rpm)	Preheating skim milk can increase the capacity of the machine by 18%	Combs & Hubbard (1932)

uniform dehydration of food products. Products with uniform moisture content could also be achieved in an efficient manner. The food product layer thickness can be controlled before application to the surface of the drum dryer. Drum dryers are also used in the manufacture of whey protein powder and casein powder. Although the product quality is poor compared to the spray drying method, drum drying is preferred due to its low production cost.

4.6 Advantages and disadvantages of drum/roller drying

There are several advantages associated with drum drying. Drum-dried milk used to be known as 'germ-free' milk because of high temperature applied to the milk during the drum-drying process. For this reason, drum drying has been regarded as the most appropriate method for the production of infant foods for many years. However, the high temperature may lead to denaturation of the protein content (Knipschildt & Andersen 1994) and the formation of cooked flavours and off-colours. Vacuum drum dryers were designed with the aim of addressing these problems, with dairy products undergoing drying at lower temperature and pressure. However, vacuum drum drying is expensive compared to atmospheric drum drying due to the cost involved in vacuum creation. Because of the boiling and evaporation processes, drum-dried products possess good porosity and good rehydration properties. Drum dryers can be used to dry highly viscous foods, a process that is difficult with other dryers. Equipment cleaning, maintaining hygienic conditions and set-up of operating parameters are very easy.

Some disadvantages of drum dryers are as follows:

- drum drying is not suitable for products which do not form a film
- products containing high sugar concentration are difficult to remove from the drum surface after drying
- lower productivity compared to spray drying
- replacing a drum is very expensive.

4.7 Conclusion

Drum dryers are mainly used for drying viscous feed materials to obtain highly porous and flaky dried products. Drum dryers are easy to handle and economical to operate compared to freeze dryers and spray dryers. Numerous types of drum dryers are avaiable in the dairy industry to suit the nature of the feed material and the dried product characteristics. Retention of flavour, colour and micronutrients, and rapid reconstitution are the major advantages of drum-dried dairy products. Moreover, drum dryers are used for the production of enriched nutritive food. Vacuum drum dryers are extensively used for drying milk products (skim milk, whey protein, weaning foods etc.) to retain micronutrients. A recent development in drum drying involves increasing energy and process efficiency. Optimization of process parameters is vital to the wider application of drum dryers.

References

Abchir, R. Vasseur, J. & Trystum, G. (1988). Modelisation and simulation of drum drying. *Proceedings of 6th International Drying Symposium, IDS'88*, Versailles, France, pp. 435–439.

Balch, J.T. (1975). Method and apparatus for distributing feed stock to dryer drum. US Patent No. 3,878,028, US Patent and Trademark Office, Washington, DC.

Bhandari, B., Bansal, N., Zhang, M. & Schuck, P. (2013). In: *Handbook of Food Powders: Processes and Properties*. Elsevier.

Brown, A.H., Van Arsdel, W.B., Lowe, E. & Morgan, J.R.A.I. (1973). Air drying and drum drying. In: *Food Dehydration* (eds Van Arsdel, W.B., Copley, M.J. & Morgan, A.I.), AVI Publishing Co., Westport, CN.

Buter, F. & Wiepking, J. (2010). Double drum arrangement. US Patent Application No. 13/583,661, US Patent and Trademark Office, Washington, DC.

Campanella, O.H. & Cerro, R.L. (1984). Viscous flow on the outside of a horizontal rotating cylinder: The roll coating regime with a single fluid. *Chemical Engineering Science*, **39**, 1443.

Campanella, O.H., Galazzo, J.L. & Cerro, R.L. (1986). Viscous flow on the outside of a horizontal rotating cylinder-II. Dip coating with a non-Newtonian fluid. *Chemical Engineering Science*, **41**, 2707.

Campbell, C.H. (1917) Drying milk. US Patent No. 1,250,427, US Patent and Trademark Office, Washington, DC.

Combs, W.B. & Hubbard, E.F. (1932). Some factors influencing the capacity of the atmospheric drum dryer. *Journal of Dairy Science*, **15**(**2**), 147–154.

Daud, W.R.W. (2014). Drum dryers. In: *Handbook of Industrial Drying*, 4th edn (ed. Mujumdar, A.S.), pp. 249–257, CRC Press, Marcel Dekker, New York.

David, C.O. (1968). Food dehydrator. US Patent No. 3,414,982, US Patent and Trademark Office, Washington, DC.

Duncan, B.M. (1933). Vacuum drum dryer. US Patent No. 1,933,819, US Patent and Trademark Office, Washington, DC.

Durand, D. & Grandadam, Y. (1985). Milky products for manufacturing chocolate-trade products and method or preparation thereof. US Patent No. 4,532,146, US Patent and Trademark Office, Washington, DC.

Gavrielidou, M.A., Valous, N.A., Karapantsios, T.D. & Raphaelides, S.N. (2002). Heat transport to a starch slurry gelatinizing between the drums of a double drum dryer. *Journal of Food Engineering*, **54**(**1**), 45–58.

Glen, O. (1929). Roll for milk-drying machines and method of making. US Patent No. 1,718,977, US Patent and Trademark Office, Washington, DC.

Hing, F.S. (1970). Dehydrated island pudding base. US Patent No. 3,539,358, US Patent and Trademark Office, Washington, DC.

John, S.H. (1951). Rotary drum dryer. US Patent No. 2,571,778, US Patent and Trademark Office, Washington, DC.

Knipschildt, M.E. & Andersen, G.G. (1994). Drying of milk and milk products. In: *Modern Dairy Technology* (ed. Robinson, R.K.), pp. 159–254, Springer.

Laurent, S., Couture, F. & Roques, M. (1999). Vacuum drying of a multicomponent pharmaceutical product having different pseudo-polymorphic forms. *Chemical Engineering Process*, **38**, 157.

Moore, J.G. (1995) Drum dryers. In: *Handbook of Industrial Drying, Vol. 1*, 2nd edn (ed. Mujumdar, A.S.), pp. 249–262. Marcel Dekker, New York.

Okos, M.R., Narsimhan, G., Singh, R.K. & Weitnauer, A.C. (1992). Food dehydration. In: *Handbook of Food Engineering* (eds Heldman, D.R. & Lund, D.B.), pp. 437, Marcel Dekker, New York.

Overton, L.S. (1952). Evacuated drum drying machine. US Patent No. 2,602,498, US Patent and Trademark Office, Washington, DC.

Overton, G. (1959). Atmospheric and vacuum drum dryer. US Patent No. 2,880,523, US Patent and Trademark Office, Washington, DC.

Peterson, C.M. (1930). Milk-drying apparatus and process. US Patent No. 1,748,618, US Patent and Trademark Office, Washington, DC.

Robert, A.D. & Lester, M.K. (1968). Double drum dryer with removable external heating means. US Patent No. 3,363,665, US Patent and Trademark Office, Washington, DC.

Tang, J. & Yang, T. (2004). Dehydrated vegetables: principles and systems. In: *Handbook of Vegetable Preservation and Processing* (eds Hui, A.H., Ghazala, S., Graham, D.M., Murrell, K.D. & Nip, W.K.), pp. 335–372, Marcel Dekker, New York.

Tang, J., Feng, H. & Shen, G.Q. (2003). Drum drying. In: *Encyclopedia of Agricultural, Food, and Biological Engineering*, pp. 211–214, Marcel Dekker, New York.

Weinstein, S.J. & Ruschak, K.J. (2004) Coating flows. *Annual Reviews in Fluid Mechanics*, **36**, 29.

Winder, W.C. & Kielsmeier, E.W. (1959). Process of drying milk. US Patent No. 2,911,301, US Patent and Trademark Office, Washington, DC.

5

Spray Drying

S. Padma Ishwarya[1] and C. Anandharamakrishnan[1,2]

[1]*Department of Food Engineering, CSIR – Central Food Technological Research Institute, Mysore 570 020, India*
[2]*Indian Institute of Crop Processing Technology, Ministry of Food Processing Industries, Government of India, Thanjavur 613 005, Tamil Nadu*

5.1 Introduction

The association between spray drying and the dairy industry dates back to the Second World War, which necessitated mechanized production of large volumes of milk powder to meet the needs of army personnel in different parts of the world. This catalysed the full-fledged use of spray-drying technology in the dairy sector. Indeed, milk powder was the first commercial product that resulted from the industrial-scale operation of spray drying. However, the evolution of spray drying as a commercial process involved a series of alterations in its design, especially with respect to its ability to handle heat-sensitive products. The spray drying process was transformed from its inception in the 1870s to the early 1900s. Following the integration of process innovations and developments, spray drying has become well established in the dairy sector.

Spray drying is used in the production of a variety of dried dairy products, including, but not limited to, skim milk powder, whole milk powder, whey protein powder, milk-based infant foods, cheese powder, powdered cream mixes and dried caseinate preparations. The list is continuously evolving due to active research and the consequent innovations in the spray-drying process. Spray drying demonstrates a competitive edge over other liquid drying techniques owing to its versatility in handling feed with widely varying characteristics, its ability to handle large production volumes and, more importantly, the possibility of continuous and hence economical production. This chapter will examine the spray-drying process with respect to its application to the drying of milk and milk products.

5.2 Spray drying: principle of operation

Spray drying of milk or milk products can be defined as the process of simultaneous atomisation and desiccation of feed with distinctive rheological properties, in the presence of hot drying medium, to facilitate rapid moisture evaporation and result in dried particulate products. The feed can be in the form of emulsion (e.g. milk), solution

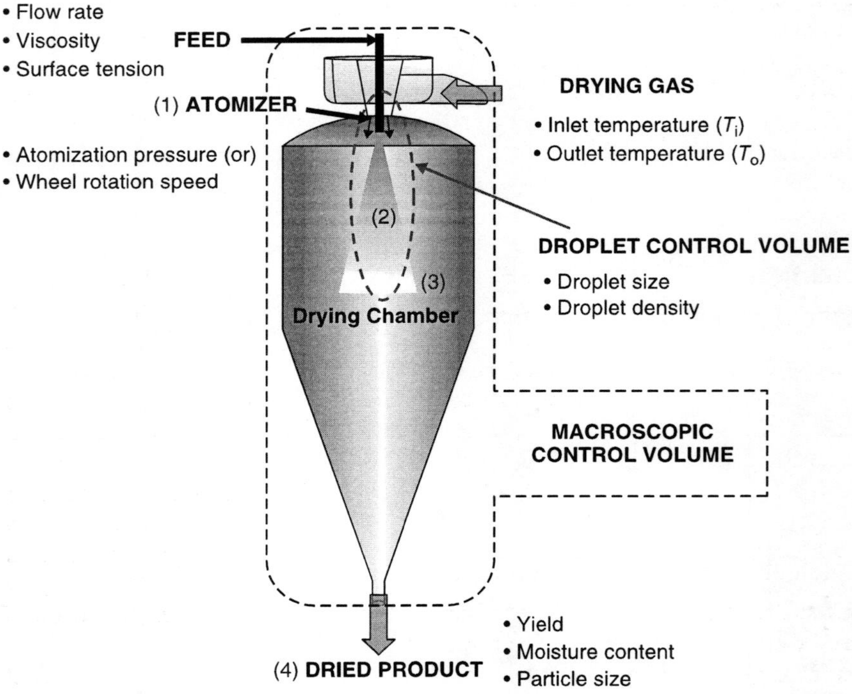

Figure 5.1 Control volumes and parameters of the spray-drying process. Adapted from Dobry *et al.* (2009).

(e.g. whey protein solution), suspension or slurry (e.g. cheese slurry). The underlying principle of spray drying is the convective mode of heat transfer between the atomized feed and hot air at atmospheric pressure.

The physical phenomena occurring during the spray-drying process can be accounted for by considering two key control volumes. The *droplet environment control volume* consists of the region around the atomizer that exists within the *macroscopic control volume*, which comprises the entire spray chamber (Figure 5.1) (Dobry *et al.* 2009). The following sequence of events occurs within the droplet environment control volume:

1. atomization of the feed solution
2. interaction between the atomized droplets and the hot drying gas
3. moisture evaporation
4. product separation at the exit of the macroscopic control volume.

The majority of the powdered product exits the spray chamber from the bottom, but the product separation stage is complete when the powder entrained in the drying gas stream is recovered using a cyclone separator, bag filter or electrostatic precipitator, depending on the particle size of product.

The timescale of these events is microseconds to milliseconds and the entire droplet-drying process is completed within an approximate residence time of 1 second. The sequential phenomena that occur during the spray-drying process and the associated parameters are shown in Figure 5.1.

Spray-drying equipment comprises a number of components: atomizer, feed pump, air heater, air disperser, aspirator, drying chamber and equipment for product discharge, transport, packaging and removing air (Devakate *et al.* 2009).

The role of the feed pump in the spray dryer is to supply the feed to the atomizer. This is especially important in the spray drying of milk, as the feed is always fed to the system at a concentration of 45–55% (multiple-effect falling-film evaporators with thermal and mechanical vapour recompression systems are commonly employed for preconcentration of milk feed prior to spray drying) in order to enhance process efficiency. The feed pumps must therefore be selected to handle such concentrated feed material of high viscosity. The feed pump selection also depends on the type of atomizer used. While low-pressure pumps are employed in combination with rotary atomizers and two-fluid nozzles, high-pressure pumps are used with pressure nozzles (Masters 1985). Peristaltic pumps are most commonly employed for spray drying. These pumps are appropriate for handling viscous feed at small flow rates and low pressure in combination with rotary wheel atomizers. On the other hand, piston pumps are suitable for high-pressure applications along with high-pressure nozzle atomizers.

The aspirator functions to supply dry air to the spray chamber by means of a motor operating under pressure conditions. The aspirator flow rate can be altered to regulate the amount of heated drying air entering the spray chamber. The air disperser is located in the roof of the spray chamber beside the atomizer to ensure equal flow of the drying gas in all directions within the spray chamber. This is facilitated by the generation of pressure drop using perforated plates or vaned channels.

The different types of atomizers, spray-drying chamber configurations and product collection systems used in the dairy industry are discussed in detail in the following sections.

5.2.1 Atomization

Atomization can be defined as the process of bringing the feed fluid into a state of minute division (Percy 1872) by virtue of the atomization energy that is supplied to the atomizer. The source of energy for atomization can be the centrifugal force of a rotating wheel, the pressure energy of the liquid feed or the kinetic energy of high-velocity air or other gaseous flow. Atomization is governed by the theory proposed by Ohnesorge in 1936. According to this theory, disintegration of the liquid jet emanating from the atomizer depends on its viscosity, density, surface tension and size. With lesser viscous and surface tension forces in the liquid, the resistance to disintegration is lower. In addition, turbulence in the liquid jet and the action of pneumatic forces causes the disintegration of liquid at the periphery or tip of the atomizer. Eventually, when the droplet is airborne, the fission proceeds as the shear stresses realign within the liquid. This is numerically expressed by the dimensionless Ohnesorge number (Oh), which is the ratio of the Weber number (We) to the Reynolds number (Re) (Eq. 5.1).

$$\mathrm{Oh} = \frac{\sqrt{\mathrm{We}}}{\mathrm{Re}} = \frac{\mu}{\sqrt{\rho\sigma L}} = \frac{\text{viscous forces}}{\sqrt{(\text{inertia} \times \text{surface tension})}} \tag{5.1}$$

where μ, ρ and σ are the viscosity, density and surface tension of the feed droplet, respectively, and L is the characteristic dimension of the feed droplet, that is, volume per unit area.

Figure 5.2 Pressure nozzle.

The atomization step exerts control over the size and size distribution of droplets, their trajectory and speed, and hence determines the final particle size (Filková *et al.* 2007). The droplet size is also important from the perspective of drying time which is proportional to the square of the particle dimension. Atomization of 1 litre of milk concentrate results in 1.5×10^{10} particles with a diameter of 50 μm and a total surface area of 120 m^2 (Westergaard 2010). The degree of aeration undergone by the feed during atomization and the consequent occlusion of air has an influence on the bulk density of the final product (Masters 1985). This in turn affects the flow properties and solubility of the powdered dairy product.

The most commonly used atomisers for the spray drying of milk and milk products are the rotary/wheel and nozzle atomizers. The use of monodisperse droplet generators is a recent approach in the atomization of milk during spray drying.

5.2.1.1 Rotary atomizers

Rotary atomizers comprise a stainless steel wheel that rotates at a speed of 10,000–30,000 rpm. The feed liquid is directed to the centre of this rotating wheel by means of centrifugal acceleration. The feed liquid spreads as a thin sheet over the entire wheel surface, then exits at the periphery of wheel in the form of droplets. The exit velocity of the droplets is in the range of 90 to 200 m/s. While feed rate and feed viscosity exhibit a direct relationship with droplet size, wheel speed and wheel diameter have an inverse relationship with it (Anandharamakrishnan 2013). Rotary atomizers can be of different types, for example cup type, wheels with or without vanes, or wheels with wear-resistant vanes. The vaneless wheels can be of bowl, cup or plate type. These

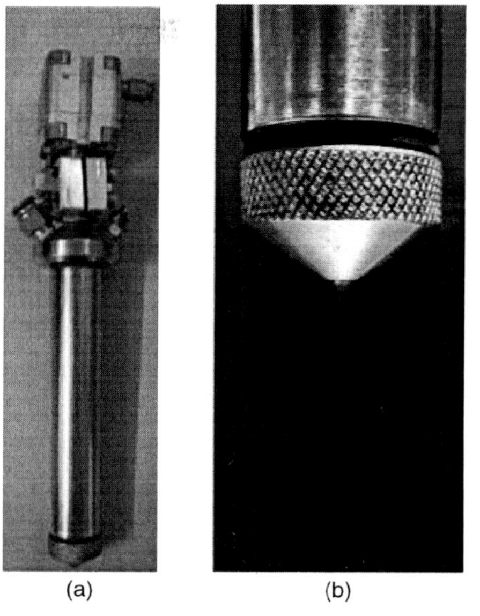

Figure 5.3 (a) Two-fluid nozzle. (b) Spray emerging from two-fluid nozzle (Anandharamakrishnan & Padma Ishwarya 2015).

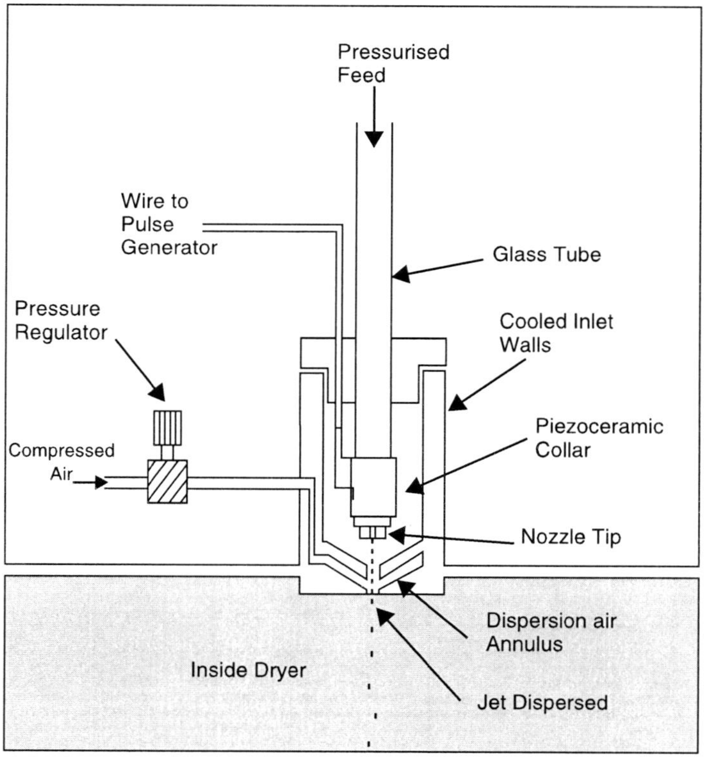

Figure 5.4 Construction of a monodisperse droplet generator used for the spray drying of whole milk (Rogers *et al.* 2012).

atomizer types are mainly used for the production of skim milk powder with high bulk density (Nasr *et al.* 2002).

Rotary atomizers produce a uniform droplet size of 30–120 μm (Anandharamakrishnan 2013), facilitating the atomisation of viscous feed. They exhibit negligible clogging and have the ability to handle a high feed rate. However, a large diameter spray chamber is required to accommodate the wide-angle spray that emerges (Masters 1985).

5.2.1.2 Pressure nozzle atomizers

Pressure nozzle atomizers are used in particular for the production of instant and re-wet skim milk powder (Nasr *et al.* 2002). Here, atomization of feed liquid is facilitated by the discharge of liquid through an orifice under pressure of 700–2000 kPa (Figure 5.2). The high-speed film of feed liquid emerging from the orifice breaks into a fine spray of droplets owing to the conversion of pressure energy into kinetic energy. The droplet diameter produced by the pressure atomizer falls in the range 120–250 μm (Anandharamakrishnan 2013). Droplet diameter decreases with an increase in atomization pressure and a decrease in feed rate and viscosity. The relationship between selected atomization parameters and droplet diameter is given by Eq. (5.2) (Masters 1985):

$$\frac{D_2}{D_1} = \left(\frac{P_2}{P_1}\right)^{-0.3} \tag{5.2}$$

where D_1 and D_2 are the initial and final droplet sizes on changing the atomization pressure from P_1 to P_2, respectively. Higher atomization pressure is required for the efficient transfer of atomization energy to bring about the fission of bulk feed liquid into droplets.

$$\frac{D_2}{D_1} = \left(\frac{\mu_2}{\mu_1}\right)^{0.2} \tag{5.3}$$

where D_1 and D_2 are the initial and final droplet sizes on changing the feed viscosity from μ_1 to μ_2, respectively.

Milk powder from nozzle atomization has higher bulk density than that from twin-fluid nozzles.

5.2.1.3 Twin-fluid atomizers

In this atomizer type the break-up of fluid into droplets is brought about by the overlap between the feed liquid and a stream of high-velocity air or other gas. The high-velocity gas stream can act externally or internally on the feed liquid (Pinder 1952). The level of shear created by the atomization gas on the liquid decides the droplet size distribution. Apart from its influence on droplet size, shear rate plays a major role in the survival of living cells in spray-dried probiotic microcapsules. The extensional and shear stresses caused by atomization leads to cell disruption and irreversible protein denaturation in the bacteria (Goldberg 2008). Furthermore, the droplet size distribution can affect bacterial survival by influencing the droplet and particle temperature. As already mentioned, fine droplets undergo more rapid drying and attain comparatively higher temperature than coarser droplets, leading to cell mortality. The shear rate (γ) during twin fluid atomization can be calculated using Eq. (5.4):

$$\gamma = \frac{2(\nu_{av} - \nu_L)}{D_L} \tag{5.4}$$

where D_L is the inner nozzle diameter (m), v_{av} is the average exit velocity of both the atomizing gas and the liquid feed (m/s), and v_L is the mean velocity of the liquid feed within the nozzle before exit (m/s) (Hede *et al.* 2008). The gas velocity is calculated from its density (kg/m^3) at absolute pressure, the annular cross-sectional area (m^2) of the twin-fluid atomizer through which the atomizing gas flows and the gas-flow rate (m^3/s). Liquid velocity is determined by measuring the density of the feed solution and its flow rate through the nozzle.

Twin-fluid nozzles (Figure 5.3a) exhibit the potential to result in highly homogeneous sprays (Figure 5.3b) of small mean droplet size irrespective of feed viscosity. The advantages over pressure nozzles include a larger orifice diameter that prevents clogging of nozzles and that the twin-fluid nozzles do not require high-pressure pumping equipment, which reduces maintenance (Masters 1985). However, the use of compressed air as the atomization medium poses certain disadvantages. The inclusion of cold air in the atomization zone reduces the temperature difference between the feed droplets and drying gas, and hence the rate of evaporation. Air occlusion also leads to lower particle density. Furthermore, the presence of oxygen in the atomization air stream leads to the susceptibility of bacterial cells to oxidation-induced damage and mortality (Ghandi *et al.* 2012).

5.2.1.4 Monodisperse droplet generators

A monodisperse droplet generator (MDG) is used for the specific purpose of obtaining droplets of uniform size distribution and non-agglomerating nature (You *et al.* 2014). The generation of monodisperse droplets is significant in obtaining uniform sized particles of the final product with optimal instantization properties. The operational principle of MDG is based on the fluid break-up theory explained above. However, the mechanism for the generation of monodisperse droplets may vary by application of pressure pulses (Brenn *et al.* 1997), strong electric fields (Meesters *et al.* 1992) or using dispersing fluids (Xu *et al.* 2006); the first two of these have been employed for the atomization of whole milk and skim milk-based feed formulations.

Monodisperse droplet creation can be achieved by two modes: inducing ejection of small droplets from an orifice or modulating the break-up of a continuous jet of fluid into droplets that are monodisperse. The former approach is known as the 'drop-on-demand' and the latter is termed 'continuous'. A single design of MDG is capable of operating in both these modes. The unique feature of MDGs is that they generate a single jet or stream of individual droplets travelling downwards. This is in contrast to conventional atomizers where droplets of poly-disperse size distribution with wider spray angle are produced (You *et al.* 2014).

A piezoelectric component-driven MDG (Figure 5.4) has been employed for the atomization of skim milk concentrate (Fang *et al.* 2012; Rogers *et al.* 2012), milk protein concentrate (Fang *et al.* 2012; Rogers *et al.* 2012) and reconstituted whole milk (You *et al.* 2014). Droplets were generated by vibrations induced by a piezoelectric pulsation device to break the liquid column that jetted out of the nozzle tip (continuous mode, as described above). The nozzle system was pressurized with compressed air at different levels to produce the required feed flow rate. For skim milk and milk protein concentrate, the feed solutions were filtered and jetted through a 75-μm orifice (Rogers *et al.* 2012a,b). The atomization of whole milk feed solution was carried out at two different orifice sizes, 75 and 90 μm (You *et al.* 2014). At the applied pulse frequency of 10 kHz, the droplet size for the milk protein concentrate of 20% solid content (w/w) (shown in Figure 5.6) was 170 ± 3 μm. The droplet break-up initiated as a continuous jet, with a short region of connected threads at the exit of the nozzle tip which eventually led to the formation of comminuted droplets. The droplets of the skim milk concentrate and reconstituted whole milk (20% solid content w/w) measured 181 ± 5 and 169 ± 1 μm, respectively. Furthermore, orifice size was found to exhibit a direct relationship with droplet size (You *et al.* 2014). Spray drying with an MDG facilitates the study of droplet/particle shrinkage during drying as both the initial droplet size and the dry particle size are known.

As mentioned above, monodisperse droplet generation can also be achieved by application of electrical charge to the spray droplets. The prerequisite for atomization using the electrostatic mode is the electrical conductivity (σ) of feed liquid. More than 80% water in the composition of milk renders it conducive for electrostatic spraying. The droplet size bears an inverse relationship with the conductivity of feed liquid. The other parameters that influence the electrospray atomisation are the flow rate (Q), density (ρ) and surface tension (γ) of the feed liquid (Jaworek 2007; Eq. 5.5):

$$d_{\mathrm{D}} = \alpha \left(\frac{Q^3 \varepsilon_0 \rho}{\pi^4 \sigma \gamma} \right)^{1/6} \tag{5.5}$$

where d_{D} is the droplet size, ε_0 is the permittivity of the vacuum and α is a constant, generally equated to 2.9. When the intensity of the electric potential applied between two oppositely charged electrodes of the nozzle system (Figure 5.5) rises to the level of kilovolts, the droplet meniscus develops into a conical shape (known as the Taylor cone) with a highly concentrated free charge. The free charge speeds up the droplets away from the needle due to the imparted electric stress. Monodisperse droplets are formed as the jet breaks into a fine mist due to varicose instabilities. In addition, the like charges imparted on the droplets prevent droplet coalescence and agglomeration of partially dried particles in the spray chamber. This is advantageous with respect to milk, which is high in fat and sugar, both of which promote the stickiness of the product to the wall of the spray chamber. However, this atomization mode demands certain modifications in the spray chamber design from a safety perspective. The stainless steel wall of the spray chamber is electrically grounded. Polytetrafluoroethylene (PTFE) (1.5 mm thickness) is used as the dielectric barrier within the spray chamber (Johnson *et al.* 1996). Monodisperse droplet generation from different milk products is shown in Figure 5.6.

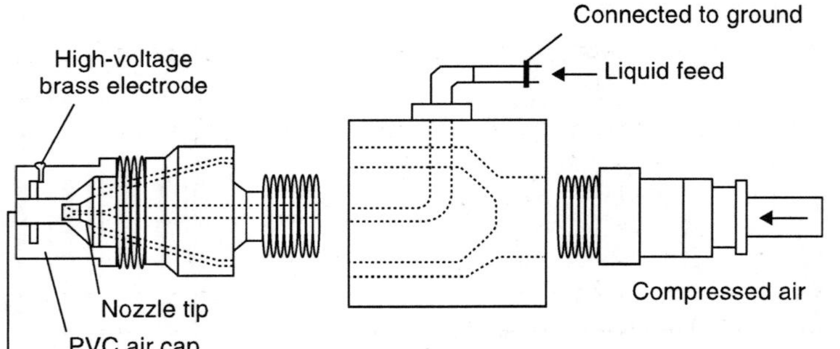

Figure 5.5 Spray-charging nozzle (Johnson *et al.* 1996).

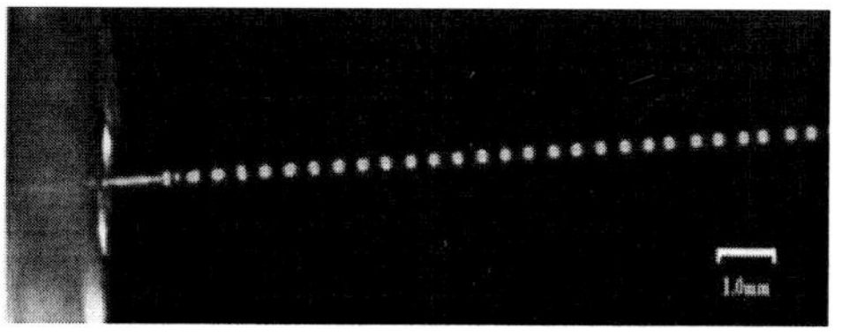

(a)

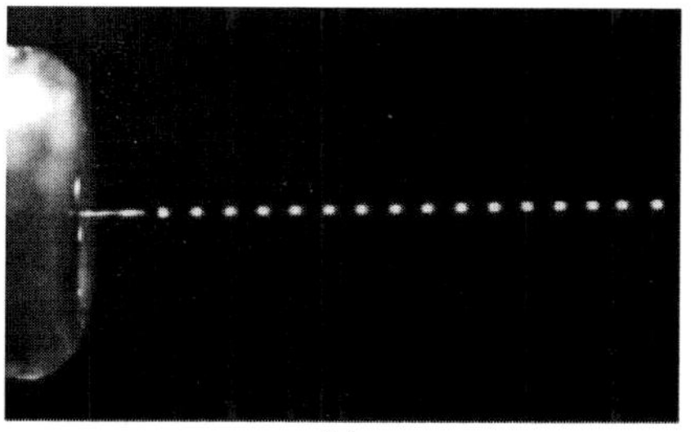

(b)

(c)

Figure 5.6 Monodisperse droplet jet of (a) reconstituted milk protein concentrate (Rogers *et al.* 2012a), (b) skim milk concentrate (Rogers *et al.* 2012b) and (c) reconstituted whole milk (You *et al.* 2014).

5.2.2 Droplet–drying air interaction and moisture evaporation

The interaction between the feed droplets and drying air occurs instantaneously after the atomization step. This stage facilitates rapid and uniform evaporation of moisture, owing to the smaller Biot numbers of the minute droplets, and marks the commencement of particle formation during spray drying. It is therefore difficult to demarcate between this step and the next stage, which is moisture evaporation, and hence they are handled together in this section. The significant process parameters in this stage are the inlet temperature (T_i) and the outlet temperature (T_o) of the drying gas. While T_i is the temperature of the heated drying air, T_o is the temperature of the air stream laden with solid particles at the exit of the spray chamber. However, T_o is not an independent parameter and is influenced by T_i and the feed flow rate. An increase in T_i and a decrease in feed flow rate would lead to high T_o values.

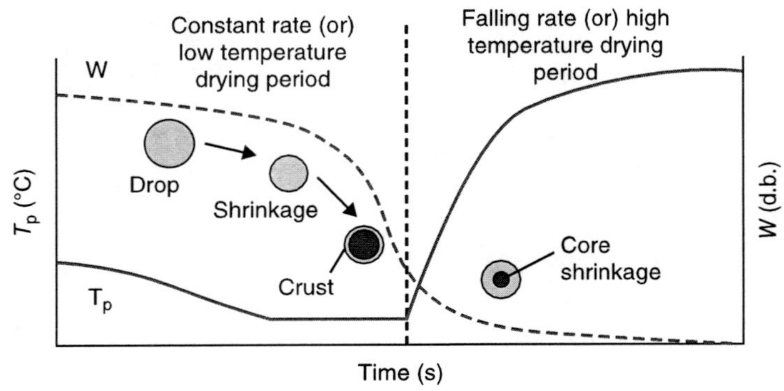

Figure 5.7 Profile of the temperature and moisture content during the spray drying of a single milk particle as a function of its residence time in the spray drier and the drying stage. Modified from Birchal *et al.* (2006) and Kim *et al.* (2009).

Once the contact is established between the droplets and the hot air, moisture evaporation occurs in two stages: (i) the constant rate period and (ii) the falling rate period (Figure 5.7). During the constant rate period, the latent heat of vaporization removed from the drying medium is used to evaporate the free water from the droplet surface. The driving force for this simultaneous heat and mass transfer is provided by the gradient in moisture concentration and the difference in vapour pressure and temperature between the interior and exterior of the droplet (Patel & Chen 2008). Owing to the lower internal resistance to mass transfer within the droplet at this stage, there is a continuous migration of moisture to the droplet surface. This facilitates maintenance of saturated conditions, promotes a constant rate of moisture evaporation from the surface and prevents the temperature of the droplet from rising above the wet-bulb temperature of drying air. This sequence of events is termed 'evaporative cooling'. This phenomenon is specifically relevant to milk and milk products due to their heat-sensitive nature and the presence of proteins that are susceptible to thermal denaturation. Thus, the constant rate period can also be termed the *low temperature phase* of spray drying. The drying time (t_c) for the constant rate drying can be calculated using Eq. (5.6) (Toledo 2007):

$$t_c = \frac{4(X_0 - X_c)r^2 \rho_L \lambda_v}{3k_f(1 + X_0)(T_a - T_s)} \tag{5.6}$$

where X_o and X_c are the initial and critical moisture contents (kg water/kg dry solids) of the droplet (on a dry basis), respectively, r is the droplet radius (m), ρ_L is the droplet density (kg/m^3), λ_v is the latent heat of vaporization of water (kJ/kg water), k_f is the convective mass-transfer coefficient (m/s), and T_a and T_s are the temperatures of the hot drying air and droplet surface, respectively (K).

The configuration of the spray chamber plays an important role in bringing about the evaporative cooling phenomenon. With respect to this, the co-current configuration is appropriate as the product and drying medium enter the spray chamber in the same direction. Because of the rapid rate of evaporation, the product temperature is always maintained below the outlet temperature of the drying air (Figure 5.8).

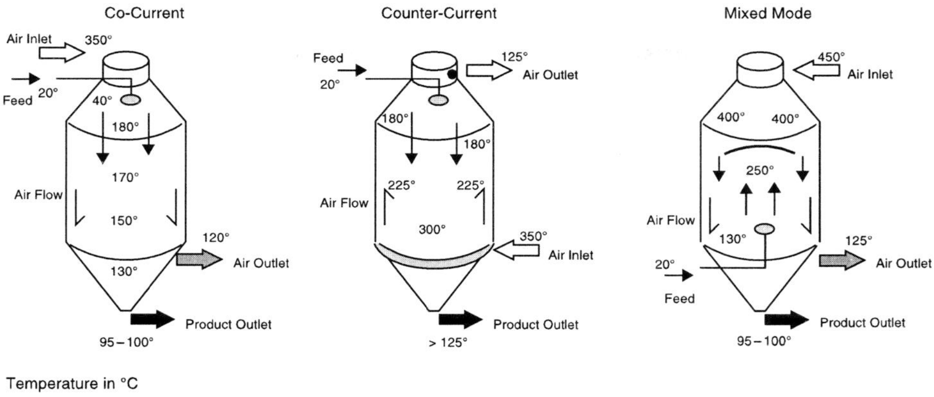

Figure 5.8 Temperature profile in different spray-dryer configurations. Modified from Vega-Mercado *et al.* (2001) and Masters (1991).

There are other configurations of spray dryers, such as counter-current and mixed-flow types, that are not suitable for the drying of milk and milk products. The product and the drying medium enter at opposite ends of the drying chamber in the counter-current configuration, leading to an outlet product temperature that is almost equal to T_i and is higher than T_o (Figure 5.8). The mixed-flow configuration is a combination of both the co-current and counter-current flow types. However, similar to the counter-current type, this mode suffers from the limitation of the higher exit temperature of the product and hence is not used for dairy spray drying. In addition to the direction of air flow, the spray dryer can also be classified on other rationale, as listed in Table 5.1.

An important phenomenon that occurs during the constant rate drying period is the redistribution of solute components within the drying droplet. The predominantly fluid state of the droplet and the absence of dense skin or membrane around it facilitates solute redistribution (Kim *et al.* 2009).

The beginning of the falling rate period is marked by a reduction in moisture content below a level that limits the maintenance of saturation conditions at the droplet surface. The dissolved solutes from the saturated feed solution form a crust of solids or skin (Figure 5.12), thus initiating the transformation of droplet to particle phase. The solid crust then continues to build up and resists the diffusion of water vapour from the droplet interior. This decreases the drying rate and elevates the particle temperature (Charlesworth & Marshall 1960; Birchal *et al.* 2006), thus initiating the transition of spray drying from a low temperature to high temperature process. The drying time for the falling rate period during the spray-drying operation time is given by the expression derived by Ranz & Marshall (1952):

$$t_f = \frac{\rho_L \lambda_v r_c^2 (X_c - X)}{3k_f(\overline{\Delta T})} \tag{5.7}$$

where r_c is the radius of the dried particle and $\overline{\Delta T}$ is the mean temperature between the surface of the particle and hot drying air during falling rate period.

Table 5.1 Classification of spray dryers

Rationale	Type of spray dryer
Aspect ratio (height to diameter)	• *Short type*: aspect ratio 2:1; suitable for accommodating flat sprays as obtained from rotary atomizers and handling of heat-sensitive products

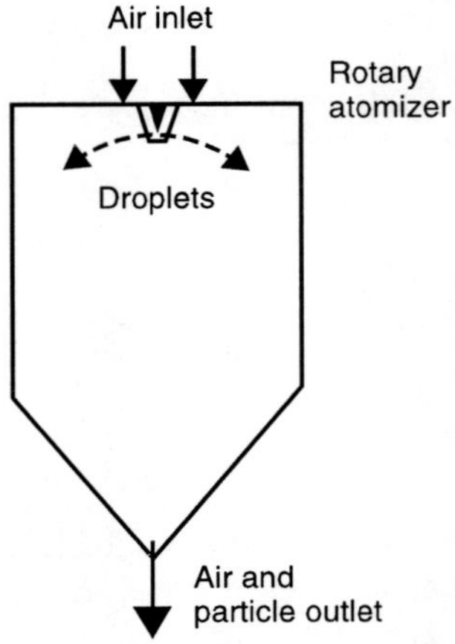

Figure 5.9 Short-type spray dryer fitted with rotary atomizer (Langrish & Fletcher 2001).

• *Tall type*: aspect ratio 5:1; longer residence time and hence not suitable for heat-sensitive products

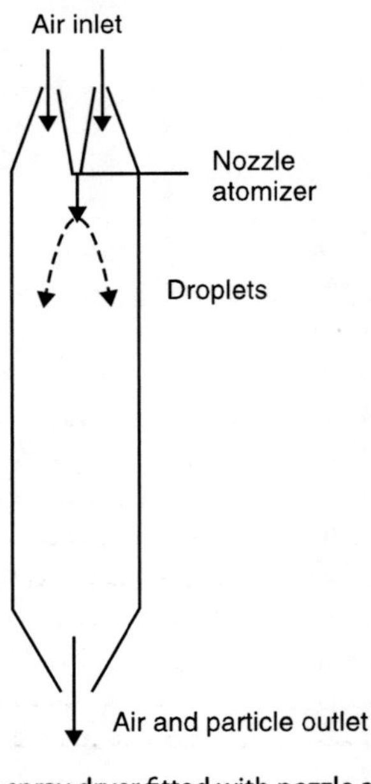

Figure 5.10 Tall-type spray dryer fitted with nozzle atomizer (Langrish & Fletcher 2001).

Table 5.1 (Continued)

Rationale	Type of spray dryer
Recycle/reuse of heating medium	• *Open cycle*: drawing the drying air from the atmosphere, heating and conveying it through the chamber once and then exhausting it back to the atmosphere.

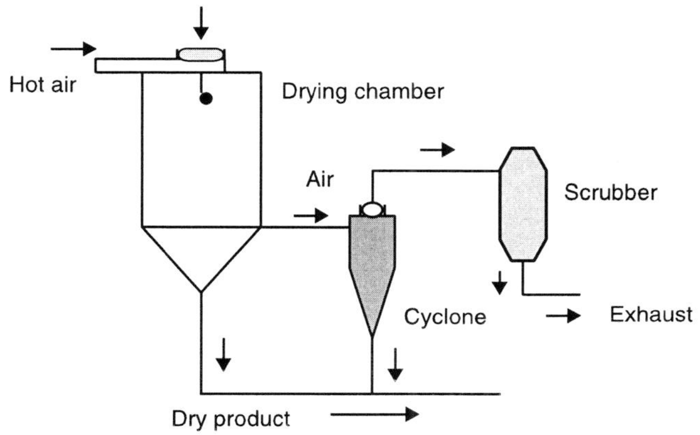

Figure 5.11 Open-cycle, co-current layout of spray dryer (Vega-Mercado *et al.* 2001; Dittman & Cook 1977).

Rationale	Type of spray dryer
	• *Closed cycle*: involves recycle and reuse of the gaseous medium; used when feed includes flammable solvents (this type is normally not encountered in the food industry as the feed material is always aqueous in nature)
Number of stages	• *Single stage*: the required final moisture content is attained in a single pass through the spray chamber; not suitable for heat-sensitive products due to high outlet temperatures and also leads to stickiness with weak bindings to form large and loose agglomerates and blow-hole formation • *Two stage*: the required moisture content is attained in two stages, the second stage being accomplished in a fluidized-bed dryer; suitable for heat-sensitive products

With respect to dairy drying, the greatest mass-transfer resistance to water flux is offered by whole milk whilst lactose solutions present the least resistance. This is due to the counter diffusion of the high fat content in whole milk because the fat component has the greatest molecular weight and the lowest diffusion coefficient (Kim *et al.* 2003). Thus, the slower moving fat molecules accumulate at the surface and the fast moving lactose molecules accumulate at the base of the powdered particles.

On the other hand, in case of milk protein concentrate and skim milk-based feed formulations that have less fat and are rich in protein and lactose, the protein content controls the rate of mass transfer. Thus, the milk protein concentrate was observed to exhibit higher mass-transfer resistances than the skim milk. However, crust or skin formation has also been observed with skim milk samples. The residual fat present in skim milk constitutes the crust, which is not of sufficient thickness to provide a major resistance to mass transfer.

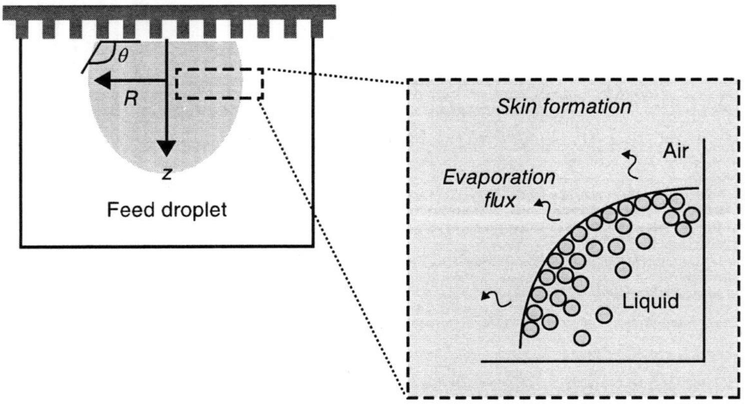

Figure 5.12 Phenomenon of skin formation. Modified from Sadek *et al.* (2015).

With whey protein isolate (WPI), the droplet demonstrated a uniform decrease in volume and retention of spherical shape during the initial stage of drying. During the falling rate period, a unique phenomenon was observed. The protein film formed exhibited distinct rheological properties due to the sol-gel transition at the interface between the surface and core of the droplet. This prevented droplet shrinkage and instead resulted in the gradual development of a porous skin layer at the air–droplet interface, which permitted continuous moisture evaporation, followed by transformation of the skin layer to a thick protein shell (Sadek *et al.* 2013).

Subsequently, the occurrence of bubble formation and droplet or core shrinkage events completes the moisture evaporation stage and decides the particle morphology. Bubble formation occurs when the partial pressure of water vapour at the centre of the droplet increases over ambient pressure with a simultaneous increase in temperature. The droplet inflates to the outer radius and ultimately results in irregular and randomly shaped particles (Etzel *et al.* 1996). When the droplet with an inflated bubble is subjected to further high temperature, it might lead to a blistered, shrivelled, hollow or inflated puffed particle. The possibility of particle collapse has also been observed (Handscomb *et al.* 2009). As explained above, drying of WPI led to the formation of an internal vacuole when the skin thickened into a protein shell during the final stage of drying (Sadek *et al.* 2013).

On the other hand, shrinkage-mediated change in droplet size has an effect on the mass and heat-transfer coefficients that determine the efficiency of moisture evaporation during spray drying (Ranz & Marshall 1952; Chen 2004). The occurrence of droplet shrinkage during spray drying has been attributed to the buckling of a viscoelastic layer of colloidal solutes that accumulated at the droplet surface during moisture evaporation. The driving force for shrinkage or deformation is the capillary forces that exist in the elastic layer of solutes on the droplet surface (Tsapis *et al.* 2005). Droplet shrinkage is in turn dependent on the temperature of the drying air and feed solid content. The shrinkage kinetics of skim milk droplets with 50% concentration at two different drying temperatures (110 and 70 °C) are shown in Figure 5.13. At the initial stage, that is, time $t = 0$, the milk droplet maintains a smooth and circular morphology irrespective of the drying temperature. At 110 °C, the higher rate of evaporation results in shell

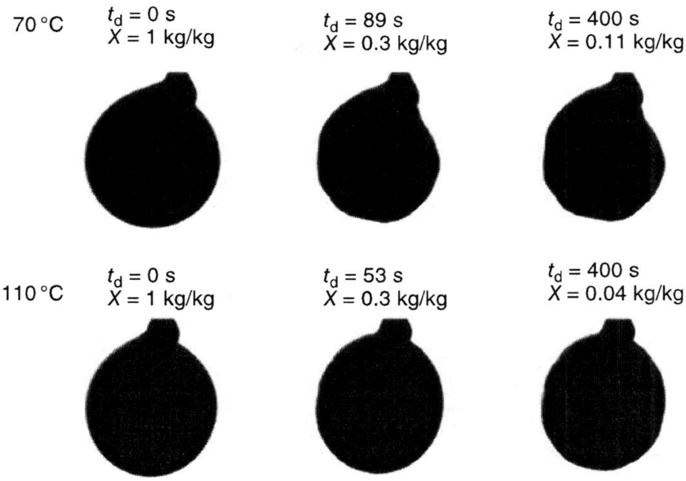

Figure 5.13 Comparison of droplet shape changes during drying of 50 wt% skim milk at 70 °C and 110 °C (Fu *et al.* 2013).

formation, which resists shrinkage and maintains the particle shape irrespective of the change in droplet solid content with progressing drying time. At 70 °C, initially the surface becomes undulated due to shell formation when the droplet reaches a moisture concentration of 0.3 kg/kg. However, the particle surface is still moist enough to allow shrinkage to occur to a lesser extent. Conversely, the effect of solid content on droplet/particle morphology (at a constant drying air temperature) can be explained in two stages. At the low initial solid level, droplet shrinkage is predominantly governed by the water fraction, hence smaller particles are produced at a higher drying temperature. With high initial solids levels (50 wt%), the solids fraction governs droplet shrinkage throughout the drying period, thus producing larger particles at a higher drying temperature (Sadek *et al.* 2015).

Quantifying the change in droplet size in real-time spray drying operation is difficult. A well-established approach is the single droplet drying (SDD) experiment. SDD can be defined as the study of the drying behavior of an isolated droplet under controlled air flow (El-Sayed *et al.* 1990; Walton & Mumford 1999) and conditions that mimic the spray chamber environment. An important prerequisite for SDD studies on a drying milk droplet is that the analysis should be carried out at appropriate feed concentration to obtain precise predictions.

The contribution of constant rate and falling rate drying periods to particle formation during spray drying can be explained by the Peclet number relationship. The Peclet number (Pe) is the ratio between the rate of droplet evaporation and the rate of solute diffusion (Eq. 5.8):

$$\mathrm{Pe} = \frac{\kappa}{D} \tag{5.8}$$

where κ is the evaporation rate and D is the diffusion rate. A low Peclet number indicates that the migration of solutes towards the centre of the droplet is fast when compared to the velocity of the diminishing droplet surface. As a result, the droplet shrinks with the migration of solutes towards the droplet centre and eventually results in small, dense and solid particles once supersaturation is attained (Vicente *et al.* 2013). A low inlet–outlet temperature combination is the major reason for a low Peclet number because at low drying temperatures a substantial amount of water is evaporated during the constant rate period when the crust or skin formation has not begun. Consequently, the dissolved solute molecules have more time to migrate within the droplet. During the subsequent falling rate period, the skin formed remains moist and flexible for longer, hence permitting the deflation and shrivelling of hollow particles as they cool (Birchal *et al.* 2005).

On the other hand, a higher inlet–outlet temperature combination results in a large Peclet number (Alamilla-Beltran *et al.* 2005), so the evaporation of the solute at the surface proceeds at a faster rate when compared to the diffusional motion of the dissolved molecules. Consequently, there is a rapid build-up of solute concentration at the surface. During the spray drying of skim milk concentrate at two different inlet air temperatures (205 and 145 °C), the higher T_i resulted in a relative increase in the lactose and protein coverage at the particle surface. This high surface concentration of the solute leads to a local increase in viscosity with subsequent skin or shell formation, which can potentially increase the particle temperature and hence lead to bubble formation, as explained above. Thus, a high Peclet number leads to hollow, light and porous particles. However, due to the formation of dry and hard skin, the hollow particle does not deflate when vapour condenses within the vacuole as the particle moves into cooler regions of the co-current spray dryer. However, damage to the particle surface, such as fissures or breakages, has been observed at high inlet air temperature (Pisecky 1978; Charm 1981). The plausible morphology patterns of spray-dried milk and milk products are depicted in Figure 5.14.

The influence of drying temperature on particle morphology also depends on the feed solid content. A high inlet temperature in combination with high feed solid content (30%) has a lesser impact on the surface composition of skim milk powders, that is, fat coverage and the ratio of surface lactose to surface protein, than when the feed solids content is low (10%). This interactive effect has been attributed to the retarded redistribution of solutes within the drying feed droplet of high viscosity at greater solid content (Kim *et al.* 2009). Furthermore, higher feed solid content reduces the diffusion coefficients of the dissolved solutes (King 1990). Thus, the solute redistribution during the constant rate drying period is prohibited by the high solid content of the feed.

5.2.3 Particle separation

The primary and secondary stages of particle separation from the gas stream involve withdrawal of dried product from the conical bottom of the spray chamber and its removal using dry collectors or wet scrubbers. While a cyclone separator is the conventional dry collection system, modern dairy spray dryers are equipped with bag filters. The construction of a cyclone separator includes a cylindrical barrel with a conical bottom. A cyclone separator utilizes centrifugal force to separate the solid particles from a carrier gas. The gas stream enters tangentially, forming an outer vortex as it travels downward. The centrifugal force imparted by the increasing air velocity causes particle separation, while an inner vortex is created in the opposite direction followed by the exit of clean gas from the top.

In the bag filter, the drying gas loaded with particles enters the hopper at the centre or bottom by virtue of suction or pressure. It then travels through the filter bag, where

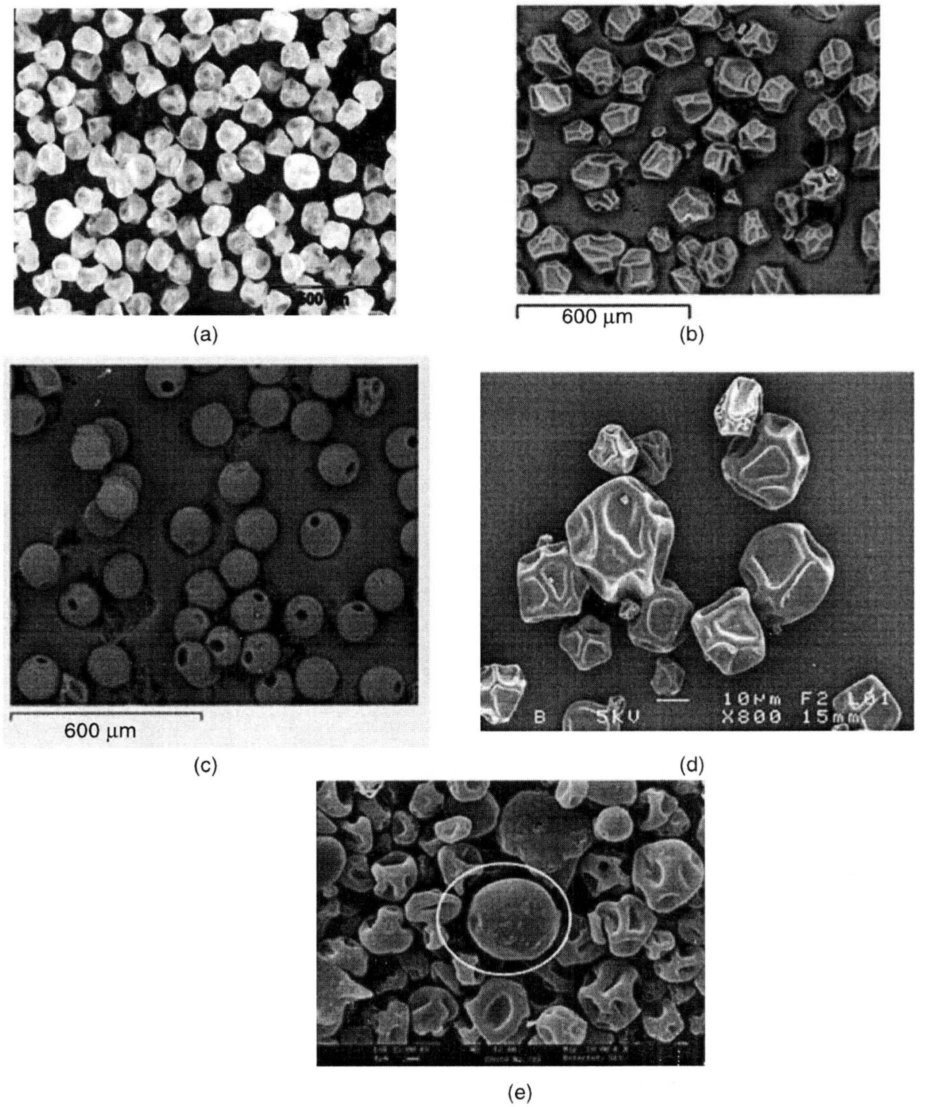

Figure 5.14 Morphology of spray-dried milk and milk products: (a) whole milk powder produced by monodisperse droplet generator (You *et al.* 2014); (b) shrivelled/wrinkled (at high T_i), (c) uniform sized, smooth and hollow particles (at low T_i) of milk protein concentrate (Fang *et al.* 2012), (d) shrinkage of casein particles (Sadek *et al.* 2016) and (e) microcapsules of *Lactobacillus plantarum* with smooth skin-forming morphology (Rajam *et al.* 2012).

the particles are retained on the surface of the bag. The clean air passes out through the bag and plenum to the outlet of the bag filter. Bag filters are also installed after the cyclones to reduce costs and collect residual powder, reducing emissions after the filter to 10 mg/N.m^3 (SPX Corporation 2013). The bag filters need to be changed frequently for sanitary reasons. The powder recovered from the bag filter or cyclone separator can be recycled either to the atomization zone or to a fluidized bed for agglomeration (SPX Corporation 2012).

5.3 Characteristics of spray-dried dairy powders

The functionality of spray-dried dairy powders is not just limited to shelf-stability, but also comprises other characteristic properties that enhance its instantization ability, application in other food products, such as bakery and confectionery, and facilitation of handling, packaging and transportation operations on a commercial scale. This section deals with the significant functional characteristics acquired by milk and milk products on spray drying.

5.3.1 Rehydration

The ability to rehydrate or reconstitute in water is an important quality attribute of dairy powders. Rehydration occurs in three stages: wettability, dispersibility and solubility (Freudig *et al.* 1999). Wettability is the ability to absorb water (IDF 1979) and can be defined as liquid penetration into a porous agglomerate system due to capillary action or the ability of agglomerates to be penetrated by the liquid (Barbosa-Cánovas *et al.* 2005). With respect to dairy powders, dispersibility is the ability of powder dispersed in water with gentle mixing to separate into individual particles that can pass through a sieve of 210 μm (IDF 1979). In other words, it is the ease with which lumps and agglomerates of powder fall apart in the water. It is expressed as the percentage of the solids dissolved (Sharma *et al.* 2012). Solubility can be defined as the dissolution of soluble agglomerates in a liquid (Barbosa-Cánovas *et al.* 2005), which is often water in case of food systems. It is an assessment of the final condition under which the powder constituents can be brought into solution or stable suspension (Sharma *et al.* 2012). With respect to spray-dried dairy powders, solubility is measured after a series of centrifugation steps (IDF 2005). The importance of solubility in dairy powders is not limited to rehydration but also indicates the impact of the drying process on the functional behavior of soluble proteins present in milk. The solubility of spray-dried dairy products is a function of the structural state of its proteins, whether in native, denatured or aggregated form (Anandharamakrishnan *et al.* 2008). Yet another important property related to dissolution of dairy powdered particles is sinkability, which is the ability of powder particles to overcome the surface tension of water and sink into the water after passing through the surface. It is expressed as milligrams of powder that sink per minute per square centimetre of surface area (Schober & Fitzpatrick 2005).

The relationship between spray-drying process parameters and the solubility of spray-dried whey protein powder was established using a reverse-phase HPLC method. The parameters considered were the outlet air temperature and feed concentration. α-lactalbumin and β-lactoglobulin, the two major whey proteins, were used as the markers for determining the loss of solubility in the reconstituted spray-dried powder. Protein solubility is measured from the concentration of proteins in a dissolved liquid phase in relation to the total amount of protein (Eq. 5.9) (Anandharamakrishnan *et al.* 2008):

$$\text{solubility} = \frac{\text{mass of soluble nitrogen}}{\text{total nitrogen in sample}} \tag{5.9}$$

Therefore,

$$\% \text{ loss of solubility} = \left(1 - \frac{S}{S_u}\right) \times 100$$

where S is the solubility of the spray-dried sample and S_u is the solubility of the untreated sample.

A study by Anandharamakrishnan *et al.* (2008) revealed that loss of solubility had a direct relationship with outlet temperature (T_o) and feed concentration (C_f), with the effect being more pronounced at high T_o and C_f. The effect of T_o was attributed to the increase in degree of protein denaturation or structural changes, which increases the formation of insoluble material with increasing temperature and holding time (Dannenberg & Kessler 1988; Law & Leaver 1997; Fang *et al.* 2012). The influence of feed concentration on solubility can be explained by two reasons, denaturation or aggregation and the different temperature histories of droplets with different initial concentrations. However, α-lactalbumin depicted a smaller drop in solubility than β-lactoglobulin owing to its heat-stable nature. The binding of α-lactalbumin with Ca^{2+} ion present in the whey increases its stability to higher processing temperatures (Permyakov & Berliner 2000).

The influence of spray-drying conditions on protein solubility can also be quantified using the SDS-PAGE technique. This is accomplished by examining the position and intensity of bands in the SDS-PAGE gel. This technique also facilitates the identification of protein involved in the formation of insoluble material. The SDS-PAGE technique detected less than 20% change in protein solubility in milk protein concentrate powders produced at an inlet temperatures of 77 and 107 °C, whereas the detected loss was greater than 20% for powders dried at 155 and 178 °C (Fang *et al.* 2012).

Rehydration properties are significantly affected by the composition of feed subjected to spray drying. For instance, amongst feed solutions comprising different fractions of milk, solubility increased in the order casein < whey protein powder < whole milk powder < semi-skim milk powder < skim milk powder (Figure 5.15b). This trend was also found to correlate with their wetting times, that is, the powders with the highest wetting times possessed least solubility (Figure 5.15c). The highest protein load and residual fat on the surface was found to deteriorate the wetting properties and hence the solubility (Gaiani *et al.* 2011).

5.3.2 Particle size and shape parameters

Particle size is a vital parameter with respect to the appearance, reconstitution properties and flow characteristics of milk powder. The relationship between the particle sizes of five categories of spray-dried milk powder (skim, semi-skim, whole milk, casein and whey protein) and rehydration properties is shown in Figure 5.15a–c. Both milk fraction type and particle size of spray-dried powder exerted influence on dispersibility, wettability and solubility. Each powder type exhibited an optimal range for rehydration properties. Compared to powders with low fat, the optimal size was found to skew towards the higher side for whole milk powders due to the presence of lipids. As a result, at an equivalent particle size, powders with low fat content showed the highest dispersibility, skim milk (95%), semi-skim milk (78%) and whole milk (71%), the reason for which can be attributed to the hydrophobicity of fat. However, the dispersibility percentages for whey and casein powders were inferior when compared to others owing to

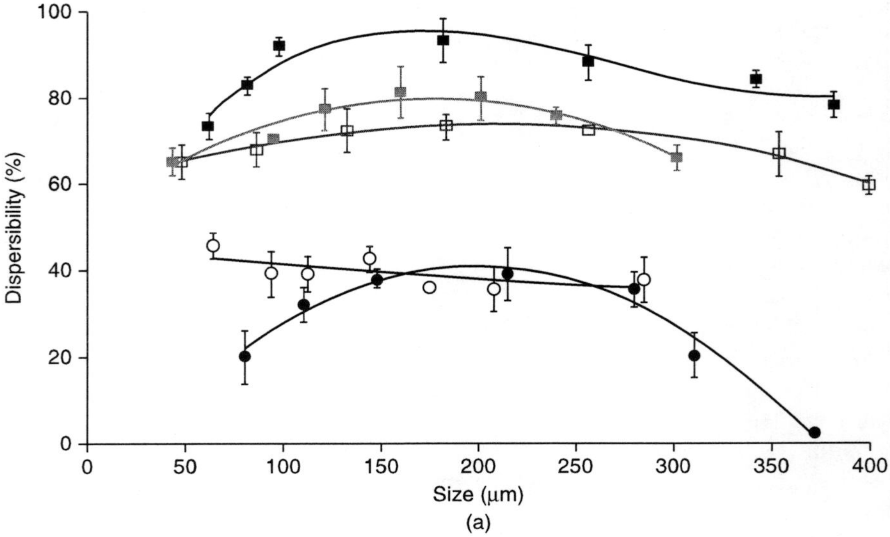

Figure 5.15 Relationship between particle size and rehydration properties: (a) dispersibility, (b) solubility and (c) wettability (Gaiani *et al.* 2011).

the high protein content. The spray-dried dairy powders exhibited a positive correlation between particle size and dispersibility up to a certain size, followed by a drop at larger particle size (Figure 5.15a). This trend was also found to hold true with the solubility of whey protein and whole milk powder (Figure 5.15b), whereas powders with low fat exhibited uniformly high solubility, irrespective of the particle size (Gaiani *et al.* 2011).

Thus, it is evident that an increase in the particle size of dairy powders promotes powder dispersibility, solubility and hence instantization properties. In general, the dispersibility of a powder decreases as the percentage of fine particles (<90 μm) increases (Singh & Newstead 1992). Agglomeration is the most commonly employed method for instant milk production. This involves wetting the fine powder particles with steam, followed by mixing and re-drying the product to the required final moisture content (Brennan 2011). This size enlargement process is usually accomplished using a fluidized bed attached to the conical bottom of the spray chamber. Agglomeration establishes interparticle contact by means of collision and binding forces such as solid bridging, liquid bridging, interparticle attraction forces and mechanical interlocking (Schubert 1980). The use of spray dryers in synergy with fluidized beds is known as a two-stage spray-drying process. However, this can also be done by re-wetting in a fluidized-bed agglomerator that is not integrated with a spray dryer. The moistening is achieved under controlled conditions of low-pressure steam, humid air or fine spray of water and then dried in a fluidized-bed dryer (Fellows 2009). In addition to agglomeration, the use of a fluidized bed overcomes the limitation of a lower diffusion coefficient for moisture removal during the final stages of drying, provides sufficient residence time for the particles to achieve the required final moisture content, and leads to higher evaporation capacity per unit weight of drying air, lower operation cost and lesser powder emission (Westergaard 2010).

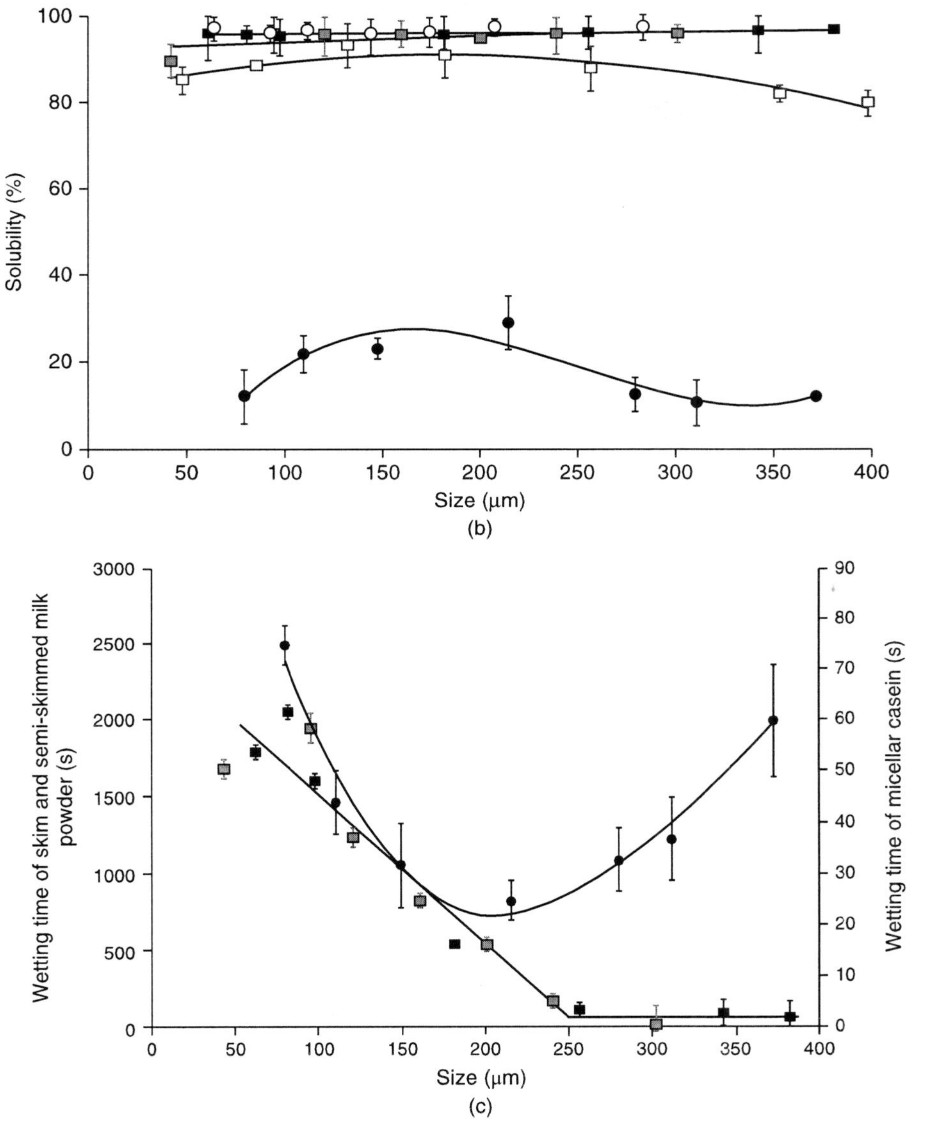

Figure 5.15 *(Continued)*

5.4 Handling spray-drying processing problems

5.4.1 Stickiness

Stickiness is a major processing problem encountered during the spray drying of sugar-rich foods, and dairy products are no exception to this. The glass transition temperature (T_g) is the major indicator of stickiness and can be defined as a temperature and time-dependent change in physical state from glassy to rubbery viscous liquid without phase change (Slade & Levine 1991; Wunderlich 1981). While the glassy state is structurally rigid, the rubbery or amorphous state is a non-equilibrium metastable state associated with hygroscopicity and stickiness (Bhandari *et al.* 1997). T_g values for milk ingredients can be estimated using the Gordon–Taylor equation (Eq. 5.10) (Gordon & Taylor 1952):

$$T_g = \frac{w_1 T_{g1} + c w_2 T_{g2}}{w_1 + c w_2} \tag{5.10}$$

where T_g is the predicted T_g of a binary system, w_1 and w_2 are the weight fractions of solids and water, respectively, T_{g1} and T_{g2} are the glass transition temperatures of dry solid and water, respectively, and k is an empirical constant for a system, for example 0.69 for the lactose–water system (Saltmarch & Labuza 1980).

When milk undergoes rapid evaporation during spray drying, an amorphous continuous matrix of lactose is formed in which proteins, fat globules and air cells are dispersed (Aguilar & Ziegler 1994). Lactose exhibits T_g in the range 97–116 °C (Roos & Karel 1990; Haque & Roos 2004). When the spray-drying temperature exceeds this temperature range, the amorphous matrix undergoes a transition from glassy to rubbery state. This transition initiates a series of events that eventually lead to stickiness (Roos & Karel 1992) owing to the increase in molecular mobility and the consequent decrease in viscosity. With respect to a thermal process like spray drying, stickiness impairs heat transfer by scale formation and fouling (Wilson 2005; Xin *et al.* 2005). The stickiness of dairy powder also poses several post-processing bottlenecks such as low product yield, difficulties in handling and storage, and poor product quality (Shrestha *et al.* 2007).

The T_g values for spray-dried powders of different dairy products (Table 5.2) vary with their composition, especially in protein and sugar content. For instance, the presence of high molecular weight glucose and galactose in skim milk powder elevated its T_g in comparison with that of hydrolysed lactose (Shrestha *et al.* 2007).

The most common approach to avoid stickiness is to operate the spray-drying process below the T_g of the amorphous matrix. However, the problem can also be alleviated by elevating the T_g by addition of high molecular weight adjuvants (e.g. glucose syrup,

Table 5.2 Glass transition temperatures of spray-dried dairy products

Spray-dried dairy product	T_g value (°C)	Reference
Skim milk powder with hydrolysed lactose	48	Shrestha *et al.* (2007)
Whole milk powder with partially hydrolysed lactose • Moisture content 0.6% • Lactose–glucose–galactose ratio 8:15:15	58	Fernandez *et al.* (2003)
Skim milk powder	93	Shrestha *et al.* (2007)
Lactose	97.8	Shrestha *et al.* (2007)

maltodextrin) as T_g increases with increasing molecular weight (Truong *et al.* 2005). Among the milk constituents, WPI exhibits this unique property due to its strong surfactant and film-forming properties. Addition of WPI to lactose solution at 0.5–1% (w/w) lowered the tensile strength of lactose solution due to the preferential migration of WPI to the droplet–air interface. As a result, a non-sticky skin was formed at high temperature that resisted droplet coalescence and also decreased the tendency of particles to adhere to the wall of spray chamber. This property enables WPI to be used as a drying aid during spray drying (Adhikari *et al.* 2007).

5.4.2 Thermal denaturation of proteins

Heat-sensitive proteins in milk are prone to thermal inactivation or denaturation due to the high-temperature operation of spray drying. A protein molecule in its native form comprises a compact globular structure with a tightly packed hydrophobic core stabilized by hydrogen and ionic bonds present in the aqueous medium. While the hydrophobic groups are buried inside the polypeptide chain, the hydrophilic groups are exposed to the surrounding aqueous milieu. When the proteins are exposed to temperatures above their denaturation temperature, the hydrophobic interactions break to change the native folded state into an inactive unfolded state. This leads to hydrophobic interaction mediated cross-linking between protein molecules, which result in aggregation, coagulation and eventually loss of solubility (Anandharamakrishnan 2008). Loss of solubility has been attributed to the formation of heat-induced protein aggregates (Bernard *et al.* 2011) during high-temperature exposure. In addition to loss in solubility, denaturation also affects the functional properties of milk proteins, including foamability and emulsifying capacity (Sourdet *et al.* 2002; Livney *et al.* 2003).

The major proteins present in milk include α-casein, β-casein, κ-casein, β-lactoglobulin and α-lactalbumin. The extent of protein denaturation during spray drying can be estimated by an increase in the amount of insoluble protein, using the SDS-PAGE method. An increase in inlet air temperature (T_i 77, 107, 155 and 178 °C) resulted in a considerable decrease in the band intensity of these proteins. Also, the protein bands were more diffused and less defined at high T_i. However, the casein proteins (α-, β- and κ-casein) were found to be more susceptible to denaturation than the whey proteins (α-lactalbumin and β-lactoglobulin) (Fang *et al.* 2012).

The potential means of alleviating protein denaturation during spray drying of milk products include generation of small droplets, use of low outlet air temperature or adding carrier or wall materials to protect these heat-sensitive components (Etzel *et al.* 1996).

5.5 Applications of spray drying for the production of dried milk and milk products

The applications of spray drying with respect to dairy products can be classified into three types according to the intended functionality of the final product: (i) to obtain a dried particulate product for extended shelf life, instantization and packaging properties, (ii) for the microencapsulation of live probiotic cells for enhancing the functional

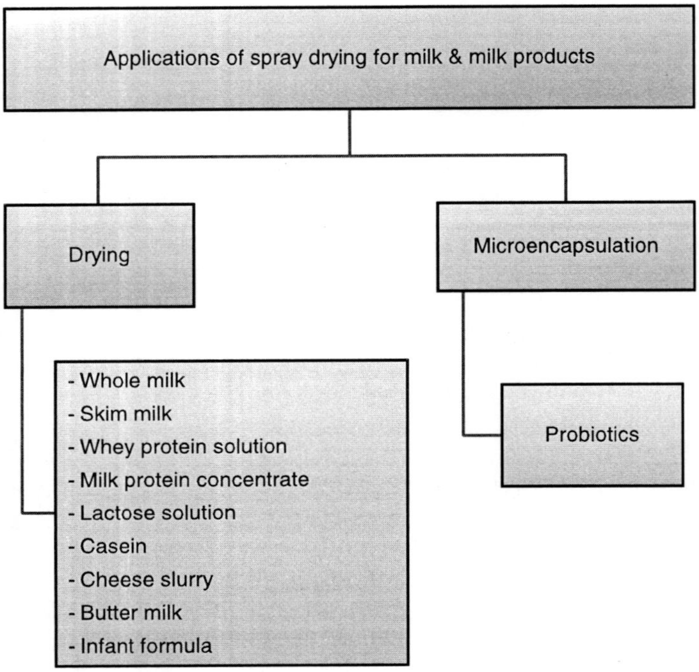

Figure 5.16 Applications of spray drying for milk and milk products.

properties of dairy products and (iii) to alleviate the sensory properties of certain dairy constituents. For instance, spray-drying encapsulation of casein hydrolysate has been used as an approach to reduce the problem of bitter taste (Molina Ortiz *et al.* 2009). The different categories of milk products handled by the spraying process are represented in Figure 5.16. The process parameters employed for the spray drying of different milk products are presented in Table 5.3. Since the applications of spray drying as dehydration technique were discussed earlier, the rationale for employing it as an encapsulation technique for ingredients of relevance to the dairy sector is briefly stated in this section.

Encapsulation by spray drying can be defined as the atomization of feed emulsions containing a bioactive core and a suitable wall material into a high-temperature drying medium, resulting in the formation of a protective crust layer due to the difference in drying characteristics between the wall and the aqueous dispersion containing the core. Thus skin formation during the falling rate drying period forms the basis of employing spray drying for encapsulation. The prerequisite for encapsulation by spray drying is that the wall material should possess inherent film-forming properties so as to form a coating around the droplet-containing core. The drying rate of most commonly used wall materials decreases rapidly as the moisture content decreases (Figure 5.17) (Matsuno & Adachi 1993). In addition to the drying rate of the wall material, the diffusion rate of the core also reduces during the falling rate period with the decrease in moisture content. This ensures the retention of core within the wall matrix after the completion of spray-drying process.

Table 5.3 Process parameters for spray drying of milk and milk products

Milk/milk product (feed material)	Atomizer type	Atomization parameters	Feed characteristics	T_i (°C)	T_o (°C)	Spray chamber configuration/ dimensions	Reference
Skim milk concentrate, whole milk concentrate	Centrifugal	Speed: 30000 rpm Feed rate: 0.8 l/h	Solid content: 10%, 20% and 30% Temperature: 50 °C	145 205	85 105	Diameter: 80 cm Height: 60 cm Conical base of 60° angle	Kim *et al.* (2009)
Standardized whole milk	Disc atomiser	Speed: 16000 rpm Feed rate: 130 l/h Feed pressure: 10 MPa	Solid content: 49% Temperature: 65 °C	180	NM	Inlet air flow: 2700 kg/h Chamber temperature: 80 °C Chamber pressure: -10^3 MPa Fluidized-bed inlet temperature: 75 °C Fluidized-bed inlet air flow: 1200 m^3/h	Ye *et al.* (2007)
Pasteurized, homogenized and condensed whole milk	Embedded-electrode, two-fluid, spray-charging nozzle	Feed rate: 25 ml/min Voltage: 300–900 V	Solid content: 50% Temperature: 55 °C	175	70	Co-current	Johnson *et al.* (1996)
Skim milk concentrate	Pressure nozzle	Nozzle pressure: 60–200 bar Internal diameter: 0.25 and 0.38 mm	Solid content: 44% and 35% Feed density: 1142, 1126 kg/m^3 Temperature: 55.7–67.1 °C	104.4–211	70–100	NM	Nikolova *et al.* (2014)

(Continued)

Table 5.3 (Continued)

Milk/milk product (feed material)	Atomizer type	Atomization parameters	Feed characteristics	T_i (°C)	T_o (°C)	Spray chamber configuration/ dimensions	Reference
Milk protein concentrate (reconstituted)	Monodisperse atomizer	Feed rate: 2 g/min Droplet size: 201–230 μm	Solid content: 20% (w/w)	178, 155, 107, 77	103, 93, 68, 54	Co-current	Fang *et al.* (2012)
Cheese slurry	Rotary atomizer	Speed: 20000–31500 rpm Diameter: 50 mm with 24 vanes Atomization pressure: 294–588 kPa	Temperature: 45 °C Solid content: 25% dry matter on cheese basis	160–230	60–100	Co-current Mass flow rate of air: 0.08 kg/s Height: 1.2 m Diameter: 0.87 m	Erbay *et al.* (2015)
Cheese bioaroma (whey permeate) (microencapsulation)	Two-fluid nozzle	Orifice diameter: 1.2×10^{-3} m Feed rate: 2.97×10^{-7} m^3/s Air pressure: 5 bar	Solid content: 25% Wall to core ratio: 4:1 Wall materials: modified corn starch in combination with maltodextrin (20DE)	163–187	NM	Inlet air flow rate: 5.8×10^{-4} m^3/s	da Costa *et al.* (2015)
Casein hydrolysate (microencapsulation to reduce bitterness)	Two-fluid nozzle	Atomization compressed air-flow rate: 50 l/min Feed rate: 3.6 l/h	Solid content: 12% Wall to core ratio: 7:3 and 8:2 Wall materials: Soy protein isolate	140	110	Co-current	Molina Ortiz *et al.* (2009)

Lactose solution	Ultrasonic atomizer	Feed rate: 14.5 and 28 g/min Atomizing air flow Rate: 10 and 20 l/min	Solid content: 9.09% w/w	170 and 230	60–100	Short-form dryer Internal diameter: 0.8 m Height: 2 m, Height of upper cylindrical part: 1.3 m Height of lower conical bottom: 0.63 m	Das *et al.* (2010)
Whey protein solution	Twin-fluid nozzle	Pressure: 45 psig Feed rate: 5–17.5 kg/h	Solid content: 20%, 30% and 40% w/w	160–255	60–120	A tall-form co-current spray dryer Height: 11 ft Diameter: 3 ft Drying gas-flow rate: 227 kg/h	Anandharamakrishnan *et al.* (2008)
Butter milk	NM	NM	Solid content: 20% (w/w)	195	85	NM	Morin *et al.* (2007)
Infant formula (lipid and protein component-containing composition)	Two-fluid nozzle	Pressure: 8 bar	Solid content: 50–60% (w/w)	180	NM	NM	Van *et al.* (2014)
Infant formula (lipid and protein component-containing composition)	Rotary atomizer	Tip speed: 65–95 m/s	Solid content: 50–60% (w/w)	175	NM	NM	Van *et al.* (2013b)

NM, not mentioned; DE, dextrose equivalent.

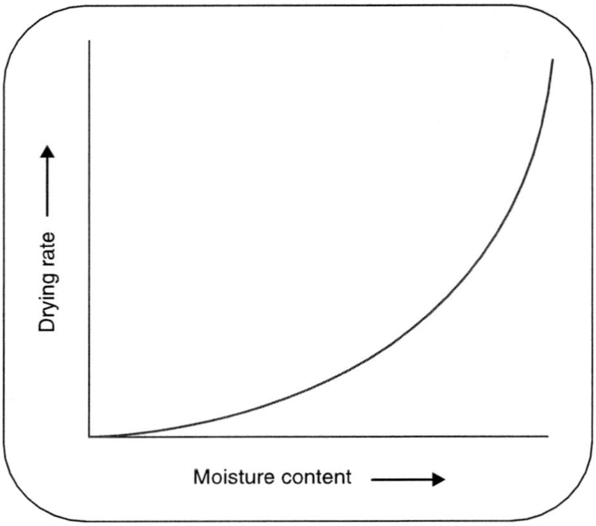

Figure 5.17 Characteristic drying curve of most often used wall materials. Modified from Anandharamakrishnan & Padma Ishwarya (2015).

A wide range of compounds can be encapsulated using spray drying, of which living entities such as probiotic bacteria and micronutrients such as vitamins A and D are of relevance with respect to dairy products. Lyophilization or freeze drying is the common approach used for the production of dried cells. However, spray drying has a competitive advantage in terms of product structure that offers the best protection to cells and also in terms of lower cost of storage and refrigeration (Paez *et al.* 2012). Moreover, the smooth skin-forming morphology of the spray-dried microcapsules (Figure 5.14e) is more favourable to cell stability during the transit through the acidic environment of the upper gastrointestinal tract. In the freeze-dried microcapsules, the cells are entrapped closer to the surface of the microcapsules and hence are susceptible to cell death under acidic conditions. However, as spray drying is a high-temperature operation it involves careful selection of the drying air temperature and wall material to obtain the optimal viability and functionality of the encapsulated cells. Various studies on the encapsulation of probiotic cells by spray drying and the associated process parameters are presented in Table 5.4.

5.6 Conclusion

Spray drying continues to be an enduring process in the field of dairy drying. The spray-drying process is highly viable in terms of its economic and industry-friendly operation. This chapter has described the applications of spray drying that extend the shelf life and improve the functionality of dairy products by its encapsulation effect. The association between spray drying and dairy industry will continue to evolve as a result of research and development activities, consumer demand and the current inclination towards food products with the synergistic benefits of healthiness and convenience.

Table 5.4 Application of spray drying for microencapsulation of probiotics

Probiotic organism	Wall material	Feed composition	Atomizer type	Feed flow rate	T_i (°C)	T_o (°C)	Maximum achieved viability/survival loss after spray drying	Reference
Lactobacillus plantarum	WPI + SA Denatured WPI (DWPI) + SA	Core-to-wall ratio of 1:1 and total solid content of 15% w/v	Twin fluid nozzle	40 ml/min	135	90	WPI + SA 80% survival DWPI + SA 87% survival	Rajam *et al.* (2012)
Lactobacillus acidophilus NCIMB 701748	Blend of whey protein concentrate, maltodextrin, glucose	MD:WPC:Glu in the ratio of 3:1:1 and balanced with 100 ml distilled water	NM	6–9 ml/min	120–160	60–92	Maximum of 84% at T_o of 60 °C	Behboudi-Jobbehdar *et al.* (2013)
Bifidobacterium Bb-12	Whey	NM	NM	6 ml/min	150	50–60	ND	De Castro-Cislaghi *et al.* (2012)
Lactobacillus rhamnosus	WPI-MD; WPI-MD-Glu, WPI-inulin, WPI-inulin-Glu	WPI:MD 1:2 WPI:MD:Glu 1:1 WPI:inulin 1:2 WPI:inulin:Glu 1:1	NM	NM	160	65	ND	Ying *et al.* (2012)
Bifidobacterium BB-12	RSM partially replaced with inulin, oligofructose and oligofructose-enriched inulin	The different combinations of RSM with inulin, oligofructose and oligofructose-enriched inulin in the ratio of 1:1	NM	6 ml/min	150	55	ND	Fritzen-Freire *et al.* (2012)

(*Continued*)

Table 5.4 (Continued)

Probiotic organism	Wall material	Feed composition	Atomizer type	Feed flow rate	T_i (°C)	T_o (°C)	Maximum achieved viability/survival loss after spray drying	Reference
Bifidobacterium bifidum BB-12	Gum arabic, gelatin, maltodextrin, pectin, SMP	Cells suspended in 50 ml phosphate buffer (pH 6.8) to gain an approximate cell number of 10^{12} 10%, 50%, or 100% of each protectant (based on the wet cell mass) was added to the cell suspension to form the feed	NM	1.5 g/min	115	70–80	Determined in terms of loss of membrane integrity and membrane potential, as explained in the text	Salar-Behzadi *et al.* (2013)
Lactobacillus acidophilus, Lactobacillus rhamnosus	MD	Ratio of total solids in raspberry juice to MD (MD ratio) set at 1:1, 1:1.5, 1:2	NM	40–60 ml/min	100–130	67–94	Maximum survival of 81.17% at T_i = 100 °C; MD ratio of 1:1 and inlet feed rate of 40 ml/min	Anekella & Orsat (2013)
Lactobacillus casei subsp. *paracasei* LMG P-21380	Trehalose and maltodextrin	Cell concentration (0.75–10 g/100 g), solids concentration (10–30 g/100 g), MD (DE5 and DE19) and MD:trehalose ratios (1:0, 2:1, 1:1)	Ultrasonic atomiser	9 ml/min	NM	NM	70.6 ± 6.2%	Semyonov *et al.* (2011)

Lactobacillus acidophilus NCIMB 701748	SA, CS, HPMC along with MD, Glu and WPC	5 g of SA, 0.5 g of CS and 5 g of HPMC in each of the 100 ml carrier solution variant with the total solid content ranging between 19% and 24%	NM	7.14 ml/min	134 ± 1	76 ± 2	Maximum survival after spray drying was obtained with microencapsulates obtained from HPMC, which was 8.73 log cfu/g with respect to 8.99 log cfu/g in the feed suspension before spray drying	Yonekura *et al.* (2014)
Lactobacillus rhamnosus	Native rice starch, inulin	Total solid content of wall material solution ranging between 10% and 20%	NM	14 g/min	135–155	NM	ND	Avila-Reyes *et al.* (2014)
Lactobacillus reuteri	Whey	20% (w/v) whey solution (with 0.5% (w/v) yeast extract as supplement) with culture of *L. reuteri* at the concentration of at least 10^8 cfu/ml	NM	2–4 ml/min	89–100	55–65	Survival of approx. >10^6–10^7 cfu/g compared to >10^8–10^9 cfu/g before spray drying	Jantzen *et al.* (2013)
Lactobacillus paracasei subsp. *paracasei*	SA, FOS	Feed solution containing 2.5% w/w SA and 1.5% w/w FOS	NM	NM	120	60	Increased survival by 4 logs after spray drying	Ivanovska *et al.* (2012)
L. casei, L. plantarum, L. paracasei, L. plantarum *L. acidophilus*	Skim milk	20% w/v	Two-fluid nozzle	NM	170	85	*L. casei* 98% *L. plantarum* 94% *L. paracasei* 95% *L. plantarum* 96% *L. acidophilus* 97%	Paez *et al.* (2012)

WPI, whey protein isolate; SA, sodium alginate; DWPI, denatured WPI; MD, maltodextrin; WPC, whey protein concentrate; Glu, glucose; NM, not mentioned; ND, not determined; RSM, reconstituted skim milk; SMP, skimmed milk powder; DE5, dextrose equivalent 5; DE19, dextrose equivalent 19; CS, chitosan; HPMC, hydroxypropyl methylcellulose; FOS, fructooligosaccharide.

References

Adhikari, B., Howes, T., Shrestha, A.K. & Bhandari, B.R. (2007). Development of stickiness of whey protein isolate and lactose droplets during convective drying, *Chemical Engineering and Processing: Process Intensification*, **46**(**5**), 420–428.

Aguilar, C.A. & Ziegler, G.R. (1994). Physical and microscopic characterization of dry whole milk with altered lactose content. 2. Effect of Lactose Crystallization. *Journal of Dairy Science*, **77**(**5**), 1198–1204.

Alamilla-Beltran, L., Canona-Perez, J., Jimenez-Aparicio, A. & Gutierrez-Lopez, G. (2005). Description of morphological changes of particles along spray drying, *Journal of Food Engineering*, **67**, 179–184.

Anandharamakrishnan, C. (2008). *Experimental and computational fluid dynamics studies on spray-freeze-drying and spray-drying of proteins*. Ph.D. thesis, Loughborough University.

Anandharamakrishnan, C. (2013). Computational fluid dynamics applications in food processing. In: *Briefs in Food, Health, and Nutrition*, Springer, New York.

Anandharamakrishnan, C. & Padma Ishwarya, S. (2015). Introduction to spray drying, In: *Spray Drying Techniques for Food Ingredient Encapsulation*, pp. 1–36, John Wiley & Sons, Chichester.

Anandharamakrishnan, C., Rielly, C.D. & Stapley, A.G.F. (2008). Loss of solubility of α-lactalbumin and β-lactoglobulin during the spray drying of whey proteins. *LWT – Food Science and Technology*, **41**, 270–277.

Anekella, K. & Orsat, V. (2013). Optimization of microencapsulation of probiotics in raspberry juice by spray drying. *LWT – Food Science and Technology*, **50**, 17–24.

Avila-Reyes, S.V., Garcia-Suarez, F.J., Jimenez, M.T., Martin-Gonzalez, M.F.S. & Bello-Perez, L.A. (2014). Protection of *L. rhamnosus* by spray-drying using two prebiotics colloids to enhance the viability. *Carbohydrate Polymers*, **102**, 423–430.

Barbosa-Canovas, G.V., Ortega-Rivas, E., Juliano, P. & Yan, H. (2005). Bulk properties. In: *Food Powders Physical Properties, Processing, and Functionality*, pp. 55-90, Kluwer Academic /Plenum Publishers, New York.

Behboudi-Jobbehdar, S., Soukoulis, C., Yonekura, L. & Fisk, I. (2013). Optimization of spray-drying process conditions for the production of maximally viable microencapsulated *L. acidophilus* NCIMB 701748. *Drying Technology*, **31**, 1274–1283.

Bernard, C., Regnault, S., Gendreau, S., Charbonneau, S. & Relkin, P. (2011). Enhancement of emulsifying properties of whey proteins by controlling spray-drying parameters. *Food Hydrocolloids*, **25**(**4**), 758–763.

Bhandari, B.R., Datta, N. & Howes, T. (1997). Problems associated with spray drying of sugar-rich foods. *Drying Technology*, **15**(**2**), 671–684.

Birchal, V.S., Passos, M.L., Wildhagen, G.R.S. & Mujumdar, A.S. (2005). Effect of spray-dryer operating variables on the whole milk powder quality. *Drying Technology*, 2005, **23**(**3**), 611–636.

Birchal, V.S., Huang, L., Mujumdar, A.S. & Passos, M.L. (2006). Spray dryers: modeling and simulation. *Drying Technology*, **24**, 359–371.

Brenn, G., Helpio, T. & Durst, F. (1997). A new apparatus for the production of monodisperse sprays at high flow rates. *Chemical Engineering Science*, **52**, 237–244.

Brennan, J.G. (2011). Evaporation and dehydration. In: *Food Processing Handbook* (eds Brennan, J.G. & Grandison, A.S.), pp. 77–152, Wiley-VCH Verlag GmBH & Co. KGaA, Boschstr, Weinheim.

Charlesworth, D.H. & Marshall, W.R. (1960). Evaporation from drops containing dissolved solids. *AIChE Journal,* **6(1)**, 9–23.

Charm, S.E. (1981). The fundamental of food engineering. In: *Products from Milk* (ed. Webb, B.H.), pp. 294, The Air Publishing Company, Westport.

Chen, X.D. (2004). Heat-mass transfer and structure formation during drying of single food droplets. *Drying Technology,* **22(1–2)**, 179–190.

da Costa, J.M.G., Silva, E.K., Hijo, A.A.C.T., Azevedo, V.M., Malta, M.R., Alves, J.G.L.F. & Borges, S.V. (2015). Microencapsulation of Swiss cheese bioaroma by spray-drying: Process optimization and characterization of particles. *Powder Technology,* **274**, 296–304.

Dannenberg, F. & Kessler, H. (1988). Reaction kinetics of the denaturation of whey proteins in milk. *Journal of Food Science,* **53(1)**, 258–263.

Das, D., Husni, H.A. & Langrish, T.A.G. (2010). The effects of operating conditions on lactose crystallization in a pilot-scale spray dryer. *Journal of Food Engineering,* **100**, 551–556.

De Castro-Cislaghi, F.P., Silva, C.D.R., Fritzen-Freire, C.B., Lorenz, J.G. & Sant'Anna, E.S. (2012). *Bifidobacterium* Bb-12 microencapsulated by spray drying with whey: Survival under simulated gastrointestinal conditions, tolerance to NaCl, and viability during storage. *Journal of Food Engineering,* **113**, 186–193.

Devakate, R.V., Patil, V.V., Waje, S.S. & Thorat, B.N. (2009). Purification and drying of bromelain. *Separation and Purification Technology,* **64**, 259–264.

Dittman, F.W. & Cook, E.M. (1977). Establishing the parameters for a spray dryer. *Chemical Engineering,* **84(2)**, 108–112.

Dobry, D.E., Setell, D.M., Baumann, J.M., Ray, R.J., Graham, L.J., & Beyerinck, R.A. (2009). A model-based methodology for spray-drying process development. *Journal of Pharmaceutical Innovation,* **4(3)**, 133–142.

El-Sayed, T.M., Wallack, D.A. & King, C.J. (1990). Changes in particle morphology during drying of drops of carbohydrate solutions and food liquids. 1. Effects of composition and drying conditions. *Industry and Engineering Chemistry Research,* **29**, 2346–2354.

Erbay, Z., Koca, N., Kaymak-Ertekin, F. & Ucuncu, M. (2015). Optimization of spray drying process in cheese powder production. *Food and Bioproducts Processing,* **93**, 156–165.

Etzel, M.R., Suen, S.Y., Halverson, S.L. & Budijono, S. (1996). Enzyme inactivation in a droplet forming a bubble during drying. *Journal of Food Engineering,* **27**, 17–34.

Fang, Y., Rogers, S., Selomulya, C. & Chen, X.D. (2012). Functionality of milk protein concentrate: Effect of spray drying temperature. *Biochemical Engineering Journal,* **62**, 101–105.

Fellows, P.J. (2009). Dehydration. In: *Food Processing: Technology-Principles and practice,* 3rd edn, pp. 4481–4512, Woodhead Publishing Limited, Cambridge.

Fernandez, E., Schebor, C. & Chirife, J. (2003). Glass transition temperature of regular and lactose hydrolysed milk powders. *Lebensmittel-Wissenschaft & Technologie,* **36**, 547–551.

Filková, I., Huang, L.X. & Mujumdar, A. (2007). Industrial spray drying systems. In: *Handbook of Industrial Drying,* 3rd edn, pp. 215–256, Boca Raton, CRC Press.

Freudig, B., Hogekamp, S. & Schubert, H. (1999). Dispersion of powders in liquids in a stirred vessel. *Chemical Engineering and Processing: Process Intensification,* **38**, 525–532.

Fritzen-Freire, C.B., Prudêncio, E.S., Amboni, R.D.M.C., Pinto, S.S., Negrao-Murakami, A.N. & Murakami, F.S. (2012). Microencapsulation of bifidobacteria by spray drying in the presence of prebiotics. *Food Research International*, **45**, 306–312.

Fu, N., Woo, M.W., Selomulya, C. & Chen, X.D. (2013). Shrinkage behaviour of skim milk droplets during air drying. *Journal of Food Engineering*, **116**, 37–44.

Gaiani, C., Boyanova, P., Hussain, R., Pazos, I.M., Karam, M.C., Burgain, J. & Scher, J. (2011). Morphological descriptors and colour as a tool to better understand rehydration properties of dairy powders. *International Dairy Journal*, **21**, 462–469.

Ghandi, A., Powell, I.B., Howes, T., Chen, X.D. & Adhikari, B. (2012). Effect of shear rate and oxygen stresses on the survival of *Lactococcus lactis* during the atomisation and drying stages of spray drying: A laboratory and pilot scale study. *Journal of Food Engineering*, **113**, 194–200.

Goldberg, S. (2008). Mechanical/physical methods of cell disruption and tissue homogenization. *Methods in Molecular Biology, 2D PAGE: Sample Preparation and Fractionation*, **424**, 3–22.

Gordon, M. & Taylor, J. S. (1952). Ideal copolymers and the second-order transitions of synthetic rubbers. I. Non-crystalline copolymers. *Journal of Applied Chemistry*, **2**, 493–500.

Handscomb, C.S., Kraft, M. & Bayly, A.E. (2009). A new model for the drying of droplets containing suspended solids. *Chemical Engineering Science*, **64**, 628–637.

Haque, M.K. & Roos, Y.H. (2004). Water plasticization and crystallization of lactose in spray-dried lactose/protein mixtures. *Journal of Food Science*, **69(1)**, FEP23–FEP29.

Hede, P.D., Bach, P. & Jensen, A.D. (2008). Two-fluid spray atomisation and pneumatic nozzles for fluid bed coating/agglomeration purposes: a review. *Chemical Engineering Science*, **63(14)**, 3821–3842.

IDF (1979). *Instant dried milk – determination of the dispersibility and wettability: IDF 087*, International Dairy Federation, Brussels.

IDF (2005). *Dried milk and dried milk products. Determination of insolubility index: IDF 129*, International Dairy Federation, Brussels.

Ivanovska, T.P., Mladenovska, K., Kavrakovski, Z., Bogdanovska, L., Grozdanov, A., Popovski, E. & Petrushevska-Tozi, L. (2012). *Macedonian pharmaceutical bulletin*, **58(1, 2)**, 45–52.

Jantzen, M., Göpel, A. & Beermann, C. (2013). Direct spray drying and microencapsulation of probiotic Lactobacillus reuteri from slurry fermentation with whey. *Journal of Applied Microbiology*, **115**, 1029–1036.

Jaworek, A. (2007). Micro- and nanoparticle production by electrospraying. *Powder Technology*, **176(1)**, 18–35.

Johnson, R.K., Anantheswaran, R.C. & Law, S.E. (1996). Electrostatic-enhanced atomization for spray drying of milk. *LWT – Food Science and Technology*, **29(1–2)**, 71–81.

Kim, E.H.J., Chen, X.D. & Pearce, D. (2003). On the mechanisms of surface formation and the surface compositions of industrial milk powders. *Drying Technology*, **21(2)**, 265–278.

Kim, E.H.J., Chen, X.D. & Pearce, D. (2009). Surface composition of industrial spray-dried milk powders. 2. Effects of spray drying conditions on the surface composition, *Journal of Food Engineering*, **94(2)**, 169–181.

King, C.J. (1990). Spray drying food liquids and the retention of volatiles. *Chemical Engineering Progress*, **86(6)**, 33–39.

Langrish, T.A.G. & Fletcher, D.F. (2001). Spray drying of food ingredients and applications of CFD in spray drying. *Chemical Engineering and Processing: Process Intensification,* **40(4)**, 345–354.

Law, A.J.R. & Leaver, J. (1997). Effects of protein concentration on rates of thermal denaturation of whey proteins in milk. *Journal of Agricultural and Food Chemistry,* **45(11)**, 4255–4261.

Livney, Y.D., Corredig, M. & Dalgleish, D.G. (2003). Influence of thermal processing on the properties of dairy colloids. *Current Opinion in Colloid and Interface Science,* **8 (4–5)**, 359–364.

Masters, K. (1991). Spray drying fundamentals: process stages and layouts. In: *Spray Drying Handbook*, 5th edn, pp. 23–64, Longman Scientific & Technical, New York.

Masters, K. (1985). Atomization. In: *Spray Drying Handbook*, 4th edn, pp. 171–297, Longman Scientific & Technical, New York.

Matsuno, R. & Adachi, S. (1993). Lipid encapsulation technology – techniques and applications to food. *Trends in Food Science & Technology*, **4(8)**, 256–261.

Meesters, G.M.H., Vercoulen, P.H.W., Marijnissen, J.C.M. & Scarlett, B. (1992). Generation of micron-sized droplets from the Taylor cone. *Journal of Aerosol Science,* **23**, 37–49.

Molina Ortiz, S.E., Mauri, A., Monterrey-Quintero, E.S., Trindadeb, M.A., Santana, A.S. & Favaro-Trindadeb, C.S. (2009). Production and properties of casein hydrolysate microencapsulated by spray drying with soybean protein isolate. *LWT – Food Science and Technology,* **42(5)**, 919–923.

Morin, P., Jimenez-Flores, R. & Pouliot, Y. (2007). Effect of processing on the composition and microstructure of buttermilk and its milk fat globule membranes. *International Dairy Journal,* **17(10)**, 1179–1187.

Nasr, G.G., Yule, A.J. & Bendig, L. (2002). Sprays in industrial production processes. In: *Industrial Sprays and Atomization – Design, Analysis and Applications,* pp. 35–118, Springer-Verlag, London.

Nikolova, Y., Petit, J., Sanders, C., Gianfrancesco, A., Desbenoit, N., Frache, G., Francius, G., Scher, J. & Gaiani, C. (2014). Is it possible to modulate the structure of skim milk particle through drying process and parameters? *Journal of Food Engineering,* **142**, 179–189.

Ohnesorge, W.V. (1936). Die Bildung von Tropfen an Düsen und die Auflösung flüssiger Strahlen, *ZAMM – Journal of Applied Mathematics and Mechanics/Zeitschrift für Angewandte Mathematik und Mechanik,* **16**, 355–358.

Paez, R., Lavari, L., Vinderola, G., Audero, G., Cuatrin, A., Zaritzky, N. & Reinheimer, J. (2012). Effect of heat treatment and spray drying on lactobacilli viability and resistance to simulated gastrointestinal digestion. *Food Research International,* **48**, 748–754.

Patel, K.C. & Chen, X.D. (2008). Surface-center temperature differences within milk droplets during convective drying and drying-based Biot number analysis. *AIChE Journal,* **54**, 3273–3290.

Percy, S.R. (1872). Improvement in drying and concentrating liquid substances by atomising. US Patent No. 125,406, US Patent and Trademark Office, Washington, DC.

Permyakov, E.A. & Berliner, L.J. (2000). α-lactalbumin: Structure and function. *FEBS Letters,* **473(3)**, 269–274.

Pinder, K.L. (1952). *Fundamentals of spray drying.* Master's thesis, Faculty of Graduate Studies and Research of McGill University, Montreal.

Pisecky, J. (1978). Bulk density of milk powders. *Dairy Industries International,* **2**, 4–11.

Rajam, R., Karthik, P., Parthasarathi, S. Joseph, G.S. & Anandharamakrishnan, C. (2012). Effect of whey protein – alginate wall systems on survival of microencapsulated *Lactobacillus plantarum* in simulated gastrointestinal conditions. *Journal of Functional Foods*, **4**, 891–898.

Ranz, W.E. & Marshall, J.M. (1952). Evaporation from drops, Part 1. *Chemical Engineering Progress*, **48**, 173.

Rogers, S., Fang, Y., Lin, S.X., Selomulya, C. & Chen, X.D. (2012a). A monodisperse spray dryer for milk powder: Modelling the formation of insoluble material. *Chemical Engineering Science*, **71**, 75–84.

Rogers, S., Wu, W.D., Lin, S.X.Q. & Chen, X.D. (2012b). Particle shrinkage and morphology of milk powder made with a monodisperse spray dryer. *Biochemical Engineering Journal*, **62**, 92–100.

Roos, Y. & Karel, M. (1990). Differential scanning calorimetry study of the phase transitions affecting the quality of dehydrated materials. *Biotechnology Progress*, **6**, 159–163.

Roos, Y. & Karel, M. (1992). Crystallization of amorphous lactose. *Journal of Food Science*, **57**(**3**), 775–777.

Sadek, C., Tabuteau, H., Schuck, P., Fallourd, Y., Pradeau, N., Le Floch-Fouere, C. & Jeantet, R. (2013). Shape, shell, and vacuole formation during the drying of a single concentrated whey protein droplet. *Langmuir*, **29**, 15606–15613.

Sadek, C., Pauchard, L., Schuck, P., Fallourd, Y., Pradeau, N., Le Floch-Fouéré, C. & Jeantet, R. (2015). Mechanical properties of milk protein skin layers after drying: Understanding the mechanisms of particle formation from whey protein isolate and native phosphocaseinate. *Food Hydrocolloids*, **48**, 8–16.

Sadek, C., Schuck, P., Fallourd, Y., Pradeau, N., Jeantet, R. & Floch-Fouéré, C.L. (2016). Buckling and collapse during drying of a single aqueous dispersion of casein micelle droplet. *Food Hydrocolloids*, **52**, 161–166.

Salar-Behzadi, S., Wu, S., Toegel, S., Hofritcher M., Altenburger, I., Unger, F.M., Wirth, M. & Viernstein, H. (2013). Impact of heat treatment and spray drying on cellular properties and culturability of *Bifidobacterium bifidum* BB-12. *Food Research International*, **54**, 93–101.

Saltmarch, M. & Labuza, T.P. (1980). Influence of relative humidity on the physicochemical state of lactose in spray-dried sweet whey powders. *Journal of Food Science*, **45**, **1231–1236**, 1242.

Schober, C. & Fitzpatrick, J.J. (2005). Effect of vortex formation on powder sinkability for reconstituting milk powders in water to high solids content in a stirred tank. *Journal of Food Engineering*, **71**, 1–8.

Schubert, H. (1980). Processing and properties of instant powdered food. In: Food Process Engineering, Vol. 1 (eds Linko, P., Mälkki, Y., Olkku, J. & Larinkari, J.), pp. 675–684, Applied Science Publishers, London.

Semyonov, D., Ramon, O. & Shimoni, E. (2011). Using ultrasonic vacuum spray dryer to produce highly viable dry probiotics. *LWT – Food Science and Technology*, **44**, 1844–1852.

Sharma, A., Jana, A.H. & Chavan, R.S. (2012). Functionality of milk powders and milk-based powders for end use applications – A review. *Comprehensive Reviews in Food Science and Food Safety*, **11**, 518–528.

Shrestha, A.K., Howes, T., Adhikari, B.P. & Bhandari, B.R. (2007). Water sorption and glass transition properties of spray dried lactose hydrolysed skim milk powder. *LWT – Food Science and Technology*, **40**(**9**), 1593–1600.

Singh, H. & Newstead, D.F. (1992). Aspects of proteins in milk powder manufacture. In: *Advanced Dairy Chemistry, Vol. 1, Proteins*, 2nd edn (ed. Fox, P.F.), pp. 735–765, Elsevier Applied Science, New York.

Slade, L. & Levine, H. (1991). Beyond water activity: Recent advances based on an alternative approach to the assessment of food quality and safety. *Critical Reviews in Food Science and Nutrition*, **30**(**2–3**), 115–360.

Sourdet, S., Relkin, P., Fosseux, P.Y. & Aubry, V. (2002). Composition of fat protein layer in complex food emulsions at various weight ratios of casein-to-whey proteins. *Lait*, **82**(**5**), 567–578.

SPX Corporation (2012). Anhydro Evaporation and Drying Solutions: For the chemical industry. SPX Flow Technology, Denmark (version 01/2012; issued 05/2012).

SPX Corporation (2013). Upgrading of Anhydro Spray Drying Plants: For extra capacity and energy savings. SPX Flow Technology, Denmark (version 01/2013; issued 07/2013).

Toledo, R.T. (2007). Dehydration. In: *Fundamentals of Food Process Engineering*, pp. 431–473, Springer Science + Business Media, New York.

Truong, V., Bhandari, B.R. & Howes, T. (2005). Optimization of co-current spray drying process of sugar-rich foods. Part I – Moisture and glass transition temperature profile during drying. *Journal of Food Engineering*, **71**, 55–65.

Tsapis, N., Dufresne, E.R., Sinha, S.S., Riera, C.S., Hutchinson, J.W., Mahadevan, L. & Weitz, D.A. (2005). Onset of buckling in drying droplets of colloidal suspensions. *Physical Review Letters*, **94**, 018302(1)–018302(4).

Van, D.B.J., Van, D.K.C., Der Steen, A.M.L.V., Moonen, R.C.J. & Van, B.A. (2013a). Process for preparing infant formula. US Patent Application No. 14/385,044, US Patent and Trademark Office, Washington, DC.

Van, D.B.J., Van, D.K.C., Moonen, R.C.J. (2014). Improved process for preparing infant formula using a rotary atomizer. US Patent Application No. 14/917,984, US Patent and Trademark Office, Washington, DC.

Vega-Mercado, H., Gongora-Nieto, M.M. & Barbosa-Canovas, G.V. (2001). Advances in dehydration of foods. *Journal of Food Engineering*, **49**, 271–289.

Vicente, J., Pinto, J., Menezes, J. & Gaspar, F. (2013). Fundamental analysis of particle formation in spray drying. *Powder Technology*, **247**, 1–7.

Walton, D.E. & Mumford, C.J. (1999). The morphology of spray-dried particles: the effect of process variables upon the morphology of spray-dried particles. *Transactions of IChemE*, **77** (**Part A**), 442–460.

Westergaard, V. (2010). Milk powder technology-evaporation and spray drying, p. 93, GEA-NIRO, Copenhagen.

Wilson, D.I. (2005). Challenges in cleaning: Recent developments and future prospects. *Heat Transfer Engineering*, **26**(**1**), 54–59.

Wunderlich, B. (1981). The basis of thermal analysis. In: *Thermal Characterization of Polymeric Materials* (ed. Turi, E.A.), pp. 91–234, Academic Press, Orlando, FL.

Xin, H., Chen, X.D. & Ozkan, N. (2005). Removal of model foulant from metal surfaces. *AIChE Journal*, **5**(**8**), 1961–1973.

Xu, J.H., Li, S.W., Tan, J., Wang, Y.J. & Luo, G.S. (2006). Preparation of highly monodisperse droplet in a T-junction microfluidic device. *AIChE Journal*, **52**, 3005–3010.

Ye, A., Anema, S.G. & Singh, H. (2007). Behaviour of homogenized fat globules during the spray drying of whole milk. *International Dairy Journal,* **17**, 374–382.

Ying, D., Sun, J., Sanguansri, L., Weerakkody, R. & Augustin, M.A. (2012). Enhanced survival of spray-dried microencapsulated *Lactobacillus rhamnosus* GG in the presence of glucose. *Journal of Food Engineering*, **109**, 597–602.

Yonekura, L., Sun, H., Soukoulis, C. & Fisk, I. (2014). Microencapsulation of *Lactobacillus acidophilus* NCIMB 701748 in matrices containing soluble fibre by spray drying: Technological characterization, storage stability and survival after in vitro digestion. *Journal of Functional Foods,* **6**, 205–214.

You, X., Zhou, Z., Liao, Z., Che, L., Chen, X.D., Wu, W.D., Woo, M. & Selomulya, C. (2014). Dairy milk particles made with a mono-disperse droplet spray dryer (MDDSD) investigated for the effect of fat. *Drying Technology*, **32**, 528–542.

6

Freeze Drying

A. Bhushani[1] *and C. Anandharamakrishnan*[1,2]

[1] *Department of Food Engineering, CSIR – Central Food Technological Research Institute, Mysore, Karnataka, India*
[2] *Indian Institute of Crop Processing Technology, Ministry of Food Processing Industries, Government of India, Thanjavur, Tamil Nadu, India*

6.1 Introduction

Freeze drying or lyophilization is the process of removing the solvent or dispersion medium from a product by sublimation and desorption. The freeze-drying process involves three stages: (i) freezing, (ii) primary drying and (iii) secondary drying (Oetjen & Haseley 2004). Briefly, the solvent in the product (water in the case of food applications) is crystallized at low temperatures (freezing) and sublimation, that is, direct transition from the solid to the gas phase, occurs without passing through the intermediate liquid phase (primary drying). After the sublimation process, the absorbed or bound water in the product is removed by desorption (secondary drying). Conventionally, the process is performed under reduced pressure or at vacuum to aid in the removal of air or non-condensable gases from the drying chamber, which might otherwise reduce the rate of water vapour transfer from the product to the condenser (Stapley 2008). This chapter discusses in detail the steps involved in freeze drying, and heat and mass transfer, and the factors influencing the freeze-drying process. Furthermore, the applications of freeze drying in the dairy industrys and the microencapsulation of probiotics are highlighted.

6.2 Steps in freeze drying

The primary solvent in the freeze drying of dairy products is water. The phenomenon of freeze drying is represented in Figure 6.1b. Water exists in three states: solid, liquid and gas (vapour). The curved lines represents six processes which define the passage of water from one state to another: *sublimation* (solid to vapour), *deposition* (vapour to solid), *melting* (solid to liquid), *freezing* (liquid to solid), *vaporization* (liquid to gas) and *condensation* (gas to liquid) (Figure 6.1a). The three phases of water coexist at the triple point, that is, at a pressure and temperature of approximately 4.58 Torr and 0.01 °C, respectively, and the critical point of water is at a pressure of 218 atm and temperature of 374 °C. The principle of freeze drying is based on the phenomena of sublimation and desorption. In the process of phase change in water, if the pressure inside

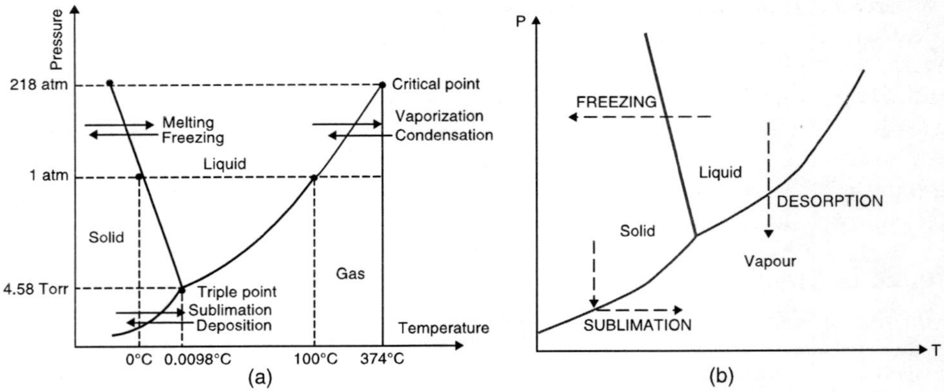

Figure 6.1 (a) The triple phases of water (Yu et al. 2011). (b) Freeze-drying phenomenon represented on a water phase diagram (Lopez-Quiroga *et al.* 2012).

the freeze-drying chamber is maintained below the triple point of water, the ice in the product can directly sublime to the vapour state. Furthermore, during the sublimation process, the moisture absorbs a large quantity of latent heat and hence heat supplied to the dryer shelf will aid in the drying of the product. Formerly it was assumed that lowering the vacuum and the condenser temperature to a minimum would speed up the freeze-drying process. However, it was later understood that the sublimation process in freeze drying can be accelerated by allowing the condenser temperature to rise and introducing dry air or inert gas to the freeze-drying chamber (Rowe 1976).

As previously mentioned, the three major steps in freeze drying are freezing and primary and secondary drying. However, in certain applications an additional pretreatment step is involved. For solid foods, pretreatment involves increasing the surface area of food to be dried by slicing, cutting or grinding. In case of liquid foods, the pretreatment process includes concentration or addition of stabilizing agents. The concentration step helps in reducing the duration of freeze drying and improving the bulk density of the dried foods.

6.2.1 Freezing

Freezing is the first step for freeze drying and converts 65–90% of the initial moisture content in the food product to the frozen state. The rest remains in the bound or adsorbed state. Most liquid dairy products freeze by the process of crystallization. Crystallization is a two-step process involving nucleation and crystal growth. The nucleation process defines the crystal structure and crystal growth defines the growth of the nuclei and critical cluster size. Once the solution is cooled below the equilibrium freezing temperature, the rate of nucleation increases exponentially with the degree of supercooling, whereas the rate of crystal growth occurs in a slow linear fashion. Hence, the degree of supercooling determines the size and number of ice crystals formed. In other words, with a high degree of supercooling, the rate of nucleation is more than the rate of crystal growth, thus favouring the formation of a large number of small crystals. On the other hand, in case of a low degree of supercooling, crystal growth is favoured and in turn fewer large crystals will be formed (Oetjen 1999). However, irrespective of the degree of supercooling, the rate of cooling is fastest at the surface and slowest at the centre, thereby giving rise to smaller crystals at the surface and larger crystals at the centre (Stapley 2008). Furthermore, the presence of solutes has a profound influence on the nucleation process. In Figure 6.2 the large circle indicates the critical nucleation radius, the arrows represent the process of diffusion and the length of the arrows represents the speed of diffusion. In the absence of solute, the nucleation process is dependant only on the water molecules to arrange themselves in a regular ice-like structure, followed by crystal growth. In the presence of solute, diffusion is reduced (represented by smaller arrows in Figure 6.2b) due to increased viscosity and the nucleation is reduced (Wolfe & Bryant 1999).

Freezing is one of the critical stages of freeze drying as the microstructure, reconstitution and storage properties of the final dried product are determined by this process. The size and structure of the ice crystals formed during freezing reflects the size and structure of the pores formed in the freeze-dried product. Furthermore, the drying rate in the primary and secondary drying stages is influenced by the properties of the frozen product matrix. When a connected network of crystals is achieved, it creates a relatively easy path for the water molecules to escape through the pore network formed. Therefore, formation of larger crystal matrices is more favourable for faster freeze drying (Kochs *et al.* 1993; Wang *et al.* 2012).

Aqueous solutions have two distinct freezing behaviours, as shown in Figure 6.3. The freezing point of milk depends on the concentration of water-soluble components such as sugars or protein. Similarly for freezing of other dairy products such as lactose, cheese, yoghurt and probiotics, the substance should be cooled to a temperature at

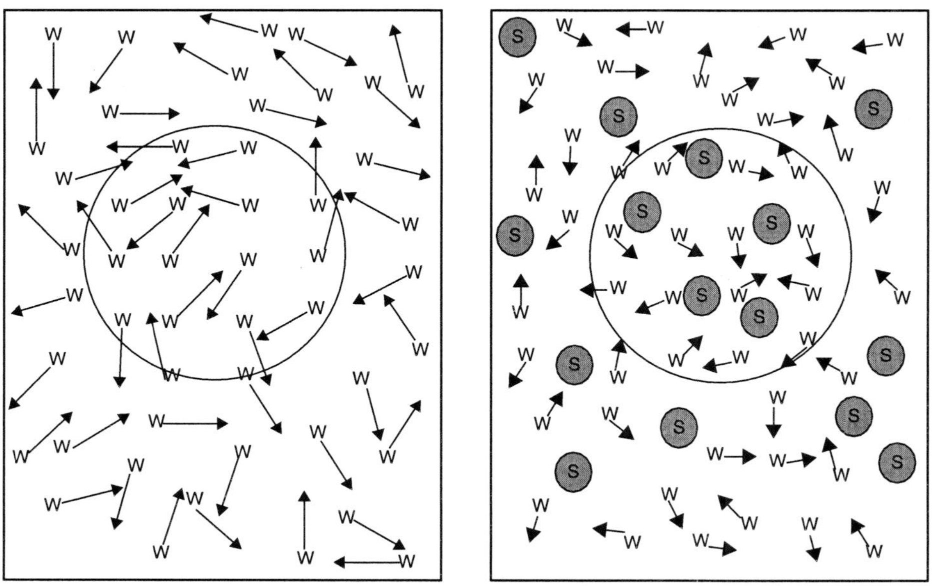

(a) (b)

Figure 6.2 The effects of solutes on the nucleation of freezing: (a) pure water and (b) water with solute at a solute-to-water molar ratio of 1:4. Symbols w and s represent water molecules and solute molecules, respectively (Wolfe & Bryant 1999).

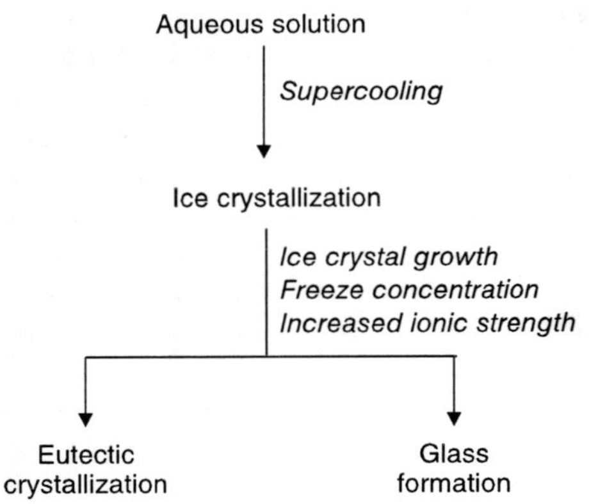

Figure 6.3 Freezing behaviours of aqueous solution (Wei *et al.* 2012).

which the water and the solids are completely crystallized or at which the solids and the crystallized ice are enclosed in zones where they remain in a mechanically solid state (Oetjen & Haseley 2004). Depending on the nature of the solids in the solution, the liquid phase either suddenly solidifies (eutectic formation) or does not solidify and becomes a highly viscous liquid (glass formation) (Liapis & Bruttini 1995). The major factors that determine the freezing process are the freezing temperature, initial concentration, rate and supercooling degree.

A notable instance could be the production of amorphous lactose using freeze drying. The freezing of a 20% lactose solution can be observed in Figure 6.4. Initially, removing

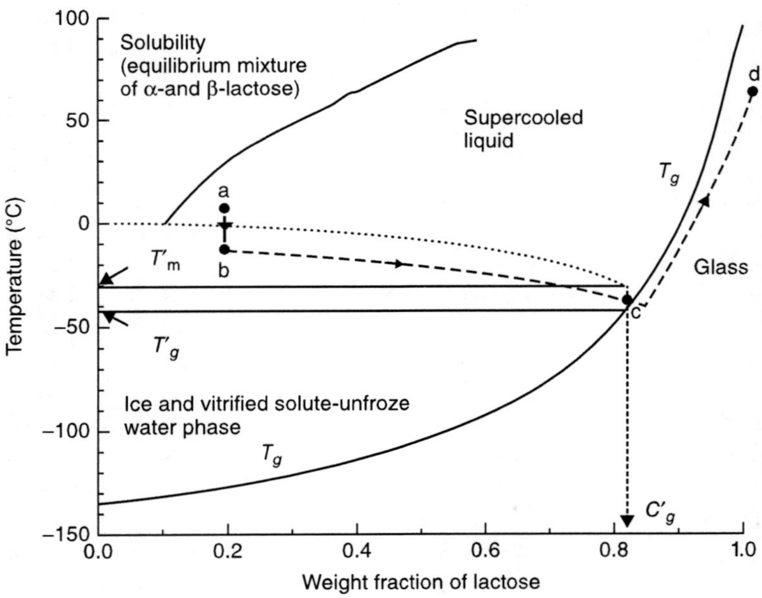

Figure 6.4 State diagram of amorphous lactose showing lactose solubility, the glass transition temperature (T_g), glass transition of maximally freeze-concentrated lactose solutions (T'_g) and the corresponding solids content (C'_g), and the onset of ice melting in the maximally freeze-concentrated solutions (T'_m). The typical path of a 20% lactose solution undergoing freezing (path a to c) followed by freeze drying (path c to d) is shown by the dotted line. Modified from Roos (2002).

heat from the solution causes it to cool to the point b, which is below the equilibrium freezing zone. Freezing is initiated with nucleation and formation of ice crystals during subcooling. The formation of ice crystals in turn reduces the water in the lactose solution and hence the solution becomes more concentrated with respected to lactose. The freezing stage ends at the point c in Figure 6.4, where the solution reaches the glass transition line and the ice crystallization ceases. The glass transition temperature of this lactose mixture is designated as T'_g, and C'_g is the corresponding solid content. This stage (point c) is known as the maximally freeze-concentrated solution (i.e. maximum ice formation) and further concentration is not possible. This point occurs in the region above T'_g but below the equilibrium melting temperature of ice (T'_m). In general, the T'_g of the product should be considered an adequate level to reach during the freezing stage. For dairy products, T'_g values range from −23 °C to −41 °C (Slade & Levine 1995).

6.2.2 Primary or sublimation drying

The next stage after freezing is the primary drying stage, in which the control of product temperature is critical for obtaining high-quality dried product. However, the product temperature is controlled indirectly by two critical factors: the shelf temperature and the chamber pressure (Rambhatla *et al.* 2006). Generally, after the freezing stage, the drying chamber is evacuated and the pressure is reduced to a point where ice sublimation takes place. It is well known that the sublimation process takes place only when the chamber pressure or the partial pressure of vapour in the chamber is lower than the vapour pressure of ice at the drying front of the product. However, if the chamber pressure is far below the vapour pressure of ice, the process will slow down due to poor convection. On the other hand, if the chamber pressure is too close to the vapour pressure of ice, the product might collapse or melt due to enhanced heat transfer. In this case, the freeze dryer would be ineffective in handling the vapour load, leading to a condition termed 'choked flow'. Searles (2004) described this phenomenon of choked flow as a primary cause for the loss of pressure control during freeze drying. This is due to the gas-flow dynamics in the duct connecting the freeze-drying chamber and the condenser, which imposes an upper limit to the maximum sublimation rate that a freeze dryer can handle. Thus, the flow is said to be choked and any increase in the vapour flow rate causes an increase in the chamber pressure above the set point and the process ultimately runs out of control (Patel *et al.* 2010). Because of these issues, it is recommended that the chamber pressure for primary drying should be one-quarter to one-half of the vapour pressure of ice at the desired product temperature (Nail *et al.* 2002; Tang & Pikal 2004). This total pressure can be controlled by two routes: (i) passage of a dry inert gas (e.g. nitrogen) into the chamber by a needle valve or (ii) closing the valve between the condenser and the vacuum pump set, which opens when the condenser set pressure exceeds. Of these two methods, the latter has advantages over the former, as only the gas of the product is used and the need for inert gas is avoided. In addition, at the end of primary drying the chamber pressure drops automatically, as required for the secondary drying process (Oetjen 1999).

In the primary drying process, the ice crystals sublimate and a sublimation front or interface is formed. During this, the temperature of the ice at the sublimation front has to be maintained below the collapse temperature. The sublimation front starting on the surface of the product gradually recedes to the material and a porous layer of the material remains after front advancing. This dried layer conducts the heat used for sublimation to the sublimation front. The sublimated vapour travels through the porous layer by diffusion and convective flow. This is an important process that affects the mass-transfer rate. As discussed earlier, this diffusivity is dependent on the pore size of the ice crystals achieved during the freezing step. Moreover, when the sublimation process continues, the moisture from the product is reduced and consequently the glass transition value changes. If the temperature at the ice front is higher than the glass transition temperature, it is likely that part of the water will be removed by evaporation and other part by sublimation (Fellows 2009).

6.2.3 Secondary or desorption drying

In this stage of freeze drying the unfrozen water is removed from the product by desorption from the dried layer. This water can be bound to the surface of the crystals in a crystallized product or embedded in the highly viscous amorphous product matrix. In the case of crystalline products, higher drying temperatures can be used and hence the secondary drying time can be reduced. However, in the case of amorphous products, high temperature cannot be used and drying is usually achieved by slow molecular diffusion within the dried cake (Pikal & Shah 1990; Oetjen & Haseley 2004; Wang *et al.* 2012).

Secondary drying is carried out by increasing the temperature and reducing the vapour pressure inside the drying chamber (Pikal 1990). Even though less water is removed during secondary drying, the time required is 30–50% of that required for primary drying (Millman *et al.* 1985). According to Pikal *et al.* (1990), at the beginning of the secondary drying the removal of water is rapid, but later it plateaus. This plateau moves with increasing temperature in the direction of smaller water contents. Hence, the drying rate increases with increasing specific surface of the product. The freeze-drying process is said to be complete when most of the water is removed, yielding a structurally stable dried product with preferably less than 1–3% residual water (Karel 1975; Liapis *et al.* 1996).

6.3 Merits of freeze drying over other drying techniques

Freeze drying is one of the most sophisticated dehydration methods and its application in the food industry is limited to high-value foods as the investment, operating and maintenance costs of freeze dryers are higher than for other thermal and non-thermal drying techniques. The chief uses are production of freeze-dried coffee, tea, meat, selected fruits and vegetables, herbs and pet foods. In the dairy industry, freeze drying is predominantly used for the preservation of starter cultures and encapsulation of probiotic organisms for usage as functional ingredients. Other dairy products that are suitable for freeze drying are milk, cheese, cream, yoghurt and ice-cream. These freeze-dried dairy ingredients are used in the baking industry, restaurants (freeze-dried mozzarella or cheese for pizza) and the biotechnology industry, and have potential as space foods, and military and emergency foods. The application of freeze drying in dairy industry is mainly due to the superior quality attributes of freeze-dried dairy products. In this context, it is essential to understand the properties and advantages of freeze-dried dairy products.

The foremost advantage of freeze drying is the *low temperature operation*. This property aids in drying of heat-sensitive materials and eliminates the possibility of reactions occurring in high-temperature drying. For instance, ascorbic acid, a heat-sensitive vitamin, can be retained in freeze-dried milk as against drum- or spray-dried milk; denaturation of whey proteins and the Maillard reaction between lactose and lysine-rich proteins in milk can be avoided. Furthermore, aroma and volatile compounds are retained at a higher level in freeze-dried dairy products compared to spray-dried or drum-dried products. Secondly, freeze-dried products have a *rigid structure* with reduced product shrinkage and low moisture content, making them shelf stable. For instance, non-fat freeze dried milk (USDA, grade A) commercialized by the US-based firm Legacy Premium LLC claims a shelf life of 25 years at ambient storage conditions. Thirdly, the *highly porous structure* of the freeze-dried particles confers excellent solubility and reconstitution properties. The porous structure of *Lactobacillus paracasei* microencapsulated in whole milk powder by freezing drying was studied by Poddar *et al.* (2014). Significant connected porosity was observed and bulk porosity was reported to be 77.5%, comprising 71.2% and 6.3% of connected and isolated porosity, respectively (see Figure 6.5). Finally, the *biological activity* of the freeze-dried probiotic or starter cultures and viability of freeze-dried microencapsulated probiotics confer advantages in the development of functional dairy ingredients (Fellows 2009).

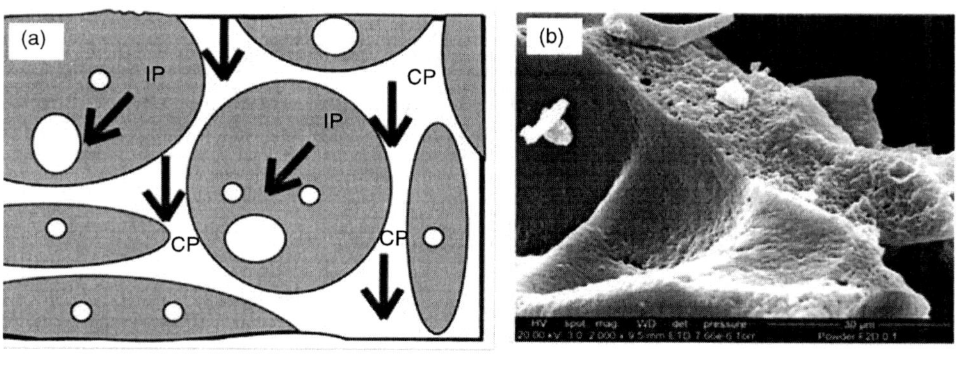

Figure 6.5 (a) Schematic representation of connected (CP) and isolated (IP) porosity in dried encapsulates. (b) SEM image of freeze-dried *Lactobacillus paracasei* powder (Poddar *et al.* 2014).

6.4 Heat and mass transfer in freeze drying

Freeze drying is a coupled heat- and mass-transfer phenomenon, wherein both the processes occur within the product, that is, energy is transported to the sublimation zone and water vapour is generated. Heat transfer can occur by conduction through the dried layer of the food product (Figure 6.6a) or through the frozen layer (Figure 6.6b) or by internal heat generation, as in the case of microwave-assisted freeze drying (Figure 6.6c). However, mass transfer occurs primarily through the dried portion of the food product, irrespective of the heat source. Figure 6.6 also shows the three distinct modes of heat transfer, which determine the drying rate: (i) heat and mass transfer occur through the dried layer, but in opposite directions (Figure 6.6a), (ii) heat transfers across the frozen

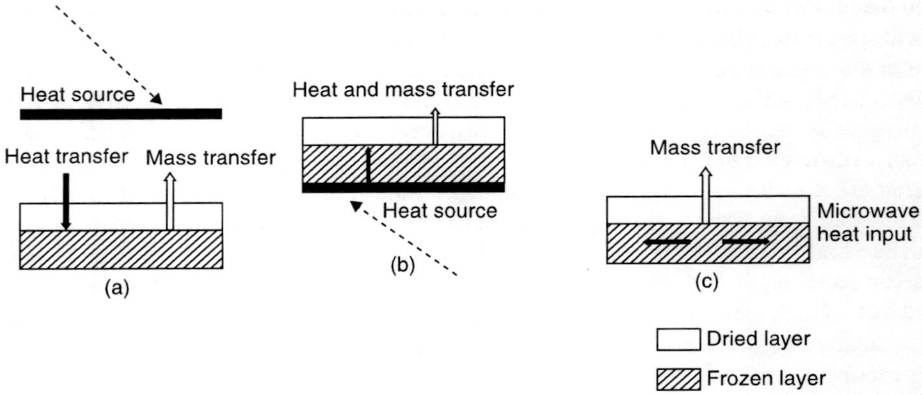

Figure 6.6 Mechanisms of heat and mass transfer in freeze drying.

layer and mass transfer occurs through the dry layer (Figure 6.6b), and (iii) heat generates within the frozen layer by irradiation or microwave heating (Figure 6.6c) (Fellows 2009; Liapis & Bruttini 2015).

Moreover, the rate of drying depends on the resistance provided by the food for heat transfer (R_d in Figure 6.7). To a lesser extent, the resistance to vapour flow, that is, mass transfer in the space between the sublimation front of the food and the condenser (R_s in Figure 6.7) also determines the drying rate (Karel & Lund 2003). The coupled heat- and mass-transfer phenomenon can be summarized using the following equation:

$$\frac{G}{A} = \frac{(p_i - p_c)}{\Sigma R} = \frac{q}{\Delta H_s}$$

where G/A is the sublimation rate per unit area (kg/m^2), q is the heat transfer per unit area (kJ/m^2), p_i is the partial pressure of water at the sublimation interface (Pa), p_c is the partial pressure of water at the condenser surface (Pa), $\sum R$ is the total resistance to vapour transfer between sublimation interface and condenser (Pa m^2 s)/kg and ΔH_s is the latent heat of sublimation (kJ/kg).

Numerous studies have focused on developing models for predicting the heat and mass transfer during freeze drying of liquid and solid foods. A study by Jafar & Farid

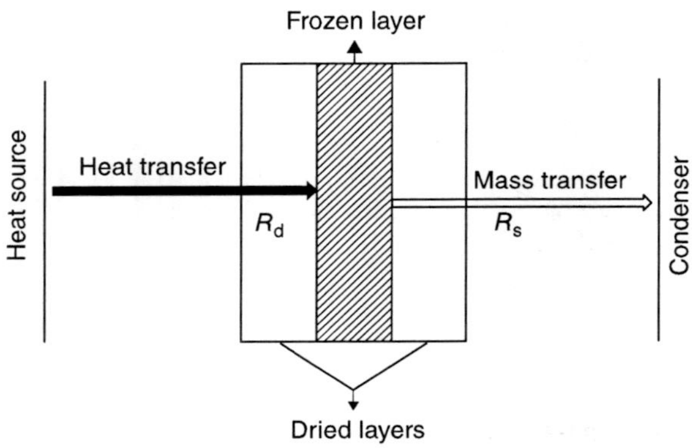

Figure 6.7 Resistances in freeze drying. R_s is the resistance of the space for mass transfer and R_d is the resistance of the dry layer for heat transfer.

(2003) deals with building a mathematical model for describing the heat and mass transfer during freeze drying of milk when both plate and radiation heating are applied. Two possible mechanisms of freeze drying with plate heating were explained in their study, one with the dried layer close to the plate heater and other with the dried layer away from the plate heater. Furthermore, the measurement of the temperature distribution during freeze drying of milk indicated that drying occurs even from the surface close to the heat source, as against the concept that drying occurs only from the top surface.

In the three possible modes of heat transfer to the sublimation front (Figure 6.6), there are few critical factors which affect the rate of heat transfer. For instance, in the condition where heat is transferred through the frozen layer (Figure 6.6a), the thickness and thermal conductivity of the ice layer determine the rate of heat transfer. On the other hand, in the case of heat transfer through the dried layer (Figure 6.6b), along with the thickness, surface area and thermal conductivity of the dry layer, the temperature difference existing between the food and the ice at the sublimation front also determines the heat-transfer rate. Similarly, the pressure of the drying chamber, the temperature of the condenser and the temperature of the ice at the sublimation front together influence the rate of mass transfer (Beaty 2006; Fellows 2009; Smith 2011).

The drying rate is highly dependent on the heat- and mass-transfer rate, which in turn is influenced to an extent by the freeze dryer instrumental setup. The following session describes in brief the significance of the major units of the freeze-drying equipment.

6.5 Freeze-drying equipment

Essentially, the selection and installation of freeze-drying equipment is dependent on the characteristics of the products to be dried. A typical freeze dryer consists of (i) a drying chamber, (ii) a condenser, (iii) a vacuum pump and (iv) a heat source (Gatlin & Nail 1993; Berk 2013) (see Figure 6.8.).

The first step in freeze drying is the freezing of products. Generally, freezing of milk and other dairy products is carried out at an industrial scale in trays by contacting the tray with a cooled surface or by freezing in contact with cryogenic liquids such as liquid

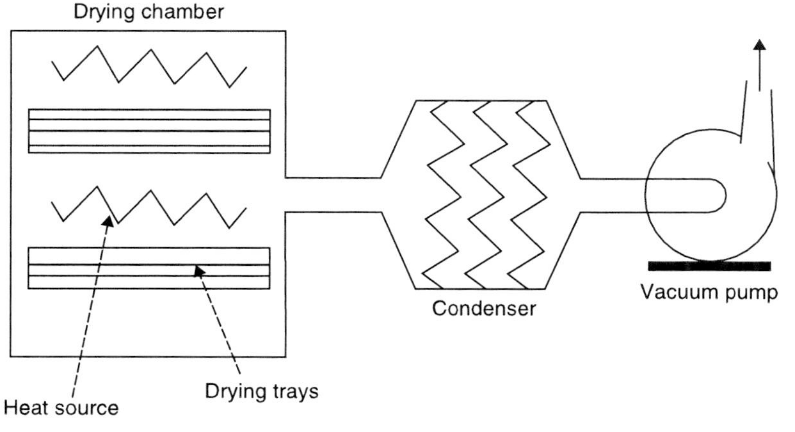

Figure 6.8 The major components of a freeze dryer (Berk 2013).

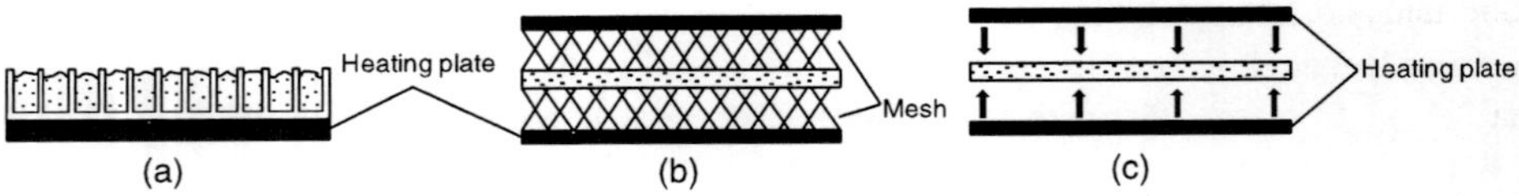

Figure 6.9 Freeze drying methods: (a) conduction through a ribbed tray, (b) expanded mesh for accelerated freeze drying and (c) radiant heating of flat trays (Fellows 2009).

nitrogen or by air-blast freezing (Stapley 2008). As discussed before, this stage has an influence on the microstructure of the product and also on the duration of the subsequent drying stages. In a study by Radaeva *et al.* (1975) on the freeze drying of yoghurt, the optimum regime of freezing was found to be −25 °C as against −15 and −40 °C for the highest survival of bacteria. Furthermore, it was established that the different freezing temperatures had no significant effect on the nutritional properties of the yoghurt, but there was a notable effect on the duration of subsequent drying stages which influenced the refrigeration and electricity requirements. As the drying process proceeds to the next stage, the role of the drying chamber, condenser, vacuum pump and heat source become vital.

The *drying chamber* in which the sample is placed has to be vacuum-tight and contain temperature-controlled shelves. The *heat source* for the shelves provides the latent heat of sublimation during the primary drying stage. Based on the freeze dryer type, the kind of tray and the heat source provided can be varied. For instance, in case of *contact or conduction freeze dryers*, ribbed trays which lay on heater plates (Figure 6.9a) are used. Here, the rate of drying is slow as the heat transfer is primarily by conduction to one side of the food. In an *accelerated freeze dryer*, heat is conducted to the foods through expanded metal mesh (Figure 6.9b) thereby enabling enhanced heat transfer compared to contact dryers. In a *radiation dryer*, the material is heated through infrared radiation from radiant heaters (Figure 6.9c) (Rolfgaard 1987; Fellows 2009).

Alternative heating methods, such as microwave heating, can also be employed to enhance the heat transfer and reduce the drying time. Microwave energy causes rapid internal heating of the material via dipole rotation and ionic conduction in polar materials. This energy, as a source of heat, can be applied under vacuum in freeze drying and the resultant process is termed *microwave-assisted freeze drying* (Wang *et al.* 2009). This microwave energy can be applied concurrently with freeze drying or can form a separate unit operation following freeze drying under vacuum. The product quality is same as for conventional freeze drying, but the operating costs can be lowered due to faster drying rates (Duan *et al.* 2010). In a study by Wang *et al.* (2005) on the microwave freeze drying of skim milk, it was found that the drying time can be significantly reduced by use of dielectric material such as silicon carbide (SiC). Using the SiC with 4 mm diameter at the spherical core of an initially unsaturated frozen skim milk, the drying time was reduced to 288.2 min, as against 380.1 min in a microwave freeze dryer and 455.0 min in a conventional vacuum freeze dryer. Therefore, it is clear that the heating system in a freeze dryer has a major effect on the overall operating costs. It is also known that, for the freeze drying of frozen liquids, slurries or suspensions, as in the case of milk and milk products, heated shelves rather than unheated shelves are recommended for the same reason of reduction in drying time. This concept is clearly illustrated in Figure 6.10, which shows the difference in drying times for the freeze drying of pure water with use of heated shelves as against unheated shelves (Christ 2010).

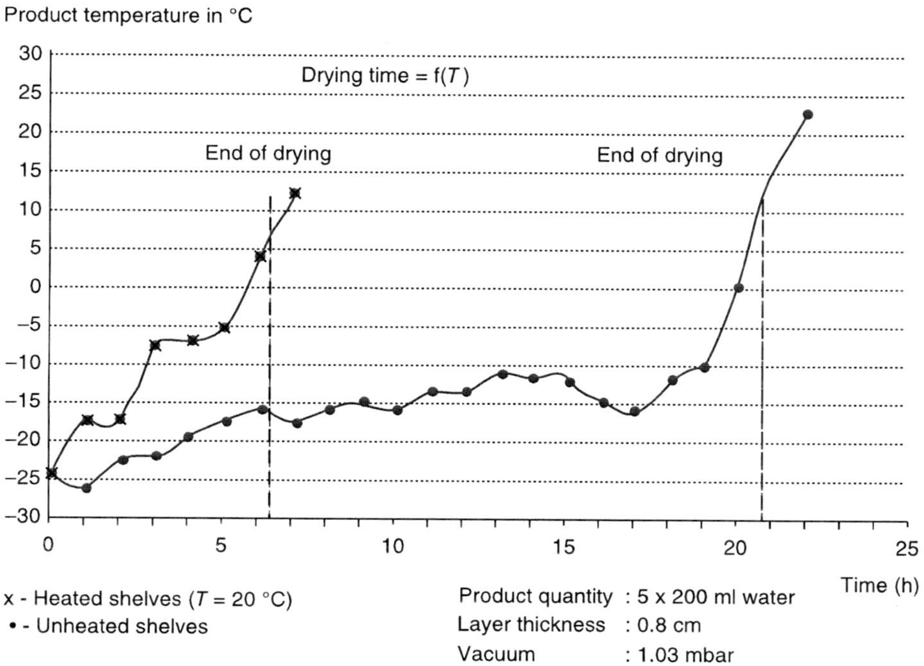

Figure 6.10 Influence of shelf heating on the drying rate of pure water (Christ 2010).

The *ice condenser* in most commercial freeze dryers operates at temperatures from −55 to −105 °C, with a capacity of 20–10,000 kg. According to Oetjen & Haseley (2004), an ideal condenser must meet the following four criteria: (i) have large surface area to condense the ice at a maximum thickness, (ii) have uniform condensation on the total coil with minimum ($\Delta T < 1$ °C) difference between inlet and outlet temperature, (iii) have less deviation in the flow of water vapour before the first condenser surface and (iv) pump off the permanent gases at the lowest position in the condenser. In industrial-scale freeze dryers, the condenser can be placed either inside or outside the drying chamber. However, it is essential that the distance between the two units is less to aid in effective reduction in partial pressure of the drying chamber.

The *vacuum pump* is used to remove non-condensable gases from the drying chamber to achieve high vacuum levels (Mellor 1978). The vacuum system can be designed based on the total volume of the dryer system that has to be evacuated and the size, type and number of vacuum pumps fitted to the system.

Thus, selection of an appropriate freeze dryer is essential for laboratory-scale manufacture or industrial-scale production of dairy powders. The functions of the four major units of the equipment, which state the significance of each, have been discussed. Furthermore, it is critical to consider these parameters during scale up of dryers from laboratory to manufacturing scale. The major problems that occur during scale up of the process are due to differences in degree of supercooling, variations in shelf temperature, and differences in the condenser system between laboratory and production units (Rambhatla & Pikal 2003). The following section will elaborate on the properties that influence the freeze drying of dairy products.

6.6 Properties influencing the freeze drying of dairy products

6.6.1 Milk

Even though spray drying is an economical process for the production of milk powder, the limitations associated with spray-dried whole milk warrant an alternative drying process for the production of whole milk powders. The major disadvantage of spray-dried whole milk is its reconstitution property, wherein, in beverage applications, the reconstituted powders leave greasy films on the containers. Also, as spray drying is a high-temperature operation, it imparts a typical cooked and somewhat astringent flavour to the whole milk powder with reduction in the milk's original aroma and flavour characteristics.

The properties that influence the freeze drying of whole milk or skim milk have been studied by various researchers. The *thermal conductivity* of the sample is important during the primary and secondary drying stages of freeze drying, which take place under vacuum. The bulk thermal conductivity of a product is dependent on the surrounding pressure and temperature, the presence of gas in the void space and, most importantly, the porosity and total solid concentration of the material. Also, the thermal conductivity of the material is essential to determine the rate of heat transfer, which in turn helps in calculating the total time required for the freeze-drying process. For instance, Fito *et al.* (1984) studied the thermal conductivity of freeze-dried milk at low pressure and proposed a mathematical model for predicting the thermal conductivity of freeze-dried products at any pressure and temperature. Based on the thermal conductivity values obtained for freeze-dried milk at different porosities, the typical curves for thermal conductivity as a function of total pressure are represented in Figure 6.11. At low pressure (i.e. 0.001–1 mmHg), there was a constant thermal

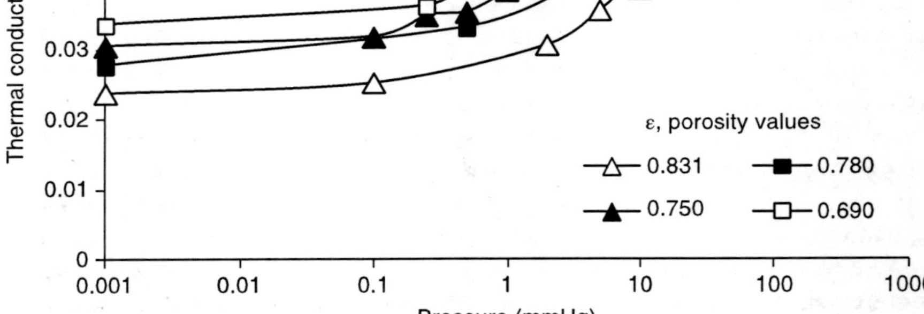

Figure 6.11 Thermal conductivity of freeze-dried milk at different porosities (based on data provided by Fito *et al.* (1984)).

conductivity, which was lower than that at higher pressure (i.e. 1–760 mmHg). Thus, it should be noted that the vacuum applied during freeze drying operation (between 0.01 and 1 mmHg) is in the range of the thermal conductivity transition (Ratti 2009). Furthermore, the influence of *porosity* on the thermal conductivity is evident from Figure 6.12. For a slight change in porosity from 0.831 to 0.690, distinct curves of thermal conductivity were observed. In general, an increase in porosity decreases the thermal conductivity (as can be observed from Figure 6.11) because the gas/air present in the pores impart insulation to the bulk material studied.

Nickerson *et al.* (1952) published the first report on the comparison of the physical characteristics and keeping quality of freeze-dried milk with spray-dried milk. The quality characteristics of freeze-dried milk were similar to those of spray-dried milk, but the former did not excel the latter in overall keeping quality. Later, Ben Souda *et al.* (1989) developed a microwave-assisted freeze-drying process for milk and demonstrated its feasibility for producing high-quality milk powders in less time than conventional freeze dryers. In a study by Fyfe *et al.* (2011), the surface characteristics and composition of freeze-dried milk was found to be better than those for spray-dried milk. Furthermore, the skim milk powder obtained by freeze drying was found to be similar in surface protein to its bulk protein composition, thereby indicating homogenous particle composition.

The physicochemical properties of milk powders, such as stickiness, caking, oxidation and non-enzymatic browning, are largely influenced by their water sorption and crystallization behaviour (Fitzpatrick *et al.* 2007). Furthermore, the amorphous lactose contents of skim milk powder cause caking and thereby affect the flowability of the dried powders during storage. Here, the critical parameters are relative humidity (RH) and temperature. A series of works conducted by Jouppila & Roos (1994a,1994b) formed the basis for understanding the crystallization and water absorption properties of milk powders. Table 6.1 lists the water contents of dehydrated milk products stored at different relative humidities for 24 h. It can be seen that the water contents of the dried powders differs significantly at the same RH. Furthermore, the water content of dried skim milk with hydrolysed lactose was lower at low RH and higher at high RH than that of dried skim milk. This indicates that crystallization is unlikely in powders with

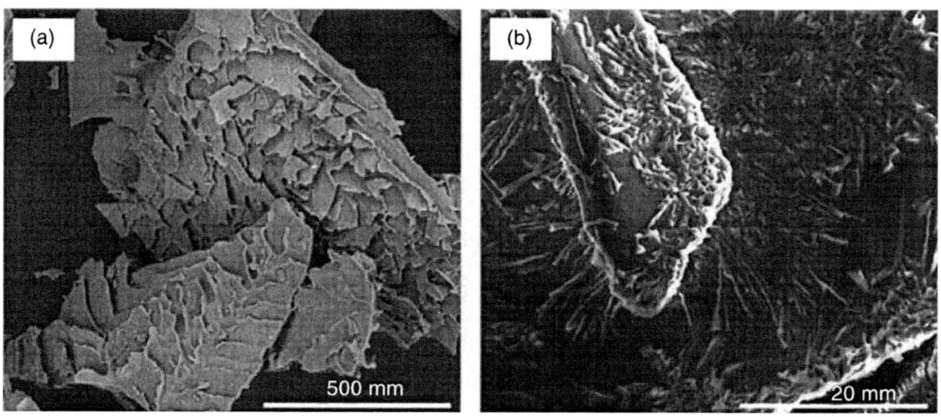

Figure 6.12 Morphology of (a) freeze-dried skim milk powder and (b) skim milk powder equilibrated at 54.4% relative vapour pressure (Marabi *et al.* 2007).

Table 6.1 Water content of dehydrated milk products stored for 24 h at various relative humidities at 24 °C (Jouppila & Roos 1994a)

Salt	Relative humidity (%)	Water content (g of H_2O/100 g)					
		Lactose	Skim milk (0% fat)	Low fat milk (10.7% fat)	Low fat milk (18.6% fat)	Whole milk (32.4% fat)	Skim milk with hydrolysed lactose (0% fat)
Phosphorus pentoxide (P_2O_5)	0	0	0	0	0	0	0
Lithium chloride (LiCl)	11.5	2.4 ± 0.5	3.7 ± 0.4	3.5 ± 0.4	3.3 ± 0.2	2.7 ± 0.2	1.5 ± 0.4
Potassium acetate (CH_3COOK)	23.9	4.3 ± 0.3	5.6 ± 0.4	5.2 ± 0.3	4.8 ± 0.3	4.0 ± 0.1	3.0 ± 0.4
Magnesium chloride ($MgCl_2$)	33.0	5.9 ± 0.4	7.1 ± 0.5	6.5 ± 0.2	5.9 ± 0.1	4.9 ± 0.2	5.2 ± 0.2
Potassium carbonate (K_2CO_3)	44.4	8.6 ± 0.7	8.8 ± 0.5	8.0 ± 0.5	7.4 ± 0.3	6.2 ± 0.3	8.1 ± 0.8
Magnesium nitrate ($Mg(NO_3)_2$)	53.8	10.7 ± 0.5	11.6 ± 0.2	10.9 ± 0.2	9.9 ± 0.5	8.3 ± 0.3	12.0 ± 0.5
Sodium nitrite ($NaNO_2$)	66.2	1.8 ± 0.9	16.9 ± 0.6	15.8 ± 0.3	14.6 ± 0.1	12.1 ± 0.1	19.3 ± 0.4
Sodium chloride (NaCl)	76.4	0.5 ± 0.2	16.7 ± 0.6	17.9 ± 1.8	18.6 ± 0.9	15.1 ± 1.1	30.9 ± 0.5

hydrolysed lactose. In addition, the fat content and the presence of other constitutents in milk powder decreases the lactose crystallization rate (Jouppila & Roos 1994a). Milk powder is considered to be a complex system as it contains carbohydrates, proteins and fats. The physical and chemical stability of milk powders is dependent not only on the glass transition temperature of their main carbohydrates but also on the other major constituents (Fernández *et al.* 2003). A study by Marabi *et al.* (2007) showed the effect of physical state on the dissolution kinetics of skim milk powder, wherein a fully amorphous sample dissolved faster than the recrystallized sample formed by equilibration at 54.5% relative vapour pressure. The scanning electron microscopy (SEM) images of the freeze-dried skim milk powder and the recrystallized skim milk, demonstrating lactose crystals on surface, are shown in Figure 6.12. With advances in freeze-drying technology, whole milk powder with superior sensory and keeping quality is being produced for various end applications.

6.6.2 Lactose

Lactose is a milk sugar composed of the monosaccharides galactose and glucose. It can be obtained in crystalline and amorphous forms. The amorphous form of lactose has no proper crystal structure and its form is expected to change with time as it is in a metastable phase. Freeze-dried milk powders contain lactose in its amorphous state as the freeze-drying process does not allow crystallization of the lactose. The presence of lactose in skim milk powder has an immense influence on its caking behaviour. As shown in Figure 6.13, above the glass transition temperature (T_g), the powder becomes sticky and this point is referred to as the sticky point temperature, which is usually about 10–20 °C above the T_g for low-molecular-weight carbohydrates. Beyond the sticky zone,

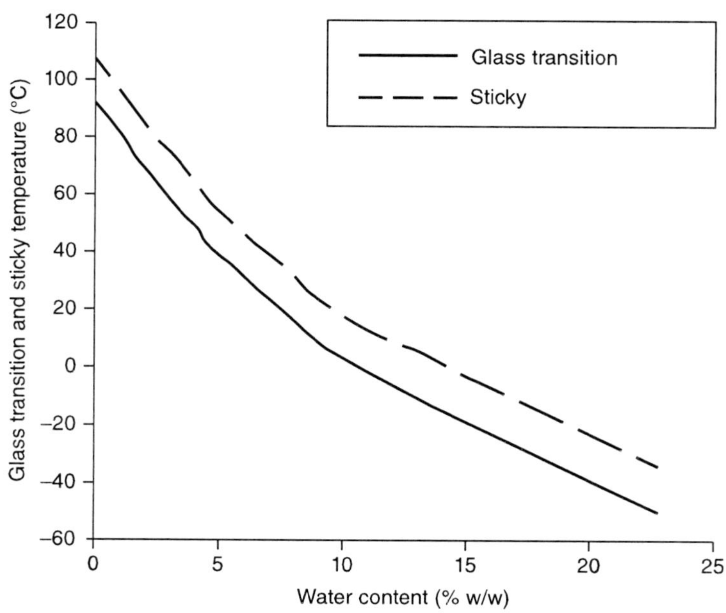

Figure 6.13 Glass transition of lactose in skim milk powder (Fitzpatrick *et al.* 2007).

the compound becomes non-sticky with reduced viscosity. Crystallization of the compound can occur only when the powder temperature is greater than its T_g, whereby the molecules have the mobility to initiate crystallization. Also, as discussed earlier, the other components of the powder interfere with the crystallization process (Kudra 2003). The caking behaviour of skim milk powder containing 51% lactose was studied after exposure to high relative humidities (76% or 100%) for a week, followed by exposure to humidity cycle (2 days exposure to high RH, 2 days exposure to low RH, i.e. 11%, 2 days exposure to original high RH). It was observed that skim milk powder containing amorphous lactose exhibited caking behaviour similar to crystalline common salt, wherein an increase in cake strength during the low RH part of humidity cycle was exhibited. The study suggested that the caking behaviour of skim milk powder might be attributed to the crystallization of lactose at high RH (Fitzpatrick *et al.* 2010). Furthermore, the glass transition temperature of freeze-dried lactose decreases with increase in water content and above the critical water content time-dependent physical changes occur (Omar & Roos 2007). Changes in the glass transition temperature of lactose can also be brought about with addition of other ingredients such as protein. A recent work by Fan & Roos (2015) demonstrated a delay in lactose crystallization by addition of whey protein isolate. The amorphous lactose–whey protein isolate mixtures were shown to have fractional water sorption properties and the protein inhibited lactose crytallization by reducing the rate of nucleation and extent of crystallization of lactose.

The pathway for production of amorphous lactose by freeze drying and spray drying is indicated in Figure 6.14, along with the phase diagram of lactose in water. Even though both the drying processes yield amorphous lactose, the water absorption behaviours

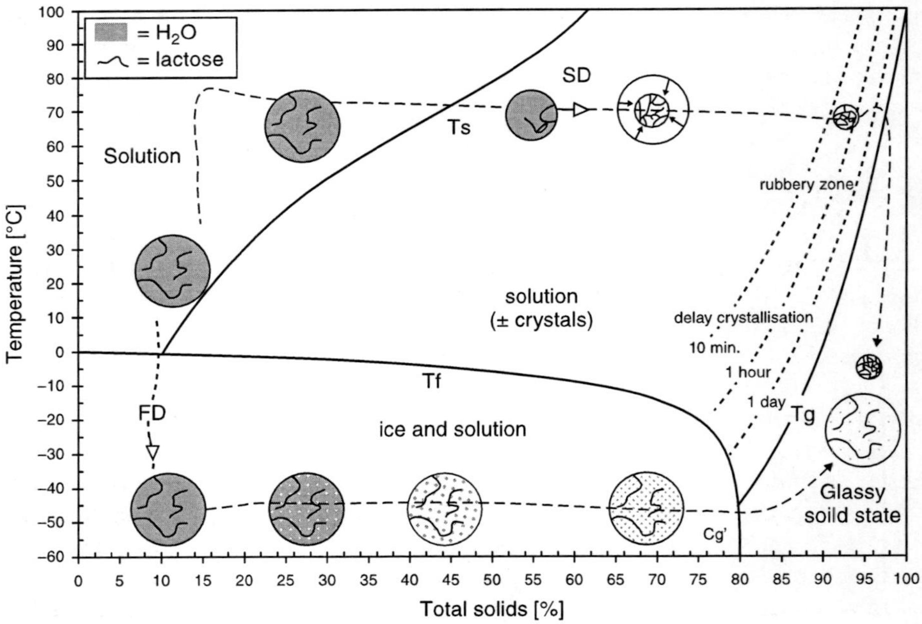

Figure 6.14 Phase diagram of lactose in water along with the pathway for production of amorphous lactose by freeze drying (FD) and spray drying (SD). Tf, freezing temperature; Tg, glass transition temperature; Ts, lactose solubility (DFE Pharma).

differ in the two kinds of lactose (Vollenbroek *et al.* 2010). However, a study by Haque & Roos (2005) revealed that there was no difference in the crystallization behaviour of spray dried and freeze dried lactose and the extent of crystallization was dependent on the storage relative vapour pressure (RVP). They found that lactose crystallized in samples stored at RVP of 44.1% and above in spray-dried and freeze-dried powders.

Furthermore, to determine the chemical stability of dairy powders, apart from water activity and glass transition concepts, molecular mobility has to be taken into consideration. The mobility of the molecules is useful in predicting many diffusion-limited properties and the chemical stability of dairy powders. A study by Schmitz-Schug *et al.* (2013) established the relation between molecular mobility and chemical stability for freeze-dried dairy powders using experimental tools such as differential scanning calorimetry and low resolution ^1H nuclear magnetic resonance spectroscopy. They concluded that the loss of available lysine (an essential amino acid) in milk powder increased with increasing molecular mobility in the glassy and rubbery state and was decelerated by crystallization. Insights developed from such works can be applied in the optimization and development of dairy powders with maximum physical and chemical stability during storage.

6.7 Preservation of kefir culture by freeze drying

Kefir is a fermented-milk-based drink produced using kefir grain. Kefir grain is a complex probiotic system, comprising a consortium of microbes, namely, yeast, various lactic acid bacteria and acetic acid bacteria. This symbiotic association of yeast and bacteria is responsible for lactic–alcoholic fermentation. This mixed culture is capable of utilizing lactose and hence whey is used as a raw material for kefir production. Nevertheless, the uses of kefir culture are manifold. They can be used as starter culture for whey beverage or cheese production and have been proposed for use as baker's yeast. However, dried cultures are preferred for commercial applications due to their advantages over conventional wet microbial cultures (Angulo *et al.* 1993; Witthuhn *et al.* 2005).

The cost of producing freeze-dried kefir culture was studied by Kourkoutas *et al.* (2007) and an industrial-scale process flow sheet (Figure 6.15) was proposed with a plant capacity for producing 300 kg of freeze-dried culture per day. As per the proposed flow, biomass slurry is produced through three bioreactors of 100, 3000 and 30000 l capacity that are connected to a heat exchanger to cool the fermenting medium. The economic analysis revealed that the production of freeze-dried cultures would cause a 3.5-fold increase in producion cost compared to the production of wet cultures. Despite the investment and production costs involved, the benefits associated with the development of freeze-dried stable culture are numerous. These freeze-dried cultures provide scope for the development of different commercial products with high market value. In a study, freeze-dried kefir culture was produced using fermented whey as a cryoprotective agent. The cultures maintained 86% survival after freeze drying, with good metabolic activity and fermentation efficiency (Papavasiliou *et al.* 2008).

These freeze-dried cultures can serve as potential value-added starter cultures in dairy technology. Various studies have shown the potential of freeze-dried kefir culture in the production of whey cheese (Dimitrellou *et al.* 2007) and feta-type cheese (Kourkoutas *et al.* 2006; Dimitrellou *et al.* 2010). The studies also suggested the use of freeze-dried

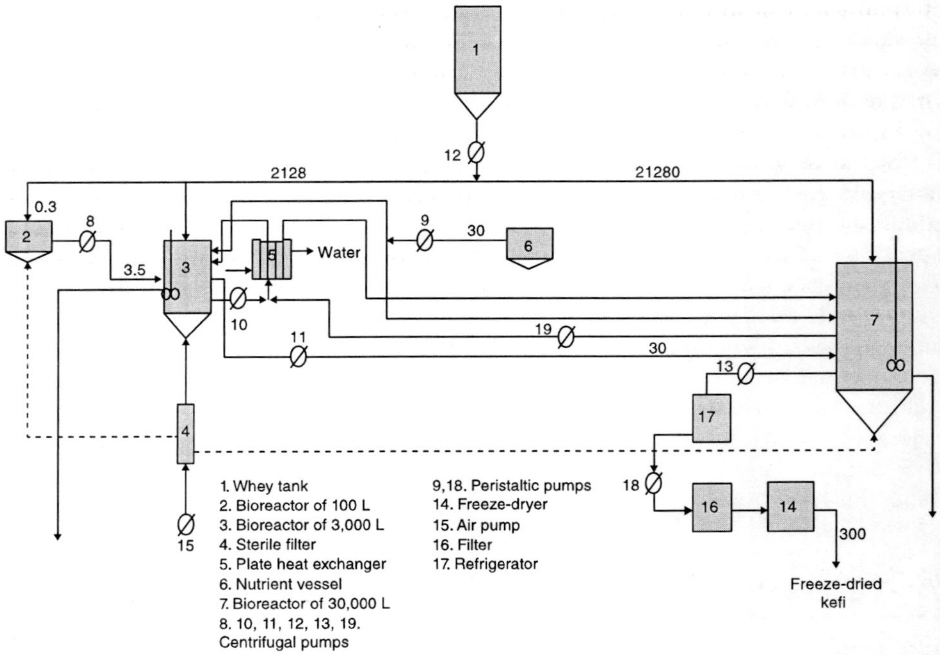

Figure 6.15 Industrial-scale process flow chart for production of freeze-dried kefir with mass balance (kg/day). Production capacity: 300 kg of dry weight/day (Kourkoutas *et al.* 2007).

kefir cultures for extending the shelf life of dairy products with less salt content. Furthermore, the cheese produced by these cultures had improved quality and sensory characteristics compared to commercial cheese.

Apart from these advantages, the use of dried probiotics is a boost for dairy ingredients in the development of dairy and non-dairy-based functional foods. Freeze drying is the most suitable technique for drying probiotics to preserve their viability and functionality. However, the use of freeze drying as a microencapsulation technique for probiotics is a relatively new concept. The following section explains in detail the basics of probiotics, the need for microencapsulation, cell viability issues during freeze drying and the characteristics of freeze-dried probiotic microcapsules.

6.8 Microencapsulation of probiotics by freeze drying

6.8.1 Probiotics

The word 'probiotics' was derived from the Greek 'pro bios' meaning 'for life'. Probiotics are live microorganisms which when administered in adequate amounts confer a health benefit on the host (FAO/WHO 2001). *Lactobacillus* and *Bifidobacterium* are the two most common genera used as probiotics. Both these genera are categorized under GRAS (generally recognized as safe) and are dominant inhabitants in the human intestinal tract (Anal & Singh 2007). *Lactobacillus* spp. is widely used in the dairy sector for the development of novel functional products. A number of food materials for probiotics incorporation have been examined recently, including yoghurt, various types of cheese or cheese-based dips, fermented milk, ice-cream, baby food, milk powder, milk chocolate, frozen dairy desserts, whey-based beverages, sour cream, normal or flavoured liquid milk and buttermilk (Capela *et al.* 2006; Possemiers *et al.* 2010; Mohammadi *et al.* 2011). Moreover, different non-dairy probiotic products like vegetable-based products, cereal-based products, fruit juices, malt-based beverages, soya-based products, oat-based desserts, confectionary products, bakery products breakfast cereals and baby foods have also been developed in recent years (Martin *et al.*, 2015). Other than these bacteria, species belonging to genera *Lactococcus, Enterococcus* (e.g. *Ent. faecalis, Ent. faecium*), some types of yeasts (e.g. *Saccharomyces cerevisiae* and *Saccharomyces boulardii*) and fungi (e.g. *Aspergillus oryzae*) have also been identified as having probiotic effects (Holzapfel & Schillinger 2002; Shah 2007).

6.8.2 Need for microencapsulation

Probiotics have to retain their active form to confer the necessary health benefits to the consumer. However, during oral delivery of probiotics in the form of fermented or functional foods, they are exposed to harsh gastrointestinal conditions such as acidic pH in the stomach and high bile salt in the intestine (Martin *et al.* 2015). Furthermore, the stability of the probiotics in the incorporated foods is subject to processing and storage conditions such as temperature, RH and exposure to light and oxygen. The various approaches suggested to improve the viability and stability of probiotic microorganisms are the selection of acid- and bile-resistant strains, two-step fermentation, stress adaptation, application of protective agents, incorporation of micronutrients and microencapsulation (Corcoran *et al.* 2004; Anal & Singh 2007; Meng *et al.* 2008). Among the various options, microencapsulation is one of the most efficient methods and has been under special consideration and investigation.

Microencapsulation is a technology that involves packing an active ingredient (core material) within sealed capsules (wall material) to improve stability, protecting the core material from adverse environmental conditions during storage and digestion, and promoting controlled-release properties (Shahidi & Han 1993). Various techniques used in probiotic microencapsulation are spray drying, freeze drying, vacuum drying, spray cooling, fluidized-bed agglomeration and coating, emulsion-based techniques, coacervation, extrusion techniques, adhesion to starch granules and compression coating. Of the microencapsulation drying methods, freeze drying is an appropriate technique for dehydration of heat labile compounds and live cells. Furthermore, freeze drying is also used as a second step in the encapsulation process after emulsion or gel entrapment methods (Chávarri *et al.* 2012). Freeze drying has been used as a preservative technology for long-term storage, easy transportation and food applications of probiotic cells. However, its use as a microencapsulation technique is a new dimension to the existing concept.

6.8.3 Cell viability issues associated with freeze drying

The viability of microencapsulated probiotic cells depends on the physiochemical properties of the microcapsules, including the type and concentration of the wall material, microcapsule size, initial cell numbers and bacterial strain (Anal & Singh 2007). Nevertheless, the major factors affecting viability are the unit operations involved in the microencapsulation process and the subsequent storage conditions. Freeze drying subjects the live probiotic cells to two kinds of stress during the freezing and drying stages. During the freezing stage there might be damage in the bacterial cell membrane due to ice-crystal formation and stress imparted by high osmolarity conditions. At a slow freezing rate water diffuses from the cells to maintain osmotic equilibrium and is frozen outside the cells (Figure 6.16a), whereas at the optimal freezing rate the rate of crystal growth and the concentration gradient are higher (Figure 6.16b), and at a high freezing rate cells reach the stabilizing temperature before water diffuses out and hence ice crystals are formed inside the cells, a condition referred to as intracellular ice formation (IIF) (Figure 6.16c) (Santivarangkna *et al.* 2008). It has to be taken into consideration that the freezing rate is not too high to cause IIF and not too low to cause solution effects (Bâati *et al.* 2000). Also, it is reported that cell size has an influence on cell viability during freeze drying, wherein spherical cells are more resistant to freezing and freeze drying than large rod-shaped lactobacilli (Meng *et al.* 2008). On the whole, the two primary mechanisms that cause damage to live cells during freeze drying are the changes in the physical state of the membrane lipids and changes in the structure of the cellular proteins. For this reason, cryoprotectants such as skim milk powder, whey protein, maltodextrin, and sugars such as trehalose, glucose and others are used to reduce the osmotic difference between the internal and external environment, thereby reducing the stress caused to the cells. The cryoprotectants have the ability to replace water during dehydration and maintain the bacteria in a hydrated condition. Sugars also have the ability to form glassy matrices, which restrict the molecular interactions (Santivarangkna *et al.* 2011; Tymczyszyn *et al.* 2012). Furthermore, to stabilize the cells during storage, protective agents like prebiotics, fibers or protein–polysaccahride blends are also employed. Despite these critical issues, freeze drying is the most widely used drying technique for the preservation of probiotic cells and starter cultures. This is attributed to the technical advantages of the freeze-drying phenomenon, wherein water phase transition and oxidation can be avoided (Berner & Viernstein 2006).

6.8.4 Characteristics of microencapsulated probiotic cells

The *morphology* of freeze-dried microencapsulated probiotic cells is largely dependent on the wall material used. In general, freeze-dried microencapsulates have flaky appearance with irregular morphology and porous structure. The SEM image of

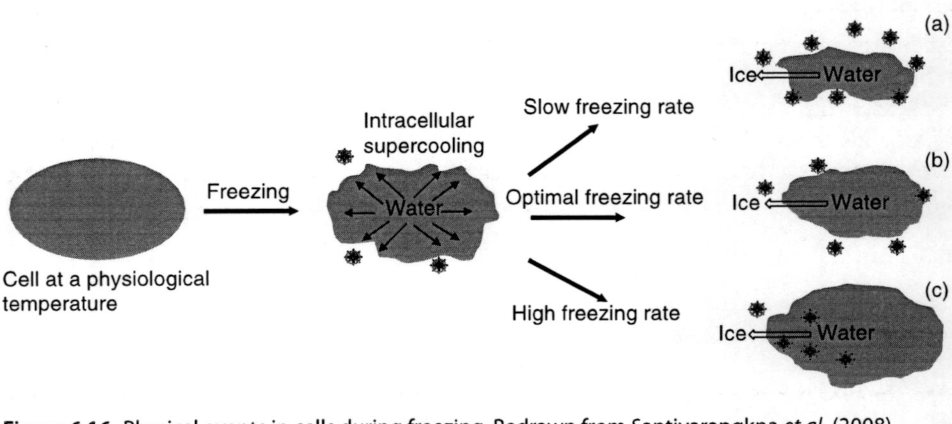

Figure 6.16 Physical events in cells during freezing. Redrawn from Santivarangkna *et al.* (2008).

microencapsulated *Lactobacillus plantarum* (MTCC 5422) is shown in Figure 6.17. The authors have used four different wall material formulations to produce probiotic (WPI + SA, whey protein isolate + sodium alginate; DWPI + SA, denatured whey protein isolate + sodium alginate) and synbiotic (WPI + FOS, whey protein isolate + fructooligosaccharide; DWPI + FOS, denatured whey protein isolate + fructooligosaccharide) microcapsules. The entrapped probiotic cells are clearly visible on the surface of the microencapsulates (Figure 6.17b,c, inset), indicating the uniform distribution of the cells in the wall matrix. The difference in the morphology with varying wall material composition is evident from the figure (Rajam & Anandharamakrishnan 2015). Similarly, Poddar *et al.* (2014) reported porous-sheet-like morphology for freeze-dried *Lactobacillus paracasei* using whole milk powder as wall material (Figure 6.17e).

The *moisture content* of freeze-dried probiotics has a profound influence on cell viability after drying and during storage. The final moisture content is determined by the process parameters and the wall materials employed. For instance, the moisture content of freeze-dried microencapsulated *L. paracasei* was as low as 1.67% when whole milk powder was used as an encapsulating material (Poddar *et al.* 2014), whereas it was 3.87% when a combination of whey protein isolate and sodium alginate was used to encapsulate *L. plantarum* (Rajam *et al.* 2012). Moreover, compared to other drying techniques, freeze drying yields powder with lower moisture content as the secondary drying process ensures drying of the microcapsules to equilibrium moisture content.

Porosity is an important physical property of freeze-dried microencapsulates. The pores in the freeze-dried powders are obtained during the sublimation process, which involves direct phase transition of the solid ice crystals to the vapour phase (Figure 6.17a,c). Porosity also determines the water absorption capacity, reconstitution ability and controlled-release properties of the microencapsulated powders (Rajam & Anandharamakrishnan 2015).

The physical properties of microencapsulates, together with the wall material attributes, determine the viability of the encapsulated probiotic cells during storage and gastrointestinal digestion. Storage temperature and RH have an effect on the glass transition temperature of microencapsulates and the rate of metabolism of the probiotic cells. Hence, increasing RH and temperature will lead to higher loss of cell viability during storage (Lapsiri *et al.* 2013). In a study by Rajam *et al.* (2015), microencapsulation enhanced the storage stability of *L. plantarum*, wherein more than 90% of the cells were viable up to 60 days after storage at refrigeration temperature, as against less than 50% viability in free unencapsulated form (control) (Figure 6.18). The stability of microencapsulated cells in simulated gastrointestinal conditions was also found to be higher than that of unencapsulated cells.

Research and development in the food sector is focused on the development of functional foods containing probiotic properties. Microencapsulated freeze-dried probiotics have numerous applications in food fortification because of their sensory acceptability. This is because the off-flavours associated with the bacterial cultures can be masked by encapsulation technology, and cell viability and storage stability can be enhanced.

6.9 Conclusion

Freeze drying is a technique for drying heat-sensitive and aroma-rich compounds with high market value. Freeze drying is used in the dairy industry for the production of

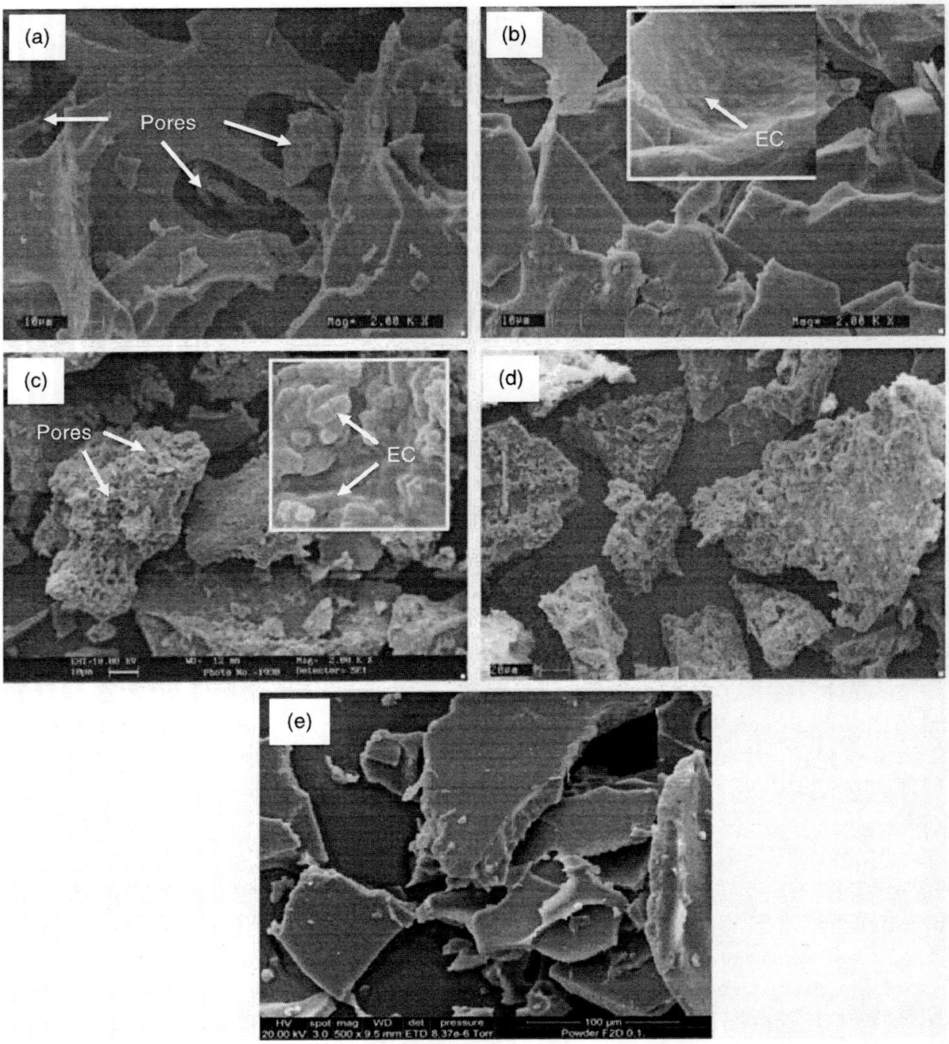

Figure 6.17 Morphology of freeze-dried microencapsulated probiotics. (a)–(d) *Lactobacillus plantarum* with WPI + SA, WPI + FOS, DWPI + SA, DWPI + FOS wall material formulations, respectively (Rajam & Anandharamakrishnan 2015). (e) *Lactobacillus paracasei* with whole milk powder as wall matrix (Poddar *et al*. 2014). EC, encapsulated cells.

quality products for a variety of other applications. Various properties of the dairy components affect the freeze-drying process and the quality of the dried product. Furthermore, freeze drying has an important role in the preservation of kefir cultures and probiotics. Recently, freeze drying has been used as a microencapsulation technique for improving the stability and release properties of probiotic bacteria. The future scope of freeze drying lies in the development of synergistic methods that reduce the drying time and in turn the production cost involved. This would herald further applications of freeze drying in the dairy sector.

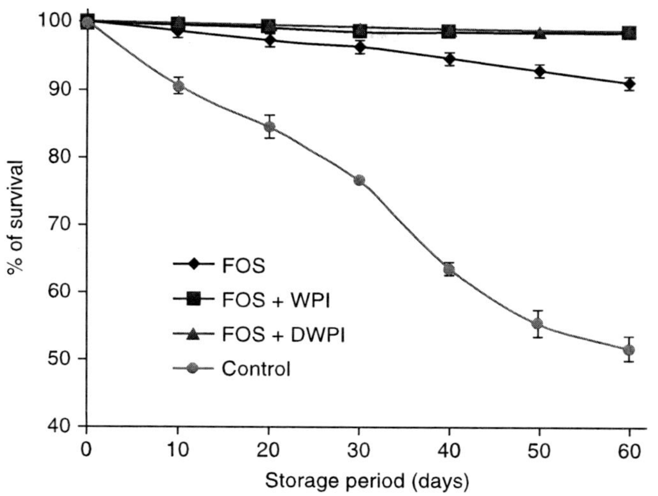

Figure 6.18 Storage stability of unencapsulated (control) and microencapsulated *L. plantarum* at 4 °C. FOS, fructooligosaccahride; WPI, whey protein isolate; DWPI, denatured whey protein isolate. From Rajam *et al.* (2014).

References

Anal, A.K. & Singh, H. (2007). Recent advances in microencapsulation of probiotics for industrial applications and targeted delivery. *Trends in Food Science and Technology*, **18**(**5**), 240–251.

Angulo, L., Lopez, E. & Lema, C. (1993). Microflora present in kefir grains of the Galician region (north-west of Spain). *Journal of Dairy Research*, **60**(**02**), 263–267.

Bâati, L., Fabre-Gea, C., Auriol, D., & Blanc P.J. (2000). Study of the cryotolerance of *Lactobacillus acidophilus*: effect of culture and freezing conditions on the viability and cellular protein levels. *International Journal of Food Microbiology*, **59**(**3**), 241–247.

Beaty, N. (2006). Lyophilization: heat and mass transfer. *American Pharmaceutical Review*, **9**, 81–83.

Ben Souda, K., Akyel, C. & Bilgen, E. (1989). Freeze dehydration of milk using microwave energy. *Journal of Microwave Power and Electromagnetic Energy*, **24**(**4**), 195–202.

Berk, Z. (2013). Freeze drying (lyophilization) and freeze concentration. In: *Food Process Engineering and Technology*, 2nd edn, pp. 518–520, Academic Press.

Berner, D. & Viernstein, H. (2006). Effect of protective agents on the viability of *Lactococcus lactis* subjected to freeze-thawing and freeze-drying. *Scientia Pharmaceutica*, **74**(**3**), 137.

Capela, P., Hay, T. & Shah, N. (2006). Effect of cryoprotectants, prebiotics and microencapsulation on survival of probiotic organisms in yoghurt and freeze-dried yoghurt. *Food Research International*, **39**(**2**), 203–211.

Chávarri, M., Marañón, I. & Villarán, M.C. (2012). *Encapsulation Technology to Protect Probiotic Bacteria*. INTECH Open Access Publisher.

Christ, M. (2010). Smart freeze drying: basic principles, optimum procedures and applications. *Germany: Martin Christ*, 51.

Corcoran, B., Ross, R., Fitzgerald, G. & Stanton, C. (2004). Comparative survival of probiotic lactobacilli spray-dried in the presence of prebiotic substances. *Journal of Applied Microbiology*, **96**(**5**), 1024–1039.

DFE Pharma. *Amorphous Lactose: Origins and Measurement*, http://www.dfepharma.com/en/knowledge-base/inhalation/technical-papers.aspx.

Dimitrellou, D., Kourkoutas, Y., Banat, I.M., Marchant, R. & Koutinas, A.A. (2007). Whey-cheese production using freeze-dried kefir culture as a starter. *Journal of Applied Microbiology*, **103**(**4**), 1170–1183.

Dimitrellou, D., Kandylis, P., Mallouchos, A., Komaitis, M., Koutinas, A.A. & Kourkoutas, Y. (2010). Effect of freeze-dried kefir culture on proteolysis in feta-type and whey-cheeses. *Food Chemistry*, **119**(**2**), 795–800.

Duan, X., Zhang, M., Mujumdar, A.S. & Wang, R. (2010). Trends in microwave-assisted freeze drying of foods. *Drying Technology*, **28**(**4**), 444–453.

Fan, F. & Roos, Y.H. (2015). X-ray diffraction analysis of lactose crystallization in freeze-dried lactose–whey protein systems. *Food Research International*, **67**, 1–11.

FAO/WHO (2001). *Report on Joint FAO/WHO Expert Consultation on Evaluation of Health and Nutritional Properties of Probiotics in Food Including Powder Milk with Live Lactic Acid Bacteria*, FAO/WHO.

Fellows, P.J. (2009). Freeze drying and freeze concentration. In: *Food Processing Technology: Principles and Practice*, Elsevier.

Fernández, E., Schebor, C. & Chirife, J. (2003). Glass transition temperature of regular and lactose hydrolyzed milk powders. *LWT – Food Science and Technology*, **36**(**5**), 547–551.

Fito, P.J., Pinaga, F. & Aranda, V. (1984). Thermal conductivity of porous bodies at low pressure: Part I. *Journal of Food Engineering*, **3**(**1**), 75–88.

Fitzpatrick, J.J., Hodnett, M., Twomey, M., Cerqueira, P.S.M., O'Flynn, J. & Roos, Y.H. (2007). Glass transition and the flowability and caking of powders containing amorphous lactose. *Powder Technology*, **178**(**2**), 119–128.

Fitzpatrick, J.J., Descamps, N., O'Meara, K., Jones, C., Walsh, D. & Spitere, M. (2010). Comparing the caking behaviours of skim milk powder, amorphous maltodextrin and crystalline common salt. *Powder Technology*, **204**(**1**), 131–137.

Fyfe, K., Kravchuk, O., Nguyen, A.V., Deeth, H. & Bhandari, B. (2011). Influence of dryer type on surface characteristics of milk powders. *Drying Technology*, **29**(**7**), 758–769.

Gatlin, L.A. & Nail, S.L. (1993). Freeze drying: A practical overview. *Protein Purification Process Engineering*, pp. 317–367, Marcel Dekker, New York.

Haque, M.K. & Roos, Y.H. (2005). Crystallization and X-ray diffraction of spray-dried and freeze-dried amorphous lactose. *Carbohydrate Research*, **340**(**2**), 293–301.

Holzapfel, W.H. & Schillinger, U. (2002). Introduction to pre- and probiotics. *Food Research International*, **35**(**2**), 109–116.

Jafar, F. & Farid, M. (2003). Analysis of heat and mass transfer in freeze drying. *Drying Technology*, **21**(**2**), 249–263.

Jouppila, K. & Roos, Y.H. (1994a). Water sorption and time-dependent phenomena of milk powders. *Journal of Dairy Science*, **77**(**7**), 1798–1808.

Jouppila, K. & Roos, Y.H. (1994b). Glass transitions and crystallization in milk powders. *Journal of Dairy Science*, **77**(**10**), 2907–2915.

Karel, M. (1975). Heat and mass transfer in freeze-drying. In: *Freeze Drying and Advanced Food Technology* (eds Goldblith, S.A., Rey, L. & Rothmayr, W.W.), pp. 177–202, Academic Press, New York.

Karel, M. & Lund, D.B. (2003). *Physical Principles of Food Preservation: Revised and Expanded, Vol. 129*, CRC Press.

Kochs, M., Körber, C.H., Heschel, I. & Nunner, B. (1993). The influence of the freezing process on vapour transport during sublimation in vacuum-freeze-drying of macroscopic samples. *International Journal of Heat and Mass Transfer*, **36**(**7**), 1727–1738.

Kourkoutas, Y., Kandylis, P., Panas, P., Dooley, J.S.G., Nigam, P. & Koutinas, A.A. (2006). Evaluation of freeze-dried kefir coculture as starter in feta-type cheese production. *Applied and Environmental Microbiology*, **72**(**9**), 6124–6135.

Kourkoutas, Y., Sipsas, V., Papavasiliou, G. & Koutinas, A.A. (2007). An economic evaluation of freeze-dried kefir starter culture production using whey. *Journal of Dairy Science*, **90**(**5**), 2175–2180.

Kudra, T. (2003). Sticky region in drying – Definition and identification. *Drying Technology*, **21**(**8**), 1457–1469.

Lapsiri, W., Bhandari, B. & Wanchaitanawong, P. (2013). Stability and probiotic properties of *Lactobacillus plantarum* spray-dried with protein and other protectants. *Drying Technology*, **31**(**13–14**), 1723–1733.

Liapis, A.I. & Bruttini, R. (1995). Freeze drying. In: *Handbook of Industrial Drying*, 2nd edn (Mujumdar, A.S. ed.), pp. 305–343, Marcel Dekker, New York.

Liapis, A.I. & Bruttini, R. (2015). Freeze drying. In: *Handbook of Industrial Drying, Vol. 1* (ed. Mujumdar A.), 4th edn, pp. 259–280, Marcel Dekker, New York.

Liapis, A.I., Pikal, M.J. & Bruttini, R (1996). Research and development needs and opportunities in freeze-drying. *Drying Technology*, **14**(**6**), 1265–1300.

Lopez-Quiroga, E., Antelo, L.T. & Alonso, A.A. (2012). Time-scale modeling and optimal control of freeze–drying. *Journal of Food Engineering*, **111**(**4**), 655–666.

Marabi, A., Mayor, G., Raemy, A., Bauwens, I., Claude, J., Burbidge, A.S. & Saguy, I.S. (2007). Solution calorimetry: a novel perspective into the dissolution process of food powders. *Food Research International*, **40**(**10**), 1286–1298.

Martin, M.J., Lara-Villoslada, F., Ruiz, M.A. & Morales, M.E., (2015). Microencapsulation of bacteria: A review of different technologies and their impact on the probiotic effects. *Innovative Food Science and Emerging Technologies*, **27**, 15–25.

Mellor, J.D. (1978). *Fundamentals of Freeze-drying*, Academic Press, London.

Meng, X., Stanton, C., Fitzgerald, G., Daly, C. & Ross, R. (2008). Anhydrobiotics: The challenges of drying probiotic cultures. *Food Chemistry*, **106**(**4**), 1406–1416.

Millman, M.J., Liapis, A.I. & Marchello, J.M. (1985). Note on the economics of batch freeze dryers. *Journal of Food Technology*, **20**, 541–551.

Mohammadi, R., Mortazavian, A.M., Khosrokhavar, R. & da Cruz, A.G. (2011). Probiotic ice cream: viability of probiotic bacteria and sensory properties. *Annals of Microbiology*, **61**(**3**), 411–424.

Nail, S.L., Jiang, S., Chongprasert, S. & Knopp, S.A. (2002). Fundamentals of freeze-drying. In: *Pharmaceutical Biotechnology 14* (eds Nail, S.L. & Akers, M.J.), pp. 281–360, Kluwer Academic/Plenum Publishers, New York.

Nickerson, T.A., Coulter, S.T. & Jenness, R. (1952). Some properties of freeze-dried milk. *Journal of Dairy Science*, **35**(**1**), 77–85.

Oetjen, G.W. (1999). *Freeze Drying*, WILEY-VCH, Weinheim.

Oetjen, G.W. & Haseley, P. (2004). *Freeze Drying*, 2nd edn, Wiley-VCH, Weinheim **978**, **2**, 76–140.

Omar, A.E. & Roos, Y.H. (2007). Water sorption and time-dependent crystallization behaviour of freeze-dried lactose–salt mixtures. *LWT – Food Science and Technology*, **40**(3), 520–528.

Papavasiliou, G., Kourkoutas, Y., Rapti, A., Sipsas, V., Soupioni, M. & Koutinas, A.A. (2008). Production of freeze-dried kefir culture using whey. *International Dairy Journal*, **18**(**3**), 247–254.

Patel, S.M., Chaudhuri, S. & Pikal, M.J. (2010). Choked flow and importance of Mach I in freeze-drying process design. *Chemical Engineering Science*, **65**(**21**), 5716–5727.

Pikal, M. (1990). Freeze-drying of proteins. Part I: Process design. *BioPharm International*, **3**(**8**), 18–27.

Pikal, M.J. & Shah, S. (1990). The collapse temperature in freeze drying: dependence on measurement methodology and rate of water removal from the glassy phase. *International Journal of Pharmaceutics*, **62**(**2**), 165–186.

Pikal, M.J., Shah, S., Roy, M.L. & Putman, R. (1990). The secondary drying stage of freeze drying: drying kinetics as a function of temperature and chamber pressure. *International Journal of Pharmaceutics*, **60**(**3**), 203–207.

Poddar, D., Das, S., Jones, G., Palmer, J., Jameson, G.B., Haverkamp, R.G. & Singh, H. (2014). Stability of probiotic *Lactobacillus paracasei* during storage as affected by the drying method. *International Dairy Journal*, **39**(**1**), 1–7.

Possemiers, S., Marzorati, M., Verstraete, W. & Van de Wiele, T. (2010). Bacteria and chocolate: a successful combination for probiotic delivery. *International Journal of Food Microbiology*, **141**(**1**), 97–103.

Radaeva, I.A., Shul'kina, S.P., Kocherga, S.I. & Efron, B.G. (1975). Effect of freezing regimes in freeze-drying on yoghurt quality. *Molochnaya Promyshlennost*, **5**, 22–23.

Rajam, R. & Anandharamakrishnan, C. (2015). Spray freeze drying method for microencapsulation of *Lactobacillus plantarum*. *Journal of Food Engineering*, **166**, 95–103.

Rajam, R., Karthik, P., Parthasarathi, S., Joseph, G.S. & Anandharamakrishnan, C. (2012). Effect of whey protein–alginate wall systems on survival of microencapsulated *Lactobacillus plantarum* in simulated gastrointestinal conditions. *Journal of Functional Foods*, **4**(**4**), 891–898.

Rajam, R., Kumar, S.B., Prabhasankar, P. & Anandharamakrishnan, C. (2015). Microencapsulation of *Lactobacillus plantarum* MTCC 5422 in fructooligosaccharide and whey protein wall systems and its impact on noodle quality. *Journal of Food Science and Technology*, **52**(7), 4029–4041.

Rambhatla, S. & Pikal, M.J. (2003). Heat and mass transfer scale-up issues during freeze-drying, I: Atypical radiation and the edge vial effect. *AAPS Pharmscitech*, **4**(**2**), 22–31.

Rambhatla, S., Tchessalov, S. & Pikal, M.J. (2006). Heat and mass transfer scale-up issues during freeze-drying, III: control and characterization of dryer differences via operational qualification tests. *AAPS PharmSciTech*, **7**(**2**), E61–E70.

Ratti, C. (2009). Freeze and vacuum drying of foods. In: *Drying Technologies in Food Processing* (eds Chen, X.D. & Mujumdar, A.S.), pp. 225–246, Blackwell.

Rolfgaard, J. (1987) Freeze drying: processing, costs and applications. In: *Food Technology International Europe* (ed. Turner, A.), pp. 47–49, Sterling Publications International, London.

Roos, Y.H. (2002). Importance of glass transition and water activity to spray drying and stability of dairy powders. *Le Lait*, **82**(**4**), 475–484.
Rowe, T.W.G. (1976) Optimization in freeze-drying. *Developments in Biological Standards*, **36**, 79–97.
Santivarangkna, C., Kulozik, U. & Foerst, P. (2008). Inactivation mechanisms of lactic acid starter cultures preserved by drying processes. *Journal of Applied Microbiology*, **105**(**1**), 1–13.
Santivarangkna, C., Aschenbrenner, M., Kulozik, U. & Foerst, P. (2011). Role of glassy state on stabilities of freeze-dried probiotics. *Journal of Food Science*, **76**(**8**), R152–R156.
Schmitz-Schug, I., Gianfrancesco, A., Kulozik, U. & Foerst, P. (2013). Physical state, molecular mobility and chemical stability of powdered dairy formulations. *Food Research International*, **53**(**1**), 268–277.
Searles, J. (2004). Observation and implications of sonic water vapor flow during freeze-drying. *American Pharmaceutical Review*, **7**, 58–69.
Shah, N.P. (2007). Functional cultures and health benefits. *International Dairy Journal*, **17**(**11**), 1262–1277.
Shahidi, F. & Han, X.Q. (1993). Encapsulation of food ingredients. *Critical Reviews in Food Science and Nutrition*, **33**(**6**), 501–547.
Slade, L. & Levine, H. (1995). Polymer science approach to water relationships in foods. In: *Food Preservation by Moisture Control. Fundamentals and Applications* (eds Barbosa-Cánovas, G.G. & Welti-Chanes, J.), Technomic Publishing Co., Inc., Lancaster, PA.
Smith, P.G. (2011). *Introduction to Food Process Engineering*, pp. 329–334, Springer.
Stapley, A. (2008) Freeze drying. In: *Frozen Food Science and Technology* (ed. Evans J.A.), Blackwell Publishing, Oxford.
Tang, X. & Pikal, M.J. (2004). Design of freeze-drying processes for pharmaceuticals: practical advice. *Pharmaceutical Research*, **21**(**2**), 191–200.
Tymczyszyn, E.E., Sosa, N., Gerbino, E., Hugo, A., Gómez-Zavaglia, A. & Schebor, C. (2012). Effect of physical properties on the stability of *Lactobacillus bulgaricus* in a freeze-dried galacto-oligosaccharides matrix. *International Journal of Food Microbiology*, **155**(**3**), 217–221.
Vollenbroek, J., Hebbink, G.A., Ziffels, S. & Steckel, H. (2010). Determination of low levels of amorphous content in inhalation grade lactose by moisture sorption isotherms. *International Journal of Pharmaceutics*, **395**(**1**), 62–70.
Wang, W., Chen, G. & Gao, F. (2005). Effect of dielectric material on microwave freeze drying of skim milk. *Drying Technology*, **23**(**1–2**), 317–340.
Wang, W., Chen, M. & Chen, G. (2012). Issues in freeze drying of aqueous solutions. *Chinese Journal of Chemical Engineering*, **20**(**3**), 551–559.
Witthuhn, R.C., Schoeman, T. & Britz, T.J. (2005). Characterisation of the microbial population at different stages of Kefir production and Kefir grain mass cultivation. *International Dairy Journal*, **15**(**4**), 383–389.
Wolfe, J. & Bryant, G. (1999). Freezing, drying, and/or vitrification of membrane–solute–water systems. *Cryobiology*, **39**(**2**), 103–129.
Yu, K.C., Chen, C.C. & Wu, P.C. (2011). Research on application and rehydration rate of vacuum freeze drying of rice. *Journal of Applied Sciences*, **11**(**3**), 535–541.

7

Spray Freeze Drying

S. Padma Ishwarya[1], *C. Anandharamakrishnan*[1,2] *and A.G.F. Stapley*[3]

[1] *Department of Food Engineering, CSIR – Central Food Technological Research Institute, Mysore 570 020, India*
[2] *Indian Institute of Crop Processing Technology, Ministry of Food Processing Industries, Government of India, Thanjavur 613 005, Tamil Nadu*
[3] *Department of Chemical Engineering, Loughborough University, Loughborough, LE11 3TU, UK*

7.1 Introduction

In this era of convenience, instant rehydration is a key aspect that determines the quality and acceptability of food powders. Powder attributes such as wettability, dispersibility and solubility are important in this regard, and are related to the surface composition and therefore hydrophilicity (Kim *et al.* 2002) and microstructure of the powder (Anandharamakrishnan *et al.* 2010). Both of these properties are greatly influenced by the choice of drying method.

Spray drying is the established process for the continuous commercial production of whole and skim milk powder. However, spray-dried dairy powders present limitations with respect to certain key physical quality attributes. Spray drying of milk results in powder surfaces with greater hydrophobicity owing to the tendency for fat to cover the particle surfaces (Millqvist-Fureby *et al.* 2001). Furthermore, the presence of sugar on the surface leads to stickiness and eventually impairs the powder solubility and flow properties (Hindmarsh *et al.* 2007). The closed internal pores (Yu *et al.* 2012) and a solid and dense surface layer of spray-dried powders hinder water penetration during reconstitution (Rogers *et al.* 2008). As a result, spray-dried whole milk powder lacks instantaneous solubility in water and leaves a fatty deposition on the container when rehydrated. Undesirable flavour quality has also been observed in fresh lots of spray dried whole milk powder, which continues to deteriorate during storage (Nickerson *et al.* 1952).

Alternatives include drum drying and freeze drying. In the case of drum drying, the main disadvantage is the scorched flavour and much lower product solubility of milk powder due to protein denaturation (Hall & Hedrick 1971). While freeze drying leads to porous particles owing to the sublimation of ice crystals, the process suffers from limitations of much greater energy expenditure and long processing times. The increase in processing cost with freeze drying is five- to eight-fold when compared to conventional drying methods (Adams 1991). In addition, freeze-dried whole milk powder shows slow wettability due to the presence of free fat, similar to that observed in spray drying (Barbosa-Canovas *et al.* 2005). However, freeze-dried skim milk powder

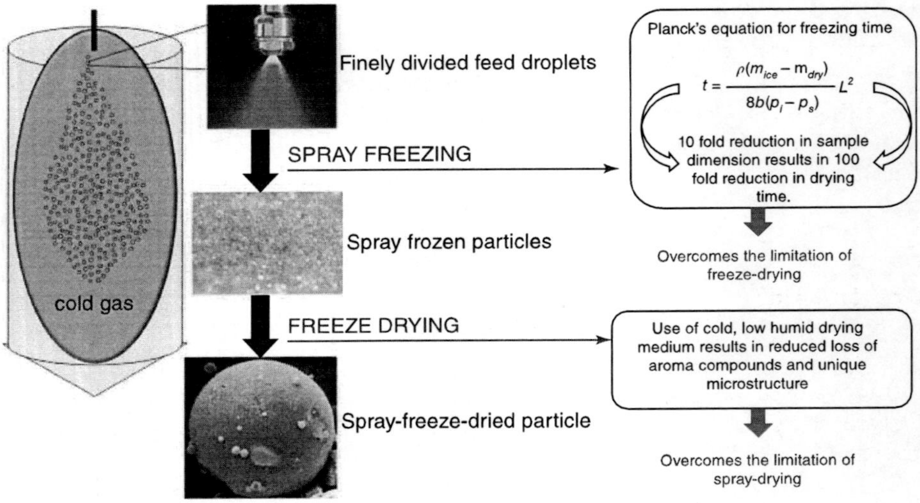

Figure 7.1 Spray-freeze-drying process.

exhibits excellent wettability and dispersibility characteristics. Thus, the current interest is to adopt a drying technique that will overcome the limitations associated with conventional dairy drying processes.

Spray freeze drying (SFD) is an unconventional freeze-drying technique that produces uniquely powdered products whilst maintaining the benefits of conventionally freeze-dried products. The applications of SFD with respect to high-value products are well established due to its edge over other drying techniques in terms of product structure, quality and the retention of bioactive compounds (Padma Ishwarya *et al.* 2015). SFD produces powdered products with higher surface-to-mass ratios and highly porous microstructures. SFD has these advantages whilst overcoming the challenges involved in both spray drying and freeze drying (Figure 7.1). The underlying principle of the different stages of SFD and its application in the production of powdered dairy products and encapsulated probiotic microcapsules will be discussed in the different sections of this chapter.

7.2 SFD process

The application of spray freezing in foods was initially proposed by Meryman (1959), with the focus on its advantages over conventional freeze drying in terms of water removal at increased mass-transfer rates. During the SFD process, the feed solution undergoes three transformation stages (Figure 7.1):

i. atomization
ii. freezing
iii. freeze drying.

Thus, SFD can be defined as a three-step process that involves a liquid or solution being atomized into droplets, solidified by contact with a cold fluid, followed by sublimation at low temperature and pressure (Leuenberger 2002).

7.2.1 Atomization

Atomization is central to the SFD process and involves spraying a feed in the form of a solution or suspension. Similar to spray drying, atomization during SFD can be accomplished using one-fluid (hydraulic), two-fluid (pneumatic) or ultrasonic nozzles. The minute droplets of feed solution resulting from atomization offer a more homogeneous temperature field for heat transfer during the freezing step, which leads to uniform nucleation and formation of fine ice crystals (MacLeod *et al.* 2006). Also, as depicted in Figure 7.1, the drying time varies approximately with the square of the sample thickness. Taking an instance of the typical conventional freeze-drying process, where the characteristic sample dimension is approximately 0.5 cm, the processing time might be 21 h according to Planck's equation for freezing time. However, with atomisation leading to droplets as fine as 50 μm, the freeze-drying time can theoretically be reduced to as little as 7 s. The reduced sample dimension also results in an increased sublimation rate (Pham 1986) and surface mass-transfer coefficient during the subsequent stages of freezing and freeze drying, respectively.

Hydraulic and twin-fluid nozzles are the most commonly used atomizers in SFD studies (Al-Hakim *et al.* 2006). With respect to the latter type, the feed flow rate (or pressure), fluid cap inner diameter, flow rate and pressure of atomization gas are found to be the key variables influencing droplet size (Costantino *et al.* 2000). While the feed flow rate and fluid cap diameter have a direct relationship with droplet size, the atomization gas flow rate and pressure depict an inverse relationship. Similar relationships have also been observed for hydraulic nozzles. In addition to these parameters, viscosity has also been found to have an influence on the size of droplets emanating from a hydraulic nozzle. While low viscosities of the feed solution (1.7–3.9 cP) led to particles less than 100 μm in diameter, viscosities of 11 cP failed to produce similar-sized particles (Barron *et al.* 2003). However, the disadvantage of hydraulic and twin-fluid nozzles is the wide range of droplet sizes, which in turn impacts on the freezing rate in the subsequent step. This limitation can be overcome by the use of ultrasonic nozzles and monodisperse droplet generators.

Ultrasonic nozzles use an electromechanical device that vibrates at a very high frequency. The fluid to be atomized passes through the surface of this vibrating device. As the frequency of vibration approaches the resonance frequency, a square wave pattern forms on the liquid surface. Further increases in the amplitude of vibration promote droplet formation and its detachment from the liquid film. This type of atomizer results in large, porous particles independent of the polarity or solubility characteristics of the active component involved. The vibration frequency of the ultrasonic nozzle is the most significant atomisation parameter which controls the droplet size. Hence, this method can exert a high degree of control over particle size by varying the frequency of the ultrasonic atomizer. Unlike twin-fluid nozzles, the operation of ultrasonic atomizers does not require an atomizing gas and enables the droplets to be captured relatively easily and effectively in a cryogenic liquid placed a short distance below the nozzle (D'Addio *et al.* 2012).

The use of four-fluid nozzles is a recent development with respect to atomization in SFD. The four-fluid nozzles utilize two separate liquid feed streams atomized by individual air streams. The streams collide with each other at the nozzle tip or the collision focal spot, generating a thin liquid film with a high-speed gas flow. The shock waves

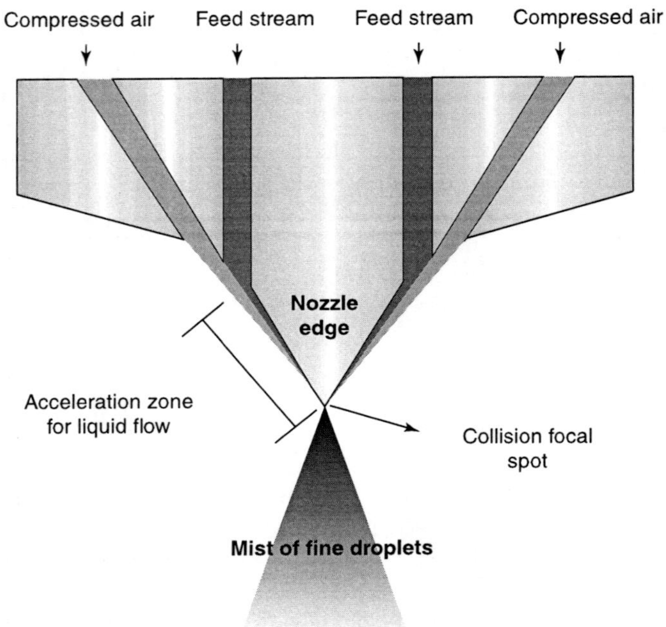

Figure 7.2 Four-fluid nozzle. Adapted from Mizoe *et al.* (2008).

generated at the focal spot create a mist of fine droplets (Figure 7.2). This is followed by transit of the fine droplets into liquid nitrogen placed at a fixed distance below the nozzle. Particles of the final product were found to have a fine and porous structure with greater specific surface area. Furthermore, this nozzle type has been found to be suitable for active ingredients that exhibit poor aqueous solubility (Niwa *et al.* 2009).

It is possible that the low temperatures used in spray freezing may cause the feed solution to freeze within and block the nozzle orifice. Hence, the nozzle housings are generally supplied with exterior air heating. An alternative approach is to use plastic nozzles (Al-Hakim *et al.* 2006; Padma Ishwarya *et al.* 2015).

7.2.2 Freezing

During this stage, the atomized droplets are subjected to rapid freezing on contact with a cryogen to form icy particles. The rapid freezing is significant because:

i. it restrains the circulation of components within the particle and therefore prevents shrinkage (Rogers *et al.* 2008)
ii. it reduces the size of the ice crystals (Costantino *et al.* 2000), which eventually leads to a highly porous structure on sublimation during the subsequent freeze drying step
iii. it achieves a brighter product colour, as the small pores resulting from the sublimation of the small ice crystals scatter more light than the larger pores formed by slow freezing (Ceballos *et al.* 2012).

Liquid nitrogen is the most commonly employed cryogen because of its low boiling point of −196 °C. The use of other cryogens such as isopentane, argon, hydrofluoroether and frozen ethanol (with an overlay of liquid nitrogen) has also been reported. Of all the cryogens listed above, liquid nitrogen tends to evaporate readily due to its much lower boiling point (−196 °C) when compared to argon (−185.8 °C), isopentane (27.7 °C) and ethanol (78.37 °C). The density and viscosity of the cryogen used for the spray-freezing step affects the formation of drops. This can either occur via Rayleigh break-up where the jet endures for some distance from the nozzle before forming droplets via Rayleigh instabilities or immediate atomization of the jet on leaving the nozzle. In the former case the drops are slightly larger in diameter than the nozzle, and in the latter case they are much smaller. This effect can be explained by the Weber (We, Eq. 7.1) and Reynolds numbers (Re, Eq. 7.2), which combine to give the Ohnesorge number (Oh, ratio of We to Re, pertaining to the cryogen used:

$$\mathrm{We} = \frac{\rho v^2 l}{\sigma} \tag{7.1}$$

$$\mathrm{Re} = \frac{l v \rho}{\mu} \tag{7.2}$$

$$\mathrm{Oh} = \frac{\mu v}{\sigma} \tag{7.3}$$

where, ρ is the density of the cryogen (kg/m^3), v is the jet velocity (m/s), l is the jet diameter (m), and σ is the interfacial tension of cryogen (N/m).

Engstrom *et al.* (2007) compared spraying into liquid nitrogen (We = 2×10^3), liquid isopentane (We = 2.8×10^3) and gaseous nitrogen (We = 11). Atomization resulted in both cases of spraying into liquid, whereas Rayleigh break-up occurred when using gaseous nitrogen. This is due to the low We number in the case of gaseous nitrogen as a consequence of the lower density. Spray freezing into liquid therefore invariably produces very small particles (Hu *et al.* 2002).

Based on the physical state of the cryogen used for freezing, SFD can be classified into three types: spray freezing into vapour (SFV), where the feed solution is sprayed into the vapour of cold dry gas, spray freezing into liquid (SFL), where the feed is sprayed directly into a cryogenic liquid, and spray freezing into vapour over liquid (SFV/L), where the feed is sprayed into a cryogenic liquid but with atomization occurring in the gaseous headspace region above the liquid.

7.2.2.1 Spray freezing into vapour

A rig for the SFV process is shown in Figure 7.3. Here, the chamber is initially purged with dry nitrogen gas from a cylinder to remove humidity from the chamber and then cooled using a liquid nitrogen supply. The chamber temperature is maintained by controlling the flows of liquid nitrogen and nitrogen cylinder gas to the chamber. This is usually done before the commencement of feed atomization into the chamber (Al-Hakim *et al.* 2006). The atomized droplets of feed solution move with respect to each other and the gas, as determined by the fluid mechanics of the spray. The resultant heat transfer between the gas and droplets depends on the temperature of the gas and droplets, and the droplet–gas slip velocity. Ice crystallization/freezing consequently occurs within the droplets (Anandharamakrishnan *et al.* 2008).

7.2.2.2 Spray freezing into vapour over liquid

Here, freezing is accomplished by atomizing the feed solution through a nozzle positioned at a short distance above a boiling cryogenic liquid (Figure 7.4). The droplets

Figure 7.3 Photograph of SFV rig during phase Doppler anemometry experiments (Al-Hakim *et al.* 2006).

start to solidify during the transit through the vapour gap and then freeze completely as contact is made with the cryogenic liquid (Adams *et al.* 1982; Buxton & Peach 1984). The distance between the nozzle tip and the surface of the cryogenic liquid should be optimized according to the required specifications of the final product. This distance influences the mean particle diameter of the dried product. With longer distance, the mean particle diameter might increase due to the gradual agglomeration and solidification of the atomized feed droplets as they pass through the vapour phase before settling onto the surface of the cryogenic liquid (Kawabata *et al.* 2011).

However, the major limitation of the SFV/L technique is the loss of protein stability during atomization. While the feed is being atomized over the ambient gas, a large gas–liquid interfacial area is created that causes protein adsorption and unfolding (Webb *et al.* 2002). In addition, the ensuing rapid cooling of the feed droplets produces a large ice–liquid interface, which denatures the proteins (Costantino *et al.* 2000). As the solvent freezes, the active ingredient begins to nucleate and grow due to its supersaturation in the unfrozen regions of the atomized droplet (Rogers *et al.* 2002b), which retards the rapid freezing process. The negative consequences are the particle growth, increase in particle size and an eventual reduction in specific surface area.

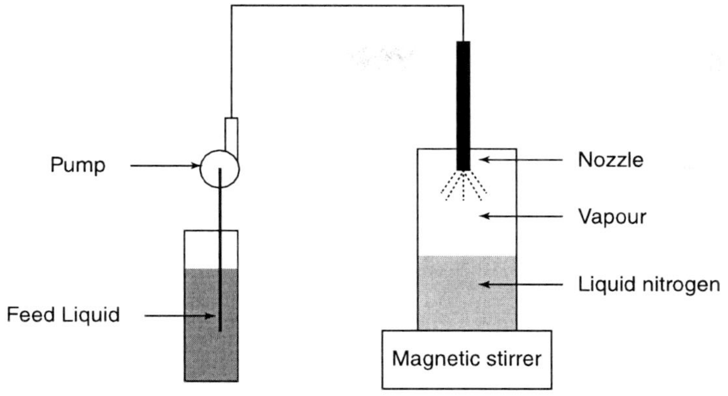

Figure 7.4 Spray freezing into vapour over liquid (Anandharamakrishnan 2008).

Formulation strategies can overcome the limitations posed by agglomeration and protein denaturation. This approach includes the use of freezing adjuvants and a choice of appropriate carrier materials that help in achieving product stability and functional integrity (Padma Ishwarya *et al.* 2015). Complexing the active ingredient (AI) with appropriate freezing adjuvants such as trehalose, mannitol and ammonium sulphate decreases the specific surface area (SSA) and limits aggregation at the interface by forming a suspension. SSA is an important factor influencing the stability of the AI, since instability is often observed at interfaces due to the high surface energy of sensitive components. The adjuvant–AI complex in the feed solution isolates the AI from its bulk water phase and prevents it from being exposed to the ice–water interface and becoming more concentrated on ice crystallization. The adjuvants provide a 'shielding' effect to the AI and also protect the structure on rehydration, thus resulting in a higher level of stability than the formulations without adjuvants with comparable SSA.

A striking anomaly is that, while a low SSA is important for obtaining interfacial stability as aforementioned, a high SSA is significant for sustained release of AI. Thus, it is important to strike an optimal balance between the level of AI stability to be achieved and the SSA of the final powdered product (Costantino *et al.* 2002).

7.2.2.3 Spray freezing into liquid

With spray freezing into liquid, the nozzle is inserted beneath the surface of the cryogenic liquid (Figure 7.5). Thus, SFL can be considered an improvement over the SFV/L process as the transit of atomized feed droplets through the vapour phase does not occur and cooling rates are much higher. As a result, the atomized droplets undergo virtually instantaneous freezing. SFL occurs by the principle of liquid/liquid impingement between the pressurized feed solution exiting the atomizer and the cryogenic liquid. Here, much smaller droplets can be produced owing to the greater viscosity and density of the cryogenic liquid when compared to the cryogenic gas of the SFV method.

Emulsification or homogenization procedures often precede the drying process for feeds containing both hydrophilic and hydrophobic components. The stability of the feed emulsion before and during the drying process is significant in terms of achieving the appropriate particle characteristics. This is especially important when the feed comprises proteinaceous material since the maintenance of emulsion stability prevents

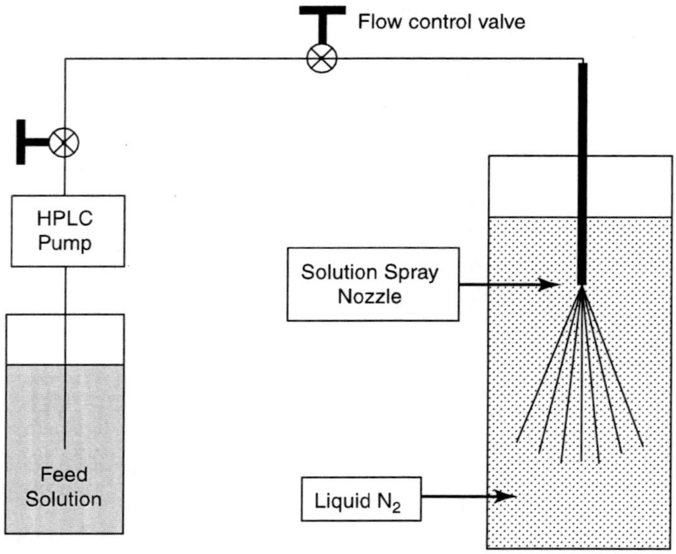

Figure 7.5 Spray freezing into liquid (Rogers *et al.* 2002a).

protein denaturation during the drying step. SFL is advantageous in this aspect as the very fast freezing rates prevent the phase separation of solutes within the feed solution and induces the rapid formation of amorphous structures (Engstrom *et al.* 2007). However, the use of appropriate stabilizing additives during the spray freezing and subsequent freeze drying is recommended in the case of a sensitive core, which will be explained later in this chapter.

7.2.3 Freeze drying

After freezing, the frozen particles are collected by sieves or separated after the cryogen is allowed to boil off, and this is then followed by freeze drying. SFD techniques can also be classified based on the method employed for freeze drying. Freeze drying can be accomplished using the vacuum drying method (Costantino *et al.* 2000, Al-Hakim & Stapley 2004, Burke *et al.* 2004, Amorij *et al.* 2007), atmospheric fluidized-bed freeze drying (Mumenthaler & Leuenberger 1991) or sub-atmospheric fluidized-bed freeze drying (Anandharamakrishnan *et al.* 2008, 2010).

7.2.3.1 Vacuum freeze drying

The principle underlying the sublimation phase of conventional freeze drying is based on the prerequisite that both the vapour pressure and the temperature should be held below that of the triple point of water (0.6 kPa and 0.01 °C) (Mumenthaler & Leuenberger 1991). As sublimation is an endothermic process, heat must be applied to the frozen food to sustain sublimation. Apart from avoiding ice crystals melting, the structural rigidity of the frozen food matrix must also be maintained to obtain an optimally freeze-dried product. It is important that the primary drying phase is performed at a temperature that is 2–5 °C below the so-called product collapse temperature (T_c) to avoid structural collapse. T_c is defined as the point at which an amorphous material in the frozen state when subjected to vacuum can no longer maintain its structure and hence undergoes collapse as the frozen solvent is removed during drying. The collapse temperature is closely related to the glass transition temperature (T_g), which can be defined as the temperature at which there is a change in physical state from glassy to rubbery viscous liquid without phase change (Slade & Levine 1991, Wunderlich 1981). Avoiding structural collapse is important as it is also associated with deteriorative reactions, loss of volatiles, shrinkage of particles and reduced rehydration properties (Stapley 2008). Product collapse generally occurs when the shelf temperature is higher than the T_c of the product (Abdul-Fattah & Truong 2010).

The sublimation of ice crystals occurs at a sublimation front which sweeps through the sample from the surface, and this is referred to as 'primary drying'. The primary drying phase is the period during which sublimation of ice occurs. At the end of the primary drying phase ~70–80% of the water has been removed in the form of ice, while the remaining water is bound to the amorphous matrix. Some of this bound water still needs to be removed to reduce the moisture content to the final value. The removal of water from the amorphous matrix is referred to as 'secondary drying'. Localized secondary drying occurs as soon as the ice sublimation front has passed, but the secondary drying phase starts at the point when the ice in the sample has sublimed.

The collapse temperature depends on the moisture content of the amorphous phase so when secondary drying starts, the collapse temperature slowly rises. This means that the sample as a whole can begin to be gradually warmed (e.g. by increasing the shelf temperature) but it should not be warmed too rapidly as this could result in localized collapse.

Although a well-established technique, the conventional freeze-drying mode suffers from the limitations of energy intensive operation due to the use of vacuum and the batch mode operation, eventually leading to high operational cost (Di Matteo *et al.* 2003).

7.2.3.2 Atmospheric SFD and atmospheric spray fluidized-bed freeze drying

Atmospheric spray freeze drying (ASFD) is based on the principle that the drying rate of a material undergoing freeze drying is driven by the difference in water vapour partial pressure between the site of water vapour formation and the chamber headspace rather than the total pressure in the drying chamber. Thus, it is thermodynamically possible to operate the freeze dryer at higher overall pressure (by the inclusion of non-condensable gases) provided the partial pressure of water remains low (Meryman 1959). Fluidization can aid in improving the ASFD process in terms of improving the heat transfer to the particles, structural quality and retention of active compounds in the final product. Here, the frozen particles are contacted with a cold, dry gas in a fluidized bed (Mumenthaler & Leuenberger 1991; Leuenberger *et al.* 2006; Anandharamakrishnan *et al.* 2010), which results in an efficient heat transfer due to good contact between particles and fluids. This concept led to the development of atmospheric spray fluidized-bed freeze drying (ASFBFD).

Although the ASFD process is advantageous in terms of product quality, it operates at relatively long residence time because of the internal resistance to mass transfer. The ASFD process demands that a very large quantity of cold dry gas is circulated through the bed. Also, the cost per kilogram of dry product depends on the drying temperature. The cost increases drastically with feed materials whose frozen solution has a low eutectic (T_e) or glass transition temperature (T_g) and hence require a low drying temperature (Padma Ishwarya *et al.* 2015).

7.2.3.3 Sub-atmospheric pressure SFD

This technique is based on the theory that a reduction in the pressure of the system reduces the mass of gas required for the process by the same factor. The reduction in pressure can also be translated as lower gas density, which in turn reduces the inertial (non-viscous) drag forces on the particles. This prevents particle elutriation from the bed (Anandharamakrishnan *et al.* 2010). The sub-atmospheric pressure also causes a reduction in drying time due to the higher fluidization velocity of the particles (Leuenberger *et al.* 2006). Furthermore, it was noticed that operating at lower pressures increased the temperature difference between the particle and gas due to its influence on the heat-transfer coefficient between the gas and the particle bed. Larger temperature differences were also seen between inlet and outlet gas temperatures (Anandharamakrishnan *et al.* 2010).

7.3 Applications of SFD in dried dairy products

SFD is mainly used for high-value food products such as dairy and instant coffee powders, apple juice powder and protein-based formulations (e.g. whey protein, egg albumen). Not limited to drying, SFD has also been established as an encapsulation technique for probiotic bacteria (Rajam & Anandharamakrishnan 2015) and bioactive nutraceutical compounds such as docosahexaenoic acid (DHA) (Karthik & Anandharamakrishnan 2013) and vanillin (Hundre *et al.* 2015). This section is intended to explain the specific applications of SFD with respect to the drying of milk and the encapsulation of probiotic organisms, the resultant product characteristics and recent process innovations in this context.

With the fundamental understanding obtained in above sections, it is important to appreciate the rationale by which the limitations of conventional dairy drying processes are addressed by SFD. The unique feature of SFD is its rapid freezing rate, which is in the order of 10^6 K/s, provided the droplet is of an appropriately small size. The role of the faster freezing rate in improving the rehydration properties of milk powders can be explained in two ways.

First, rapid freezing and the small drop size leads to a very high level of super-cooling at the point of nucleation. The number of nucleation sites is proportional to the $(size)^3$ of the droplets and hence small droplets are associated with fewer nucleation sites (MacLeod *et al.* 2006). The relative rate of nucleation compared to growth increases dramatically with supercooling and results smaller average crystal sizes in the particle (Fellows 2009).

Second, SFD products exhibit a non-porous surface layer. This could be the result of water evaporation from the droplet before freezing and/or solute redistribution during freezing (Hindmarsh *et al.* 2007). The formation of a solute surface layer has earlier been attributed to the very large cooling rates experienced by the surface regions that result in solidified glassy state ice crystals (Windhab 1999).

An alternative theory for the absence of a surface layer during spray freezing has been related to the use of a two-fluid nozzle, which introduces relatively warm air into the spray that chaotically mixes with the cold gas in the chamber. This causes the spray droplets to meet gas pockets at different temperatures, resulting in varied cooling rates. These cooling rate differences might possibly lead to different surface properties (Al-Hakim & Stapley 2004). All these reasons can potentially reduce the surface hydrophobicity and stickiness of whole milk powders due to the covering layer of fat and sugars, as observed during spray drying and freeze drying.

7.3.1 SFD of whole milk and skim milk

SFD has been successfully used to dry both whole and skim milk, with certain process modifications. A monodisperse droplet generator was used to combat the issues of different sized droplets resulting from the commonly used hydraulic and twin-fluid nozzles. A uniform droplet size distribution can be brought about by the application of an external force to the feed liquid being atomized. This can be accomplished by using the shearing force of a second fluid or with the aid of an electrostatic repulsive force applied in pulsated mode to the feed reservoir. As a result, this nozzle generates a single stream or a jet of droplets with a predefined trajectory and a narrow size distribution (Rogers *et al.* 2008). This phenomenon can be explained by Rayleigh's theory (Rayleigh & Strutt 1878). The sinusoidal perturbations cause wave disturbances on the surface of the droplet jet and the jet continues to grow until it disintegrates into droplets connected by narrow threads. The disintegration occurs when the fastest growing disturbance attains a wavelength, λ_{opt}, of $4.51d$, where d is the initial jet diameter. After break-up, the cylinder of length $4.51d$ becomes a spherical drop (Figure 7.6), which can be approximated to a sphere of equal volume (Eq. 7.3):

$$4.51\, d \times \left(\frac{\pi}{4}\right) d^2 = \left(\frac{\pi}{6}\right) D^3 \tag{7.3}$$

where D is the droplet diameter, which can be obtained as:

$$D = 1.89\, d \tag{7.4}$$

The droplets were plunged into liquid nitrogen placed in a container followed by a conventional freeze-drying process. The pressure pulse frequency and the stepping of the feed pump are the key parameters that control the droplet size.

In comparison with industrially spray-dried milk powders, the particles of SFD skim milk powder were highly porous, with a finely textured surface. The intact and uniform spherical particles (Figure 7.7) indicate the absence of droplet shrinkage during the SFD process. There was no significant difference between the mean droplet diameter (423 μm) and mean particle diameter (415 μm). The structure was uniform over the entire cross-section of each particle, with pores evenly dispersed throughout. Because of the quick freezing of water, the droplet maintains its shape and volume despite the

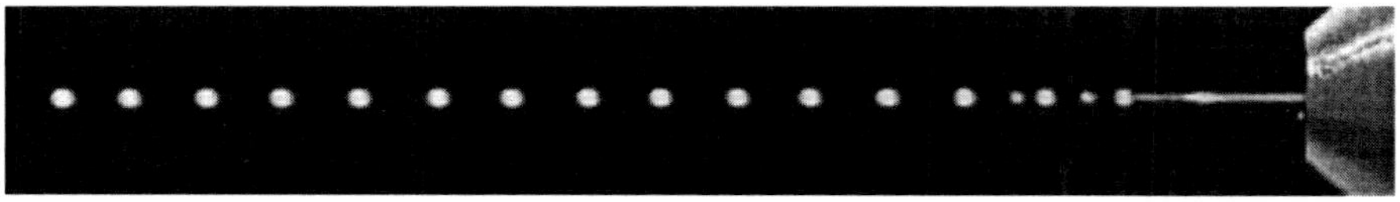

Figure 7.6 Break-up of milk jet into droplets from a monodisperse droplet generator (Rogers *et al.* 2008).

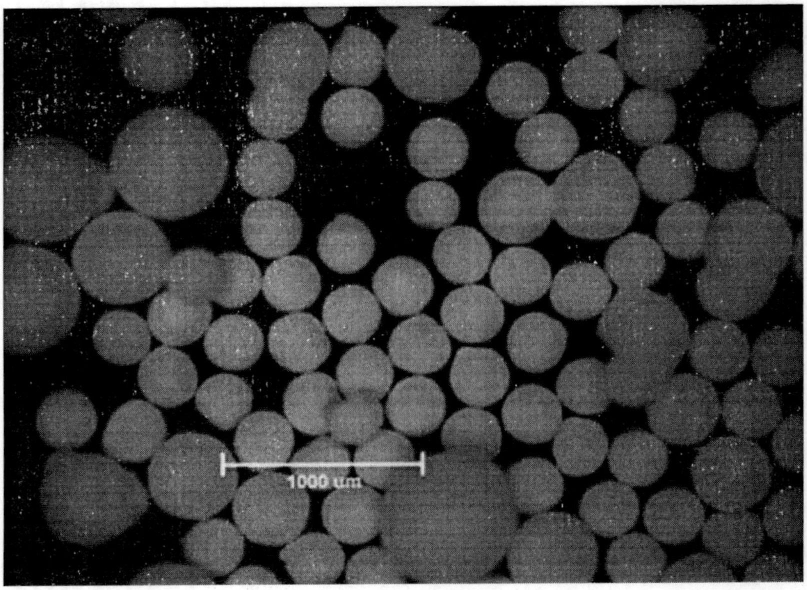

Figure 7.7 Light microscope image of SFD skim milk powder (Rogers *et al.* 2008).

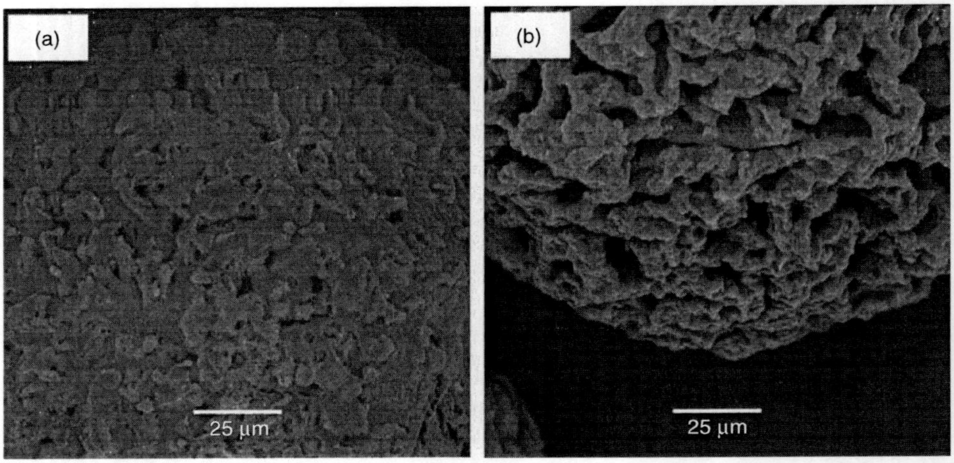

Figure 7.8 SEM micrograph of (a) SFD skim milk powder and (b) SFD whole milk powder (Rogers *et al.* 2008).

removal of water by sublimation during the subsequent freeze-drying step. This also led to a uniform and similar microstructure on the surface and within the centre of the SFD skim milk powder (SFD-SMP) particle (Figure 7.8a). Consequently, the SFD-SMP had a three-fold faster wettability (2.31 s) than industrially spray-dried skim milk powder (6.2 s). However, the rehydration time of SFD whole milk powder (Figure 7.8b) was higher when compared to that of SFD-SMP and spray-dried whole milk powders, and thus demand the addition of surfactants in the feed formulation.

7.3.2 SFD of whey protein

Whey protein is a by-product from the whey resultant from cheese production. The four principal protein fractions present in whey protein are α-lactalbumin, β-lactoglobulin, immunoglobulins and serum albumin. Whey proteins are globular in structure and can remain in solution over a wide range of pH (Morr & Ha 1993). Whey protein powder has applications in beverage mixes and infant formulas as the substrate for fortification and supplementation of micronutrients.

Spray freezing with vacuum fluidized-bed freeze drying (VFBFD) (Figures 7.9 and 7.10) was used for the drying of whey protein (isolate) solution (Anandharamakrishnan *et al.* 2010).

A fluidization vessel with volume 0.022 m^3 was employed for this trial. The VFBFD chamber was operated at three different inlet gas temperatures of −10, −15 and −30 °C at 0.1 bar pressure. The inlet gas temperature was controlled by a double-pipe heat exchanger by varying the heater temperature and the air-flow rate to the heater. The total freeze drying durations achieved at −10, −15 and −30 °C were 60, 65 and 100 min, respectively. The inlet gas temperature exerted a significant influence on the moisture content of the final product. The highest inlet gas temperature (−10 °C) resulted in the lowest moisture content of 8.1% and the shortest drying time of 1 h. The faster drying rate (Figure 7.11), greater mass transport coefficients and pure ice vapour pressure at the higher temperature were reported as the potential reasons underlying the above observations. The moisture contents at −15 and −30 °C were found to be 9.5% and 14%, respectively.

Figure 7.9 Spray-freezing into cold vapour chamber (right), with sub-atmospheric fluid bed freeze dryer (far left) (Anandharamakrishnan 2008).

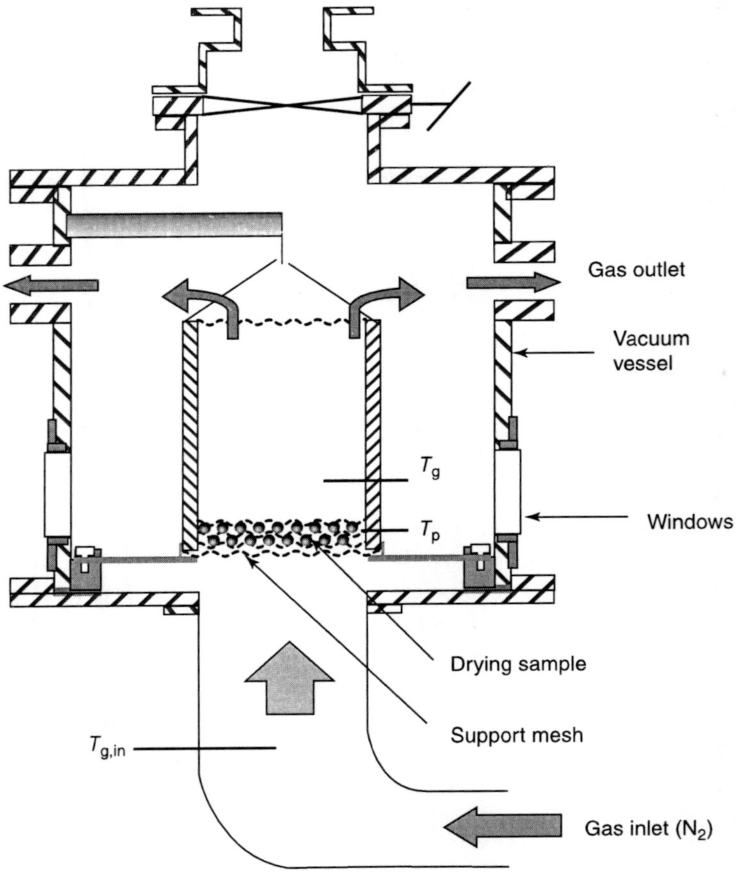

Figure 7.10 Experimental rig for sub-atmospheric fluidized-bed freeze drying (Anandharamakrishnan *et al.* 2010).

Furthermore, reverse-phase HPLC analysis of the SFD whey protein powder showed insignificant loss of solubility in α-lactalbumin and 2% solubility loss in β-lactoglobulin. With respect to particle microstructure, using an inlet gas temperature of −10 °C resulted in a large number of broken particles with irregular shapes (Figures 7.12a–d). This may be due to structural collapse, as the inlet gas temperature is greater than the glass transition temperature (T_g = −25.8 °C) of whey protein isolate. Furthermore, particle agglomeration was also observed at this temperature (Figure 7.12a,b). Thus, freeze drying of whey proteins should be carried out below −25.8 °C to avoid structural collapse.

At −15 and −30 °C the surfaces of the particles were smooth with small pores and an occasional surface blemish. Magnification of one of these blemishes revealed a porous structure inside the particle (Figures 7.13e,f), which is indicative of the successful conduct of the sublimation phase without structural collapse. Figures 7.12d, 7.13c,d and 7.14e,f show the internal pore structure created by the ice sublimation route (region B). The pore route is shown in Figure 7.14b,e with circular marks. During the freezing

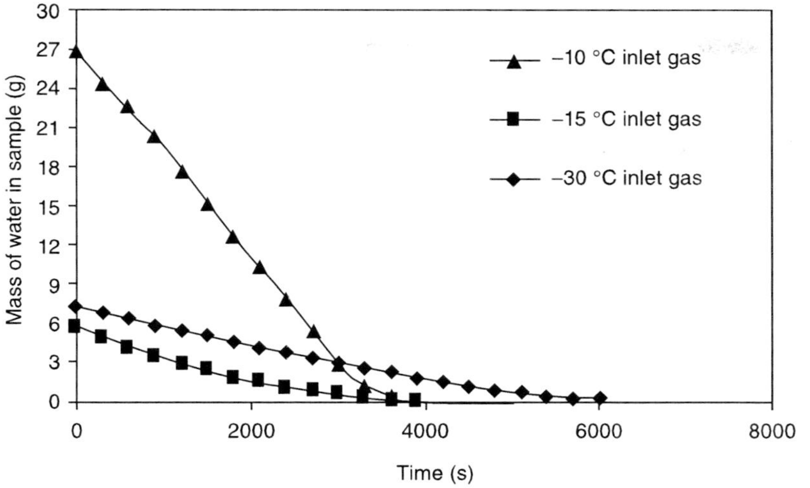

Figure 7.11 Influence of inlet gas temperature on the drying rate of whey protein solution during vacuum fluidized-bed SFD (Anandharamakrishnan *et al.* 2010).

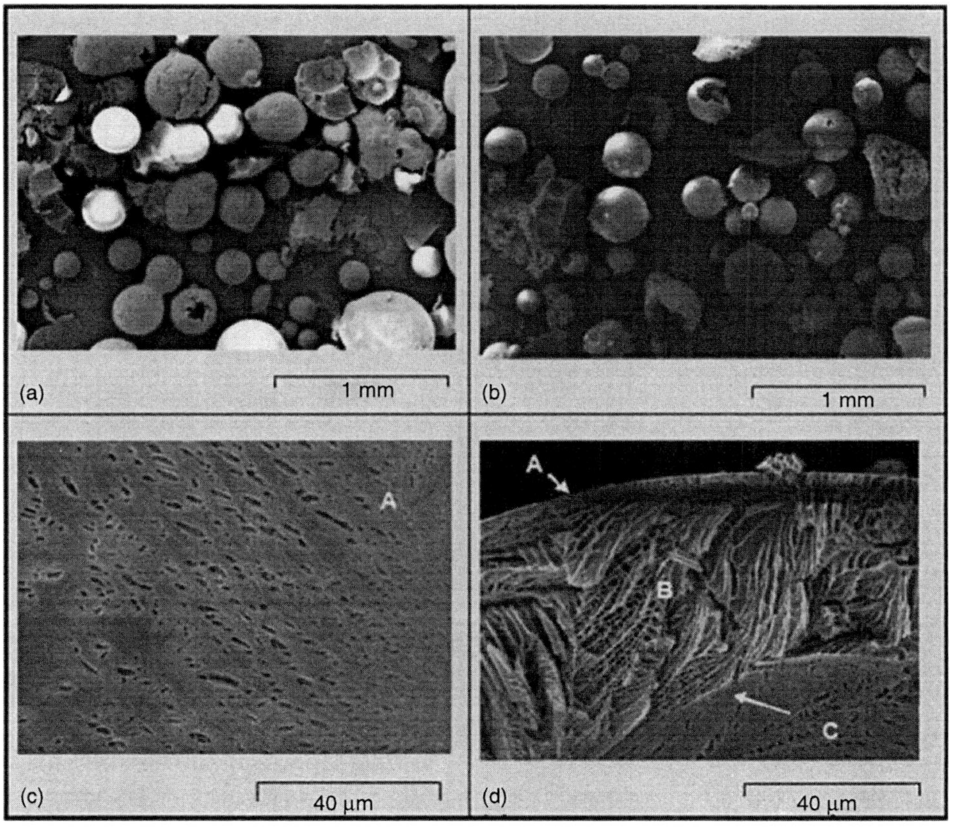

Figure 7.12 SEMs of spray-freeze-dried whey protein powders at −10 °C inlet gas temperature. A, surface of the particles; B, pore structure created following ice sublimation; C, inside the core region of the particles (Anandharamakrishnan 2008).

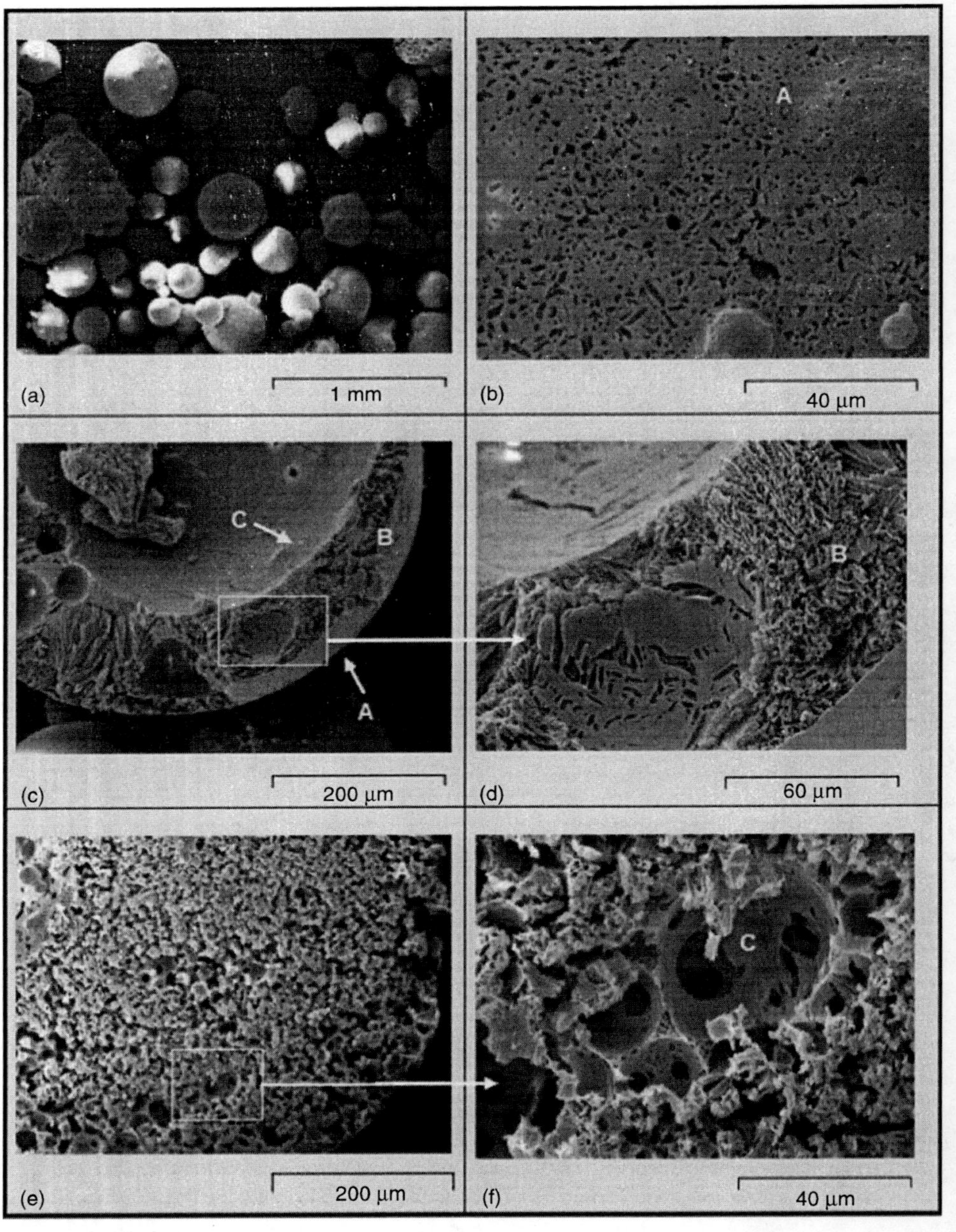

Figure 7.13 SEMs of spray-freeze-dried whey protein powders at −15 °C inlet gas temperature. A, surface of the particles; B, pore structure created following ice sublimation; C, inside the core region of the particles (Anandharamakrishnan 2008).

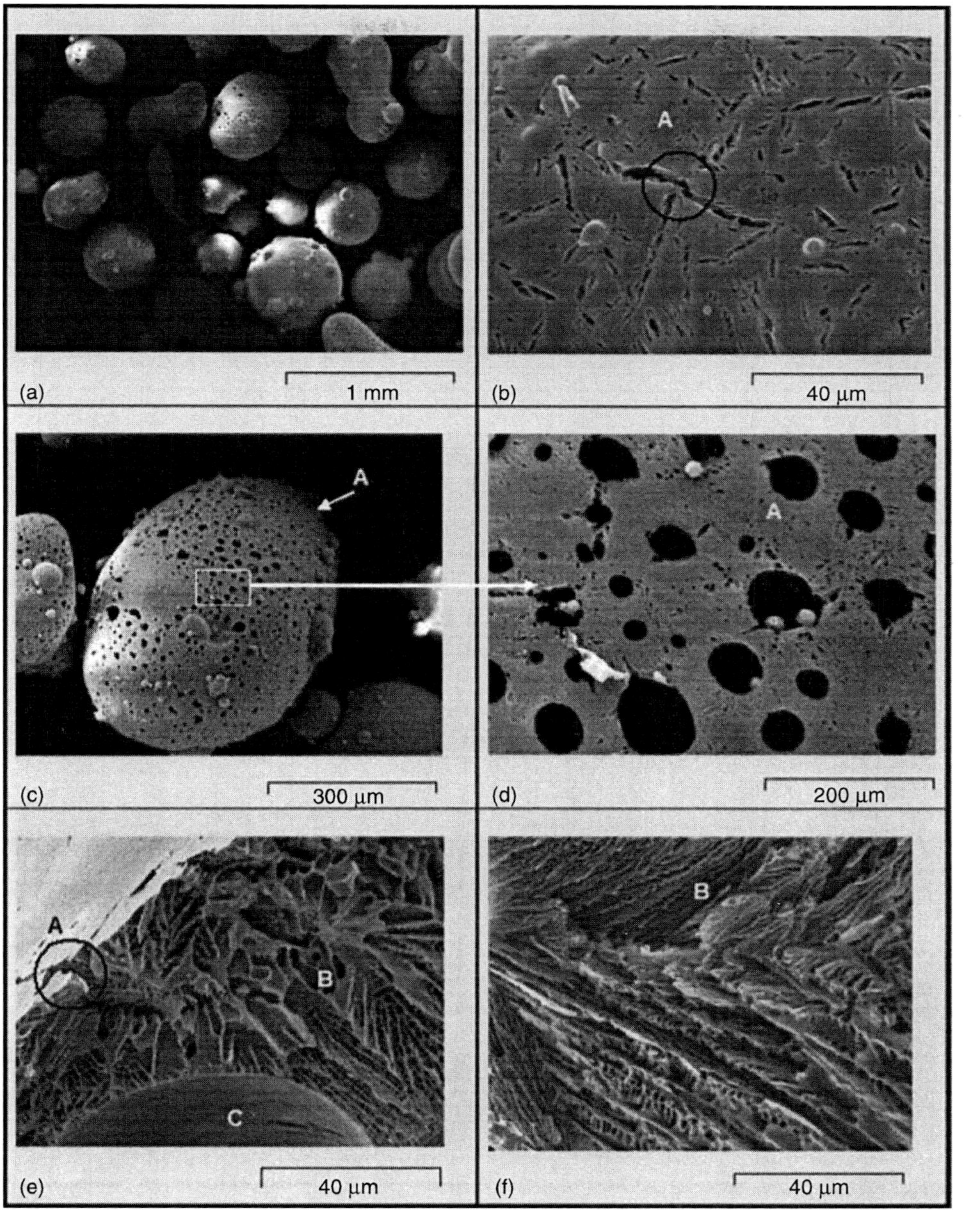

Figure 7.14 SEMs of spray-freeze-dried whey protein powders at −30 °C inlet gas temperature. A, surface of the particles; B, pore structure created following ice sublimation; C, inside the core region of the particles (Anandharamakrishnan 2008).

stage, whey proteins were absorbed on the ice–liquid interface and during the drying stage when ice was removed it left voids at the centre core, hence whey protein isolate particles appeared as a perforated shell wall (Figure 7.14c,d). The presence of hollow cavities within the particle suggests the formation of gas bubbles inside the particles (Figures 7.12d (region C), 7.13 (region C) and 7.14 (region C)). These bubbles are likely to result from dissolved gas that was released during atomization, as the feed was previously held under pressure of 6 bar using compressed nitrogen in the feed chamber. This study revealed that an inlet gas temperature of −30 °C, which is well below the T_c of whey protein isolate, is suitable for the SFD of whey protein isolate (Anandharamakrishnan 2008).

7.3.3 SFD for microencapsulation of probiotics

Apart from drying of milk and milk proteins, the SFD technique has also been used for the microencapsulation of probiotic bacteria. Probiotics can be defined as 'live microorganisms that, when administered in adequate amounts, confer a health benefit on the host' (FAO/WHO 2002). The genera of Lactobacillus and Bifidobacterium along with enterococci and yeasts comprise the category of probiotic organisms. Dairy products have been the most commonly used vehicle for the inclusion of probiotics. Milk-based infant foods, pasteurized milk, ice-cream, yoghurt and cheeses are some of the dairy product categories that are marketed as 'probiotic dairy products' (Tamime *et al.* 2005). In order to be declared as probiotic foods and to exert the intended health and functional benefits, the concentration of live probiotic bacteria should be more than 10^7 cfu per gram of the finished product at the time of consumption (FAO/WHO 2001).

Microencapsulation may be defined as a process to entrap an active compound or living entities such as the probiotic cells within a stable and protective carrier material to produce encapsulates of size 0.2–5000 μm and characteristic functional properties. The necessity for encapsulation of probiotics has emerged as knowledge of the relationship between gut health and beneficial bacteria present in the gastrointestinal tract has evolved. Microencapsulation of probiotics by spray drying has been extensively studied and well established. Thermal injury to the cells caused by the high temperature operation of spray drying is inevitable. SFD has a competitive edge over spray drying in enhancing probiotic cell viability and stability in the gastrointestinal tract, owing to its low temperature process and unique microstructure.

When the SFV/L approach was used for the microencapsulation of *Lactobacillus plantarum*, the spray-freeze-dried probiotic microcapsules showed a 20% higher cell viability than the spray-dried counterpart. In simulated gastrointestinal conditions, the cells microencapsulated by SFD demonstrated a better tolerance (up to 4 h) than those cells encapsulated by spray drying and the unencapsulated cells. Furthermore, during the storage period of 40 days at 4 °C, the SFD probiotic microcapsules showed only a 3–5% loss in survival against a 10% loss observed with spray drying. However, atomization stress on live cells during the spray-freezing stage led to a loss of viable cells when compared to the freeze-drying process. Use of a higher core-to-wall ratio was found to be favourable in obtaining a greater survival rate and better storage stability because of the enhanced protective properties (Dolly *et al.* 2011).

Rajam & Anandharamakrishnan (2015) have established the suitability and efficiency of SFD for the encapsulation and drying of live and sensitive probiotic cells. The influence of wall material formulations on the characteristics of SFD microencapsulates of *Lactobacillus plantarum* (MTCC 5422) was studied using whey protein isolate (WPI) with sodium alginate (SA), WPI with fructooligosaccharide (FOS), denatured WPI (DWPI) with SA and DWPI with FOS. Each formulation was advantageous with respect to different specific product attributes.

With reference to particle morphology, the WPI + SA formulation was advantageous, with particles exhibiting a smooth outer surface (Figure 7.15a) owing to the well-established skin-forming property of whey protein (Sheu & Rosenberg 1998; Anandharamakrishnan *et al.* 2007). This type of microstructure results when the drying curve of the polymeric wall material shows a rapid decrease in drying rate with a decrease in moisture content. This aspect is important for achieving better protection and stability of the probiotic cells as a denser wall material membrane will be formed around the live cells when wall material with a high drying rate is used. On the other hand, the WPI + FOS combination exhibited a slightly rough surface, with the smaller particles sticking to the larger particles (Figure 7.15b). The lower glass transition temperature and hygroscopic nature of FOS might be responsible for its stickiness. The presence of DWPI in the wall material formulation led to rough particle surfaces with protrusions (Figure 7.15c,d).

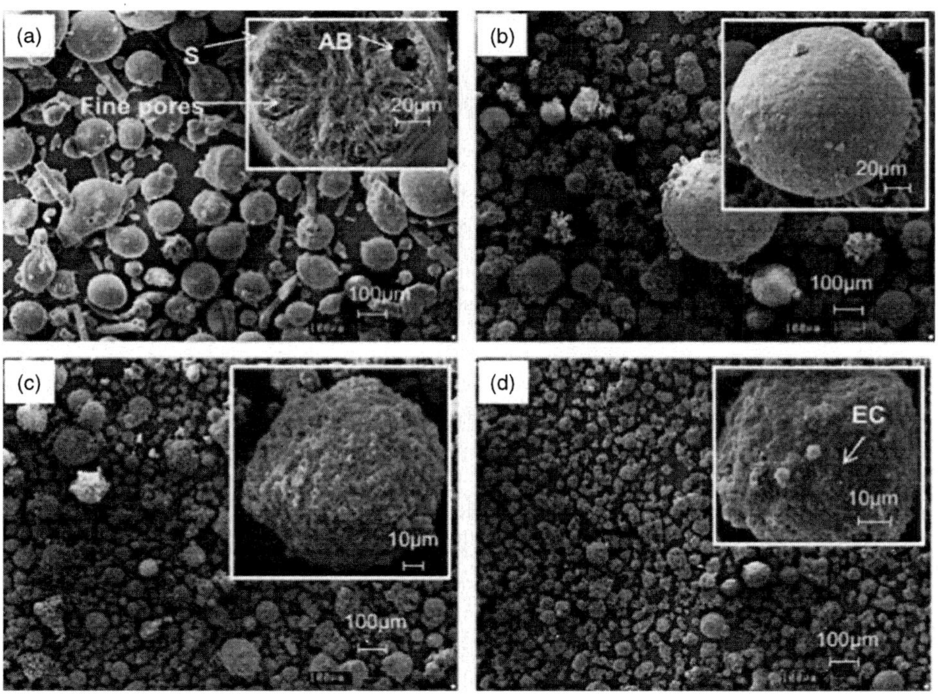

Figure 7.15 SEM images of spray freeze dried microcapsules of *L. plantarum* produced with different wall material formulations: (a) WPI + SA, (b) WPI + FOS, (c) DWPI + SA and (d) DWPI + FOS (Rajam & Anandharamakrishnan 2015). AB, air bubble; S, smooth outer skin; EC, encapsulated cells.

The SFD process using different wall material formulations resulted in encapsulation efficiencies in the range of 88–95%. In general, encapsulation efficiency can be defined as the ratio of the amount of core compound in the microencapsulates to the initial amount of core compound taken for microencapsulation. With respect to probiotic cells, estimation of encapsulation efficiency involves determining the viability of probiotic cells before and after drying. Of all the wall material combinations, the formulations that contained FOS were found to show the greatest encapsulation efficiency. FOS tends to penetrate the cell membrane and partially replace the water molecules in the cells that protect the cells from adverse effects during freezing and freeze drying (Schwab *et al.* 2007). FOS is also known to exhibit a bifidogenic effect due to its dietary fibre properties (Roberfroid 2007). The protective effect of FOS can be enhanced in the presence of WPI and DWPI due to their film-forming properties and the formation of a stronger gel network (Parthasarathi *et al.* 2013), respectively.

The DWPI + FOS formulation was found optimal with respect to stability of the SFD microencapsulates during storage. The poor water vapour permeability of DWPI and cell membrane stabilization by FOS were stated as plausible reasons (Pérez-Gago *et al.* 1999; Pérez-Ggago & Krochta 2001).

The importance of using stabilizing additives on the survival of spray-freeze-dried probiotic microcapsules was established by the study of Semyonov *et al.* (2010). The SFL method used in this study posed both thermal and osmotic stresses to the cells of *Lactobacillus paracasei*. Cell viability was also reduced by the fusion of cell membranes of *L. paracasei* and denaturation of its proteins. The use of trehalose and maltodextrin of various dextrose equivalents prevented viability loss during both spray freezing and freeze drying. The protective effect of trehalose and maltodextrin has been attributed to the theory of vitrification. Vitrification refers to the passing of the material into the glassy state, that is, below the glass transition line on the state diagram. During conventional freezing, plenty of time is allowed for ice formation to occur and freeze concentration is thus able to take place until the solution reaches close to an equilibrium state of maximal freeze concentration. At this point the unfrozen domains become so viscous that they vitrify at this concentration (Engstrom *et al.* 2007). Thus, during microencapsulation by SFD, the bacteria as well as water would be immobilized in the viscous glass, thus preventing any deteriorative effect on the cells (damage to cell membrane and protein unfolding) due to the low mobility. The use of trehalose owing to its low molecular weight can further facilitate this phenomenon by minimizing the water crystal size in the inter-membrane space. This can eventually prevent changes in the physical state of the membrane lipids and reduce the mechanical stresses in membranes (Koster *et al.* 2003, Bryant & Wolfe 1992, Wolfe & Bryant 1999).

Her *et al.* (2015) employed SFD to obtain a fine probiotic powder of *Lactobacillus casei* (IFO 15883). The most effective atomisation pressure and protective excipient during SFD were found to be 20 kPa and buffered peptone water (BPW), respectively. The mean particle diameter of optimized probiotic powder particles was 24.8 μm. The survival rate of the final SFD probiotic powder under conditions was 97.7%.

The atomization parameters and size properties of the different SFD studies discussed above are shown in Table 7.1.

Table 7.1 Atomization parameters and particle size of spray-freeze-dried milk powder and probiotic microcapsules

Product	Cryogen	Nozzle type	Feed rate	Air-flow rate/air pressure	Particle size	Reference
Whole milk and skim milk powder	Liquid nitrogen	Monodisperse droplet generator (orifice size 100 and 150 μm)	2.5 ml/min	Pulse frequency: 1500–2000 Hz Static dispersal voltage: 1.85 kV	415 μm	Rogers *et al.* (2008)
Whey protein	Nitrogen gas at inlet gas temperature: (i) −10 °C (ii) −15 °C (iii) −30 °C	Hydraulic nozzle	0.0125 kg/s	8 bar	i) 480 ± 53 μm ii) 393 ± 75 μm iii) 412 ± 4 μm	Anandharamakrishnan *et al.* (2010)
Lactobacillus plantarum microencapsulates	Liquid nitrogen	Twin-fluid nozzle (0.5 mm diameter)	6 ml/min	25 m^3/h	53.99–105.07 μm	Rajam & Anandharamakrishnan (2015)
Lactobacillus casei microencapsulates	Liquid nitrogen	Twin-fluid nozzle	5 ml/min	20 and 30 kPa	24.8 μm	Her *et al.* (2015)
Lactobacillus paracasei microencapsulates	Liquid nitrogen	Pneumatic nozzle	(a) 0.15 ml/min (b) 0.3 ml/min (c) 0.8 ml/min	2.12 l/min 3.08 l/min 4.52 l/min	400–1800 μm	Semyonov *et al.* (2010)

7.4 Advantages and limitations of SFD

SFD exhibits a competitive advantage over spray-drying and freeze-drying techniques in terms of product stability and quality. When SFD was employed for the production of soluble coffee, it resulted in a higher average retention of volatiles (93%) than either freeze drying (77%) or spray drying (57%). Also, the spray-freeze-dried particles were larger (91 μm) and more porous than those resulting from spray drying (50 μm). SFD also resulted in instantaneous rehydration, high bulk density and brighter colour, which are considered to be key quality characteristics of the soluble coffee product (Padma Ishwarya & Anandharamakrishnan 2015). SFD has been found to result in reductions in processing time by factors of 2.5 (Rajam & Anandharamakrishnan 2015) and 2.75 (Karthik & Anandharamakrishnan 2013) when compared to conventional freeze drying.

Maa *et al.* (1999) showed that the spray-freeze-dried particles could efficiently retain the spherical shape created during the atomization stage owing to the immediate freezing. During spray drying, the atomized droplets shrank from 10 to 3 μm on water removal by hot air drying. The shape of the dried particles was also independent of the drying conditions and feed formulation.

Certain practical challenges were also encountered in the SFD process, some of which are discussed above. The vacuum conditions and refrigeration required in the case of SFD with conventional freeze drying led to a higher production cost. The high energy requirement and longer processing time ultimately result in a processing cost that is 30–50 times more expensive than spray drying (Gharsallaoui *et al.* 2007). With ASFD and ASFBFD, the large amount of drying gas and particle elutriation are considered major limitations.

7.5 Conclusion

Spray-freeze-dried products have a clear competitive edge over industrial spray-dried dairy products in terms of its porous microstructure, instant solubility and superior quality. Tailoring the particle microstructure by appropriate adjustment of SFD processing conditions is an interesting research area. Nevertheless, work to transform SFD into a commercially viable drying process and establish its economic viability is still at an early stage. Approaches such as the inclusion of heat-pump systems could help to enhance the energy efficiency of the SFD process and reduce the requirement of cryogens. The use of recently evolved computational modeling tools can aid in the development of an optimised design for SFD. Future research on these perspectives could transform SFD intos an industry-friendly technique for use in the dairy sector.

References

Abdul-Fattah, A.M. & Truong, V.L. (2010). Drying process methods for biopharmaceutical products: an overview. In: *Formulation and Process Development Strategies for Manufacturing Biopharmaceuticals* (eds Jameel, F. & Hershenson, S.), pp. 705–737, John Wiley & Sons, New Jersey.

Adams, G.D.J. (1991). Freeze-drying of biological materials. *Drying Technology*, **9**(**4**), 891–925.

Adams, T.H., Beck, J.P. & Menson, R.C. (1982). Novel particulate compositions. US Patent No. 4,323,478, US Patent and Trademark Office, Washington, DC.

Al-Hakim, K. & Stapley, A.G.F. (2004). Morphology of spray-dried and spray-freeze-dried whey powders. In: *Proceedings of the 14th International Drying Symposium* (IDS 2004), pp. 1720–1726.

Al-Hakim, K., Wigley, G. & Stapley, A.G.F. (2006). Phase Doppler anemometry studies of spray freezing. *Chemical Engineering Research and Design*, **84**(**A12**), 1142–1151.

Amorij, J.P., Saluja, V., Petersen, A.H., Hinrichs, W.L.J., Huckriede, A. & Frijlink, H.W. (2007). Pulmonary delivery of an inulin-stabilized influenza subunit vaccine prepared by spray-freeze drying induces systemic, mucosal humoral as well as cell-mediated immune responses in BALB/c mice. *Vaccine*, **25**(**52**), 8707–8717.

Anandharamakrishnan, C. (2008). *Experimental and computational fluid dynamics studies on spray-freeze-drying and spray-drying of proteins*, Ph.D. thesis, Loughborough University.

Anandharamakrishnan, C., Rielly, C.D. & Stapley, A.G.F. (2007). Effects of process variables on the denaturation of whey proteins during spray drying. *Drying Technology*, **25**(**5**), 799–807.

Anandharamakrishnan, C., Gimbun, J., Stapley, A.G.F. & Rielly, C.D. (2008). Application of computational fluid dynamics (CFD) simulations to spray-freeze drying operations.*16th International Drying Symposium* (IDS 2008), 537–545.

Anandharamakrishnan, C., Rielly, C.D. & Stapley, A.G.F. (2010). Spray-freeze-drying of whey proteins at sub-atmospheric pressures. *Dairy Science and Technology*, **90**(**2–3**), 321–334.

Barbosa-Canovas, G.V., Ortega-Rivas, E., Juliano, P. & Yan, H. (2005). Bulk properties. In: *Food Powders: Physical Properties, Processing, and Functionality*, pp. 84, Kluwer Academic/Plenum Publishers, New York.

Barron, M.K., Young, T.J., Johnston, K.P. & Williams, R.O. (2003). Investigation of processing parameters of spray freezing into liquid to prepare polyethylene glycol polymeric particles for drug delivery. *AAPS Pharm Sci Tech*, **4**(**2**), 1–13.

Bryant, G. & Wolfe, J. (1992). Interfacial forces in cryobiology and anhydrobiology. *CryoLetters*, **13**, 23–36.

Burke, P.A., Klumb, L.A., Herberger, J.D., Nguyen, X.C., Harrell, R.A. & Monica Zordich. (2004). Poly(lactide-co-glycolide) microsphere formulations of darbepoetin alfa: spray drying is an alternative to encapsulation by spray-freeze drying. *Pharmaceutical Research*, **21**(**3**), 500–506.

Buxton, I.R. & Peach, J.M. (1984). Process and apparatus for freezing a liquid medium. US Patent No. 4,470,202, US Patent and Trademark Office, Washington, DC.

Ceballos, A.M., Giraldo, G.I. & Orrego, C.E. (2012). Effect of freezing rate on quality parameters of freeze dried soursop fruit pulp. *Journal of Food Engineering*, **111**(**2**), 360–365.

Costantino, H.R., Firouzabadian, L., Hogeland, K., Wu, C.C., Beganski, C., Carrasquillo, K.G., Cordova, M., Griebenow, K., Zale, S.E. & Tracy, M.A. (2000). Protein spray-freeze drying. effect of atomisation conditions on particle size and stability. *Pharmaceutical Research*, **17**(**11**), 1374–1382.

Costantino, H.R., Firouzabadian, L., Wu, C.C., Carrasquillo, K.G., Griebenow, K., Zale, S.E. & Tracy, M.A. (2002). Protein spray freeze drying. 2. Effect of formulation variables on particle size and stability. *Journal of Pharmaceutical Sciences*, **91(2)**, 388–395.

D'Addio, S.M., Chan, J.G.Y., Kwok, P.C.L., Prud'homme, R.K. & Chan, H.K. (2012). Constant size, variable density aerosol particles by ultrasonic spray freeze drying. *International Journal of Pharmaceutics*, **427(2)**, 185–191.

Di Matteo, P., Donsi, G. & Ferrari, G. (2003). The role of heat and mass transfer phenomena in atmospheric freeze-drying of foods in a fluidized bed. *Journal of Food Engineering*, **59(2)**, 267–275.

Dolly, P., Anishaparvin, A., Joseph, G.S. & Anandharamakrishnan, C. (2011). Microencapsulation of *Lactobacillus plantarum* (MTCC 5422) by spray-freeze-drying method and evaluation of survival in simulated gastrointestinal conditions. *Journal of Microencapsulation*, **28(6)**, 568–574.

Engstrom, J.D., Simpson, D.T., Lai, E.S., Williams, R.O. & Johnston, K.P. (2007). Morphology of protein particles produced by spray freezing of concentrated solutions. *European Journal of Pharmaceutics and Biopharmaceutics*, **65(2)**, 149–162.

FAO/WHO (2001). Evaluation of health and nutritional properties of powder milk and live lactic acid bacteria. *Food and Agriculture Organization of the United Nations and World Health Organization Expert Consultation Report*, 1–34.

FAO/WHO (2002). Guidelines for the evaluation of probiotics in food. *Report of a joint FAO/WHO working group on drafting guidelines for the evaluation of probiotics in food*, 1–11.

Fellows, P.J. (2009). Freezing. In: *Food Processing: Technology – Principles and Practice*, 3rd edn, pp. 650–686, CRC, New York.

Gharsallaoui, A., Roudaut, G., Chambin, O., Voilley, A. & Saurel, R. (2007). Applications of spray-drying in microencapsulation of food ingredients: An overview. *Food Research International*, **40(9)**, 1107–1121.

Hall, C.R. & Hedrick, T.I. (1971). *Drying of Milk and Milk Products*, 2nd edn, AVI Publishing, Westport.

Her, J. Y., Kim, M.S. & Lee, K.G. (2015). Preparation of probiotic powder by the spray freeze-drying method. *Journal of Food Engineering*, **150**, 70–74.

Hindmarsh, J.P., Russell, A.B. & Chen, X.D. (2007). Fundamentals of the spray freezing of foods – microstructure of frozen droplets. *Journal of Food Engineering*, **78(1)**, 136–150.

Hu, J., Rogers, T.L. Brown, J., Young, T., Johnston, K.P. & Williams R.O. (2002). Improvement of dissolution rates of poorly water soluble APIs using novel spray freezing into liquid technology. *Pharmaceutical Research*, **19(9)**, 1278–1284.

Hundre, S.Y., Karthik, P. & Anandharamakrishnan, C. (2015). Effect of whey protein isolate and β-cyclodextrin wall systems on stability of microencapsulated vanillin by spray-freeze drying method. *Food Chemistry*, **174**, 16–24.

Karthik, P. & Anandharamakrishnan, C. (2013). Microencapsulation of docosahexaenoic acid by spray-freeze-drying method and comparison of its stability with spray-drying and freeze-drying methods. *Food and Bioprocess Technology*, **6(10)**, 2780–2790.

Kawabata, Y., Wada, K., Nakatani, M., Yamada, S. & Onoue, S. (2011). Formulation design for poorly water-soluble drugs based on biopharmaceutics classification system: basic approaches and practical applications. *International Journal of Pharmaceutics*, **420(1)**, 1–10.

Kim, E.H.J., Chen, X.D. & Pearce, D. (2002). Surface characterization of four industrial spray-dried dairy powders in relation to chemical composition, structure and wetting property. *Colloids and Surfaces B: Biointerfaces*, **26**(**3**), 197–212.

Koster, K.L., Lei, Y.P., Anderson, M., Martin, S. & Bryant, G. (2000). Effects of vitrified and nonvitrified sugars on phosphatidylcholine fluid-to-gel phase transitions. *Biophysical Journal*, **78**, 1932–1946.

Leuenberger, H. (2002). Spray freeze drying – the process of choice for low water soluble drugs? *Journal of Nanoparticle Research*, **4**(**1–2**), 111–119.

Leuenberger, H., Plitzko, M. & Puchkov, M. (2006). Spray freeze drying in a fluidized bed at normal and low pressure. *Drying Technology*, **24**(**6**), 711–719.

Maa, Y.F., Nguyen, P.A., Sweeney, T., Shire, S.J. & Hsu, C.C. (1999). Protein inhalation powders: spray drying vs spray freeze drying. *Pharmaceutical Research*, **16**(**2**), 249–254.

MacLeod, C.S., McKittrick, J.A., Hindmarsh, J.P., Johns, M.L. & Wilson, D.I. (2006). Fundamentals of spray freezing of instant coffee. *Journal of Food Engineering*, **74**(**4**), 451–461.

Merymann, H.T. (1959). Sublimation freeze-drying without vacuum. *Science*, **130**(**3376**), 628–629.

Millqvist-Fureby, A., Elofsson, U. & Bergenstahl, B. (2001). Surface composition of spray-dried milk protein-stabilised emulsions in relation to pre-heat treatment of proteins. *Colloids and Surfaces B: Biointerfaces*, **21**(**1**), 47–58.

Mizoe, T., Ozeki, T. & Okada, H. (2008). Application of a four-fluid nozzle spray drier to prepare inhalable rifampicin-containing mannitol microparticles. *AAPS PharmSciTech*, **9**(**3**), 755–761.

Morr, C.V. & Ha, E.Y. (1993). Whey protein concentrates and isolates: processing and functional properties. *Critical Reviews in Food Science & Nutrition*, **33**(**6**), 431–476.

Mumenthaler, M. & Leuenberger, H. (1991). Atmospheric spray freeze-drying: a suitable alternative in freeze drying technology. *International Journal of Pharmaceutics*, **72**(**2**), 97–110.

Nickerson, T.A., Coulter, S.T. & Jenness, R. (1952). Some properties of freeze-dried milk. *Journal of Dairy Science*, **35**(**1**), 77–85.

Niwa, T., Shimabara, H., Kondo, M. & Danjo, K. (2009). Design of porous microparticles with single micron size by novel spray freeze drying technique using four fluid nozzle. *International Journal of Pharmaceutics*, **382**(**1**), 88–97.

Padma Ishwarya, S. & Anandharamakrishnan, C. (2015). Spray-freeze-drying approach for soluble coffee processing and its effect on quality characteristics. *Journal of Food Engineering*, **149**, 171–180.

Padma Ishwarya, S., Anandharamakrishnan, C. & Stapley, A.G.F. (2015). Spray-freeze-drying: A novel process for the drying of foods and bioproducts. *Trends in Food Science & Technology*, **41**(**2**), 161–181.

Parthasarathi, S., Ezhilarasi, P.N., Jena, B.S. & Anandharamakrishnan, C. (2013). A comparative study on conventional and microwave-assisted extraction for microencapsulation of Garcinia fruit extract. *Food and Bioproducts Processing*, **91**(**2**), 103–110.

Pérez-Gago, M.B. & Krochta, J.M. (2001). Denaturation time and temperature effects on solubility, tensile properties, and oxygen permeability of whey protein edible films. *Journal of Food Science*, **66**(**5**), 705–710.

Pérez-Gago, M.B., Nadaud, P. & Krochta, J.M. (1999). Water vapor permeability, solubility, and tensile properties of heat-denatured versus native whey protein films. *Journal of Food Science*, **64(6)**, 1034–1037.

Pham, Q.T. (1986). Simplified equation for predicting the freezing time of foodstuffs. *International Journal of Food Science & Technology*, **21(2)**, 209–219.

Rajam, R. & Anandharamakrishnan, C. (2015). Spray freeze drying method for microencapsulation of *Lactobacillus plantarum*. *Journal of Food Engineering*, **166**, 95–103.

Rayleigh, L. & Strutt, J.W. (1878). On the instability of jets. *Proceedings of the London Mathematical Society*, **10(4)**, 4–13.

Roberfroid, M. (2007). Prebiotics: The concept revisited. *Journal of Nutrition*, **137(3)**, 830S–837S.

Rogers, T.L., Hu, J., Hu, Z., Johnston, K.P. & Williams, R.O. (2002a). A novel particle engineering technology: spray-freezing into liquid. *International Journal of Pharmaceutics*, **242(1)**, 93–100.

Rogers, T.L., Nelsen, A.C., Hu, J., Brown, J.N., Sarkari, M., Young, T.J., Johnston, K.P. & Williams, R.O. (2002b). A novel particle engineering technology to enhance dissolution of poorly water soluble drugs: spray-freezing into liquid. *European Journal of Pharmaceutics and Biopharmaceutics*, **54(3)**, 271–280.

Rogers, S., Wu, W.D., Saunders, J. & Chen, X.D. (2008). Characteristics of milk powders produced by spray freeze drying. *Drying Technology*, **26(4)**, 404–412.

Schwab, C., Vogel, R. & Gänzle, M.G. (2007). Influence of oligosaccharides on the viability and membrane properties of *Lactobacillus reuteri* TMW1.106 during freeze-drying. *Cryobiology*, **55(2)**, 108–114.

Semyonov, D., Ramon, O., Kaplun, Z., Levin-Brener, L., Gurevich, N. & Shimoni, E. (2010). Microencapsulation of *Lactobacillus paracasei* by spray freeze drying. *Food Research International*, **43(1)**, 193–202.

Sheu, T.Y. & Rosenberg, M. (1998). Microstructure of microcapsules consisting of whey proteins and carbohydrates. *Journal of Food Science*, **63(3)**, 491–494.

Slade, L. & Levine, H. (1991). Beyond water activity: Recent advances based on an alternative approach to the assessment of food quality and safety. *Critical Reviews in Food Science and Nutrition*, **30(2–3)**, 115–360.

Stapley, A. (2008). Freeze drying. In: *Frozen Food Science and Technology* (ed. Evans, J.A.), pp. 248–275, Blackwell Publishing, Oxford.

Tamime, A.Y., Saarela, M.A.K.S., Sondergaard, A.K., Mistry, V.V. & Shah, N.P. (2005). Production and maintenance of viability of probiotic micro-organisms in dairy products. In: *Probiotic Dairy Products* (ed. Tamime, A.), pp. 39–72, Blackwell Publishing, Oxford.

Webb, S.D., Golledge, S.L., Cleland, J.L., Carpenter, J.F. & Randolph, T.W. (2002). Surface adsorption of recombinant human interferon-γ in lyophilized and spray-lyophilized formulations. *Journal of Pharmaceutical Sciences*, **91(6)**, 1474–1487.

Windhab, E. (1999). New developments in crystallization processing. *Journal of Thermal Analysis and Calorimetry*, **57(1)**, 171–180.

Wolfe, J. & Bryant, G. (1999). Freezing, drying and/or vitrification of membrane-solute-water systems. *Cryobiology*, **39**, 103–129.

Wunderlich, B. (1981). The basis of thermal analysis. Thermal characterization of polymeric materials (ed. Turi, E.A.), pp. 91–234, Academic Press, Orlando, FL.

Yu, T., Macnaughtan, B., Boyer, M., Linforth, R., Dinsdale, K. and Fisk, I.D. (2012). Aroma delivery from spray dried coffee containing pressurised internalised gas. *Food Research International*, **49(2)**, 702–709.

8

Optimization of Dairy Product Drying Processes

S. Parthasarathi[1] *and C. Anandharamakrishnan*[1,2]

[1] *Department of Food Engineering, CSIR – Central Food Technological Research Institute, Mysore 570 020, India*
[2] *Indian Institute of Crop Processing Technology, Ministry of Food Processing Industries, Government of India, Thanjavur 613 005, Tamil Nadu*

8.1 Introduction

Increasing demand for high-quality products at low production cost prompted the engineers to look for rigorous methods of optimization. Optimization techniques are used in the automobile, chemical, electrical and food manufacturing industries. In the dairy industry, engineers have to take many technological and managerial decisions during each process. However, the ultimate aim of all such decisions is either to minimize the effort required or to maximize the desired benefit. Optimization can be defined as the process of finding the conditions that give the maximum or minimum value of a function (Rao 2009).

The drying process plays a crucial role in the dairy industry. Factors affecting the drying process have been studied experimentally. However, finding the optimum conditions experimentally is difficult and time-consuming. This prompted researchers to examine the importance of optimization tools like response surface methodology, artificial neural networks, statistical optimization and computational modelling.

8.2 Experimental design tools for process optimization

8.2.1 Response surface methodology

Response surface methodology (RSM) is a collection of statistical and mathematical methods that are useful for modelling and optimizing a process in which the response of interest is influenced by several variables (Baş & Boyacı 2007a). The main objective of RSM is to optimize the responses that are influenced by various input process parameters. During the modelling and optimization of process parameters, sufficient data has to be collected through designed experimentation. Food researchers started using RSM in the early 1970s (Henika 1972; Henselman *et al.* 1974) to optimize various parameters for the production of food products such as malted milk beverages (Singh *et al.* 2008), edible whey protein films (Ozdemir & Floros 2008), kefir powder (Atalar & Dervisoglu 2015) and prebiotic beverages (Rebouças *et al.* 2014). For instance, a food engineer may wishto find the temperature (x_1) and pressure (x_2) that maximize the product yield (y) of a process. The process yield is a function of the temperature and pressure:

$$y = f(x_1, x_2) + \varepsilon \tag{8.1}$$

where ε represents the noise or error observed in the response y. If we provide the expected response as $E(y) = f(x_1, x_2) = \eta$, then the response surface is represented by:

$$\eta = f(x_1, x_2) \tag{8.2}$$

Graphical representation of the responses can be drawn in three-dimensional space or as contour plots, which helps to visualize the response surface (see Figure 8.1).

Optimization using RSM can be carried out in three stages. The first stage is the preliminary work to determine the independent parameters and their levels. The second stage is the selection of the experimental design and the prediction and verification of the model equation. The third stage is obtaining the response surface plot and contour plot of the response as a function of the independent parameters and determination of optimum points (Baş & Boyacı 2007a; Rao 2010). Two important models are commonly used in RSM (Khuri & Mukhopadhyay 2010).

i) First-degree model

$$y = \beta_0 + \sum_{i=1}^{k} \beta_i x_i + \varepsilon \tag{8.3}$$

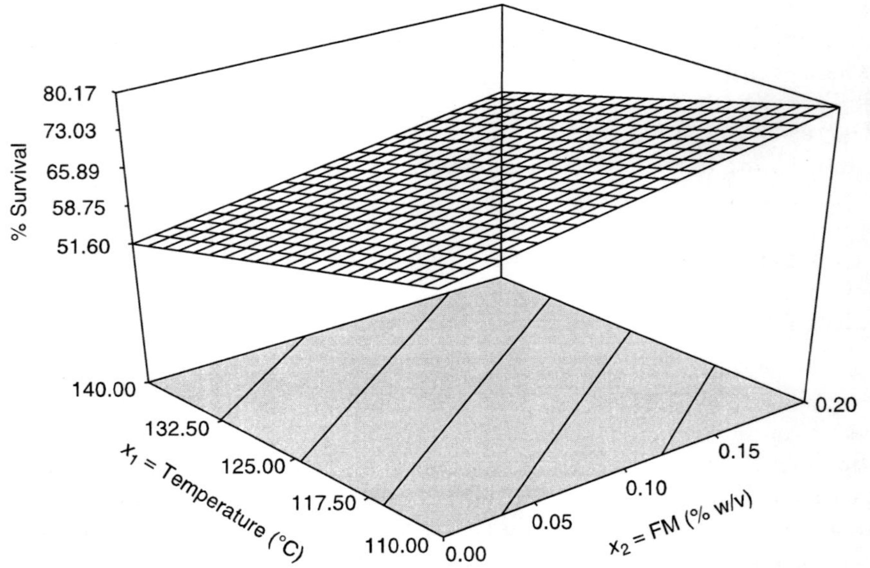

Figure 8.1 Three-dimensional responses showing the percentage survival of *Lactobacillus acidophilus* La-05 after spray drying as a function of temperature (x_1) and flaxseed mucilage (FM) (x_2) (Bustamante *et al.* 2015).

ii) Second-degree model:

$$y = \beta_0 + \sum_{i=1}^{k} \beta_i x_i + \sum_{i=1}^{k} \beta_{ii} x_i^2 + \sum_{j>1}^{k} \beta_{ij} x_i x_j + \varepsilon \tag{8.4}$$

The second-degree model is most frequently used for RSM, where y is the response and the $x_i(1, 2, \ldots k)$ are the coded levels of k quantitative variables. The coefficient β_0 is the free term, the coefficients β_i are the linear terms, the coefficients β_{ii} are the quadratic terms and the coefficients β_{ij} are the interaction terms (Rao 2010). The matrix notation of the second-degree model can be written as follows:

$$y = x\beta + \varepsilon \tag{8.5}$$

$$\begin{bmatrix} y_1 \\ y_2 \\ \cdot \\ \cdot \\ y_n \end{bmatrix} = \begin{bmatrix} 1 & x_{11} & x_{12} & \cdot & \cdot & x_{1k} \\ 1 & x_{21} & x_{22} & \cdot & \cdot & x_{2k} \\ \cdot & \cdot & \cdot & \cdot & \cdot & \cdot \\ \cdot & \cdot & \cdot & \cdot & \cdot & \cdot \\ 1 & x_{n1} & x_{n2} & \cdot & \cdot & x_{nk} \end{bmatrix} \begin{bmatrix} \beta_0 \\ \beta_1 \\ \cdot \\ \cdot \\ \beta_k \end{bmatrix} + \begin{bmatrix} \varepsilon_1 \\ \varepsilon_2 \\ \cdot \\ \cdot \\ \varepsilon_n \end{bmatrix} \tag{8.6}$$

Variables in these equations can be derived by determining the values of the coefficients using the method of least squares (MLS), a multiple regression technique.

8.2.1.1 Advantages of RSM

RSM has two advantages over classical optimization techniques: (i) it produces a large amount of information from a small number of experiments and (ii) it is possible to investigate the interaction effect of independent parameters on the response. On the other hand, classical methods are time-consuming and a large number of experiments need to be conducted to explain the behaviour of a system.

8.2.1.2 Limitations of RSM

The main limitation is that estimating the accuracy of approximation is difficult. In addition, this method is a local analysis. The developed response is not valid for the regions other than the studied ranges of factors. Furthermore, RSM is not suitable for solving highly non-linear multimodal functions or multiple objectives (Carlyle *et al.* 2000). Moreover, the objective function needs to be continuously differentiable, which may not be the case in many complex physical processes (Rao 2010).

8.2.2 Artificial neural networks

Artificial neural networks (ANNs) are a recently developed optimization approach in food process engineering that is conceptually different from traditional mathematical approaches. An ANN is a circuit composed of interconnected elements called neurons. The ANN's operating principle is based on the neural network of the brain and the way it encodes and solves problems. The brain is a highly complex and non-linear processing system that contains approximately 10^{23} neurons. Neurons are the structural constituents of the brain, which can perform computations faster than a digital computer. Neurons have four basic components: dendrites, the soma, axons and synapses

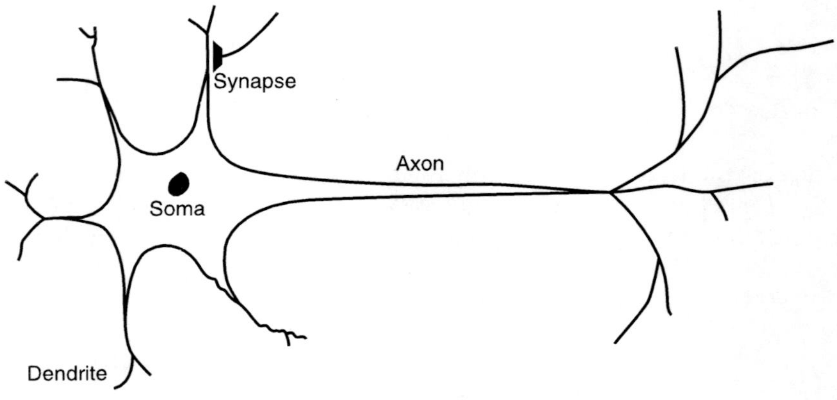

Figure 8.2 Structure of a biological neuron (Baş & Boyacı 2007b).

(Figure 8.2). Dendrites receive their input through synapses and the soma processes these inputs. The processed outputs are transferred to other neurons through axons and synapses (Baş & Boyacı 2007b). In a similar way an ANN models the optimization problem as a network consists of several neurons and is trained to solve the optimization problem efficiently (Haykin 1998).

Artificial neurons or *nodes* are the basic processing elements of an ANN. Computation is performed by a dense mesh of computing nodes and connections. Connection strength is expressed by a numerical value called *weight* (Zurada 1992). Each neuron represents a circuit with multiple input and a single output, whereas the output of the neurons is a function of the sum of the inputs (Chong & Zak 2001). Figure 8.3 is a pictorial representation of a neural network. Multiple arrows represent the output of a neuron that can be given as input to several other neurons.

8.2.2.1 Feed-forward neural network

The feed-forward neural network is the most widely applied ANN for optimizing food-processing parameters. In a feed-forward neural network the neurons are interconnected in layers that make the data to flow only in one direction. Each neuron

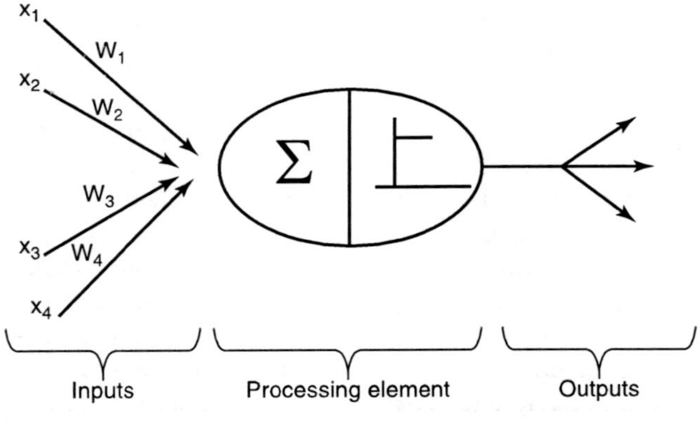

Figure 8.3 Basic structure of artificial neural network (Baş & Boyacı 2007b).

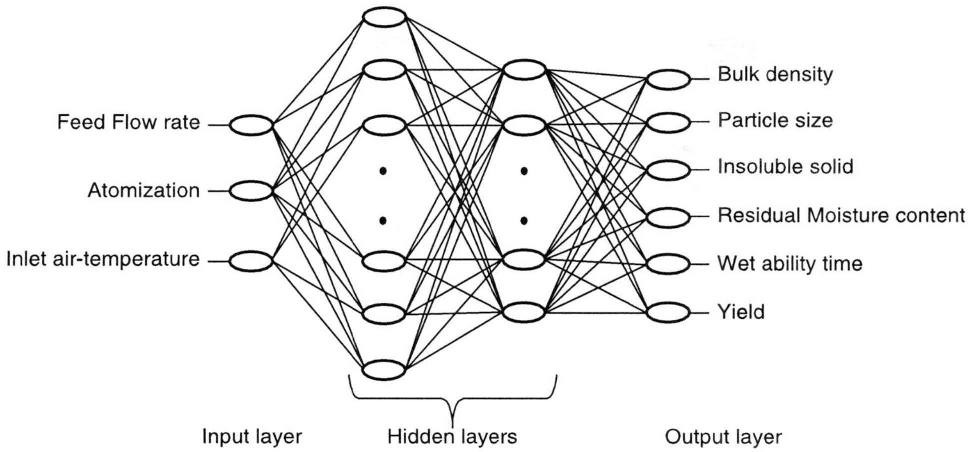

Figure 8.4 A feed-forward neural network for predicting process and product parameters in a spray dryer (Chegini *et al.* 2008).

receives information from the neurons in previous layer, that is, the input of each neuron is the weighted output of neurons in the previous layer. The first layer in the network is called the *input layer* and the last layer is called the *output layer*. The layers in between the input and output layers are called *hidden layers*. Figure 8.4 shows a typical feed-forward ANN structure containing input, hidden and output layers. More hidden layers in the neural network makes it more complex to solve. Reducing the number of neurons in the hidden layer is an effective way of improving the process (Murphy *et al.* 2014).

8.2.2.2 Learning process of an ANN

Unlike mathematical modelling, an ANN has to be trained to perform a particular function. The primary significance of an ANN is its ability to learn from its environment and to improve its performance through learning. A neural network learns about its environment through an interactive process of adjustments applied to its synaptic weights and bias levels. Haykin (2005) defines learning as 'a process by which the free parameters of a neural network are adapted through a process of stimulation by the environment in which the network is embedded. The type of learning is determined by the manner in which the parameter changes take place.' Learning procedure is based on the gradient search, with the criteria of achieving optimal least sum squared errors between the predicted and desired values. The training process is continued until the minimum error is reached (Chegini *et al.* 2008).

A prescribed set of well-defined rules for the solution of a learning problem is called a learning algorithm. It includes five basic learning rules: error-correction learning, memory-based learning, Hebbian learning, competitive learning and Boltzmann learning (Haykin 2005). A *back-propagation algorithm* is frequently used by food researchers to train a feed-forward network. *Back-propagation* is a process that computes the derivatives of network error, with respect to network weights and biases. Feed-forward neural networks can be trained in three stages (Fausett 1994): (i) the feed forward of the input training pattern, (ii) the calculation and back-propagation of the associated error and (iii) the adjustment of the weights (Zhang & Friedrich 2003). In brief, the above three stages can be explained as: (i) information from the input layer is fed forward through zero or more hidden layers to the output layer, (ii) output errors are back-propagated from the output later to the input layer through the hidden layers and (iii) the connection weights are adjusted to minimize the error at the output (Huang *et al.* 2007).

Figure 8.4 illustrates a typical multilayer feed-forward ANN constructed with three layers, that is, input, output and hidden. The hidden layer consists of one or several layers for practical applications. Increasing the number of hidden layers creates more flexibility for the network, but too many neurons in the hidden layers result in a problem called 'over-fitting'.

The sum of weighted input signals is passed to a transfer function to generate an output signal. Transfer functions for neural networks may be logarithmic sigmoid, tangential-type sigmoid, linear or radial, but the sigmoidal functions (logarithmic and tangential) are most commonly used for backpropagation neural networks:

$$\mathrm{f}(x) = \frac{1}{1 + e^{-x}} \tag{8.7}$$

The relation between the input vector $X_i^{(n)}$ and the output vector $X_j^{(n+1)}$ can be described as:

$$X_j^{(n+1)} = \mathrm{f}\left(\sum_i W_{ji}^{(n)} X_i^{(n)}\right) \tag{8.8}$$

where $X_j^{(n+1)}$ is the output of unit j in the nth layer, $W_{ji}^{(n)}$ is a weight function from unit i in the nth layer to unit j in the $(n + 1)$th layer and $\mathrm{f}(x)$ is the transfer function.

8.2.2.3 Optimization of process parameters

ANNS have been widely used to optimize dryers. ANNs trained with the feed-forward back-propagation algorithm have been studied extensively and applied successfully to various areas of food research, including drying of cassava crackers (Lertworasirikul & Tipsuwan 2008), carrot (Erenturk & Erenturk 2007) and orange juice (Chegini *et al.* 2008). Erenturk & Erenturk (2007) developed a neural network model to study the effect of drying temperature, air velocity, material geometry and drying time on the moisture content for a tray-drying operation. Pilot-scale experiments were carried out to determine the optimal number of hidden layers. A quarter of the measured experimental data were used as learning samples. The neural network represented the drying characteristics better than other mathematical drying models.

8.2.3 Finite element and finite volume methods

Fluid dynamics is an amazing, important field in many sectors, including aeronautics, automobiles and the chemical, food processing and pharmaceutical industries. Fluid-flow calculations can be done by experiment (expensive, lengthy, time-comsuming) or using modelling techniques. Mathematical and numerical modelling helps us to find qualitative and quantitative properties, but many mathematical models are described by partial differential equations (PDEs). PDEs govern the kinematic and mechanical behaviour of general bodies, which are derived with physical principles such as conservation of mass, momentum and energy. Numerical solutions of PDEs can be obtained by any of the three classical methods: (i) the finite difference method (FDM), the finite element method (FEM) and the finite volume method (FVM). The FDM is the oldest method of solving PDEs and it is based on the application of Taylor's expansion to approximate the differential equation. The main drawback of FDM over the other methods is that it cannot handle higher dimensional geometric objects. This drawback motivated researchers to use an integral form of PDEs and subsequently FEM and FVM techniques were developed.

8.2.3.1 Finite element method

The FEM is a numerical method to solve differential or integral equations. The FEM was originally developed to solve mechanical engineering problems related to fluid dynamics and structural analysis. It is one of the most powerful numerical tools for analysing complex engineering problems. The FEM is similar to the FDM: both are used to solve physical problems and both methods eventually lead to the inversion of the matrix-vector equation $Ax = b$. Although FEM was originally employed for structural analysis, it was later applied to solving Euler/Navier–Stokes equations, starting with a subdivision of the physical space into triangular (2D) or into tetrahedral (3D) elements (Figure 8.5). The FEM provides the solution for the differential equation by generating discrete algorithms (Brenner & Scott 2008). The equations are multiplied by a weight function before integrating over the entire domain and this approximation is substituted into the weighted integral of the conservation law. A set of non-linear algebraic equations can be obtained by minimizing the residuals (Ranade 2002). An important advantage of the FEM is its superior ability to deal with a solution domain having complex geometry. This method is also suitable for the treatment of non-Newtonian fluids, but it is difficult to develop computationally efficient methods for solving strongly coupled and non-linear equations.

8.2.3.2 Finite volume method

The FVM is closely related to the FEM and the FDM. However, the FVM can be derived from the integral form of the conservation law (LeVeque 2002). It is popular because it obeys the clear physical principle of conservation on the discrete scale

Figure 8.5 Finite element grid of a carrot slice (Aprajeeta *et al.* 2015).

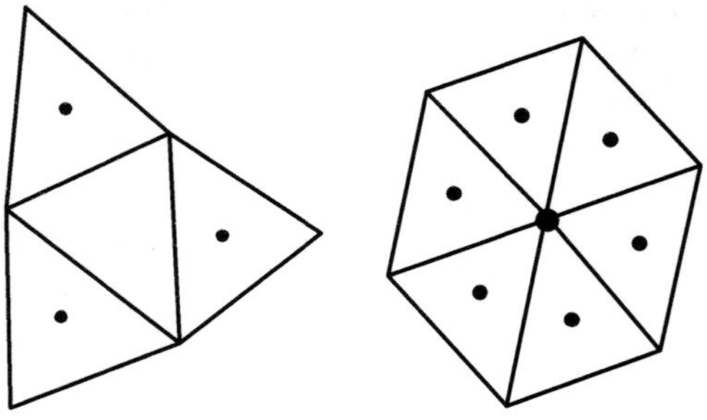

Figure 8.6 Finite volume structured grids (Manzini & Putti 2007).

(Tijskens *et al.* 2001). The numerical solution of the FVM involves integrating the governing equations over all the (finite) control volumes of the domain. The control source is in the centre of the cells as shown in Figure 8.6. The resulting statements express the conservation of relevant properties for each finite size cell. The main attraction of the FVM is the clear relationship between the numerical algorithm and the underlying physical principle, which provides more clarity than using the FEM (Versteeg & Malalasekera 1995). An algebraic equation has to be determined for each control volume by approximation of surface and volume integrals using suitable quadrature formulae. Surface integrals can be approximated either by variable values at one or more location on the cell face or on the basis of the nodal value at each cell centre (Bates *et al.* 2005). The advantages of the FVM include spatial discretization, which is carried out directly in the physical space, no complexities in the transformation between coordinate systems and easy implemention on structured as well as unstructured grids as the grid only defines the boundaries of the volume.

8.3 Drying process variables and their influence on process and product quality

Most food and bioproducts can undergo drying, which can increase the shelf life of a product at ambient temperature to months or even years (Ishwarya *et al.* 2015). Drying contributes to the reduction of the environmental impact of food distribution and consumption. Dried products have significantly reduced weight and thus require less energy for transport. The transport weight of milk can be reduced by seven or eight times by transforming liquid milk into milk powder (Palzer *et al.* 2012). Dairy-based products include skim and whole milk powder, filled milk powder, powdered creamer, infant formulas and powdered dairy food ingredients, which are manufactured by concentrating and subsequent drying of milk-based liquid blends.

In the dairy industry the following drying processes are used: spray drying, spray-belt drying, fluid-bed drying, microwave drying, freeze drying and osmotic drying. Reducing processing costs and developing specialities are considered to be the main challenges involved in the production of powders. Production capacity can be maximized by installing huge-capacity dryers, but the process conditions need to be optimized towards the minimal fouling of equipment, minimal product losses, quality and energy consumption. During drying, the limiting factor for dairy product quality is the formation of insoluble materials. Unfolding of β-lactoglobulin, followed by aggregation of casein, forms the insoluble material (De Jong 1996). Developing a drying model coupled with kinetic model for the formation of insoluble material helps the formation of insoluble materials to be understood (Straatsma *et al.* 1999).

8.3.1 Drum drying

Drum drying (or roller drying) and spray drying are the most frequently used techniques for the production of milk powders because of their low processing cost, with drum drying being cheaper (Thomas *et al.* 2004). A drum dryer is a hollow metal cylinder that rotates on horizontal axes. Heat is internally supplied by steam, hot water or another heating medium. Drum dryers are used for the dehydration of thick liquids like re-cooked breakfast cereals, dry soup mixtures, yeast and various fruit purees (Moore 1995; Gavrielidou *et al.* 2002). Viscous liquid feed is spread on the counter-rotating drums to form thin sheets on the hot cylinders in the drum dryer and the adhering thin film is rapidly dried by the high heat flux of the steam inside the drum. Energy efficiency in drum drying ranges from 70 to 90%, corresponding to a steam consumption of 1.2–1.5 kg per kg of water evaporated (Tang *et al.* 2003; Nindo & Tang 2007). Under ideal conditions, the maximum evaporation capacity of a drum dryer can be as high as 80 kg H_2O/h m^2 and it can produce dried products at the rate of 5–50 kg/h m^2, depending on the moisture content and feed material properties (Hall & Iglesias 1997).

The products of drum dryers are widely used in bakery goods, beverages, cereal and dairy foods. Double drum dryers are most frequently used for industrial purposes because of their ability to handle wider range of products, higher production rate and fewer labour requirements (Moore 1995). Designing a drum dryer with minimum energy consumption and maximum throughput remains a challenge for researchers. Extensive experimental and theoretical studies have been conducted to optimize the operating variables of drum dryers. Gavrielidou *et al.* (2002) listed the five important process variables for the operation of double drum dryer:

1. steam pressure
2. speed of rotation
3. gap between the drums
4. pool level between the drums
5. characteristics of the feed material, that is, concentration and physical characteristics.

8.3.1.1 Heat and mass transfer

In drum drying, heat is transferred from the condensing steam to the product through the drum wall. Generally, it undergoes three drying periods: initial heating, constant product temperature and rising product temperature (Tang *et al.* 2003).

Initial heating period: This is the first stage of drying, where intensive heat transfer takes place between the drum surface and the thin wet layer on the drum surface. The temperature of the product rapidly increases to reach the boiling point of free water.

Constant product temperature period: In this second stage, a large amount of free water evaporates and the temperature of the product remains constant. However, the drum surface temperature decreases due to evaporative cooling.

Rising product temperature period: Most of the free water is removed in the second stage. Heat transfer from the steam gradually exceeds the energy used for evaporation, which results in an increased drum surface temperature. Now, bound moisture in the material plays a major role in controlling the rate of evaporation. The product temperature gradually increases until it reaches the knife blade, which scrapes off the product, and the drum surface temperature continues to increase until new wet material is loaded.

The evaporation rate of free water can be calculated from Eq. (8.9) (Okos *et al.* 1992; Tang *et al.* 2003):

$$\frac{dm}{dt} = 30.94\ V^{0.8}\Delta P \tag{8.9}$$

where dm/dt is the evaporation rate per unit drum surface (kg H_2O/h m^2), V is the velocity of ambient air (m/s) and ΔP is the difference between the vapour pressure at the product surface (p_s) and in ambient air (p_a).

The water removal rate can be estimated from Eq. (8.10):

$$\frac{dm}{dt} = 3.6\frac{h(T_w - T_{evp})}{L} \tag{8.10}$$

where T_w is the drum surface temperature (°C), T_{evp} is the temperature of the evaporating surface (°C), h is the overall heat-transfer coefficient, which varies from 200 to 2000 (W/m^2 °C) and L is the latent heat (kJ/kg H_2O).

Many food researchers have used RSM to optimize the drum-drying parameters. This might be attributed to the simplicity of RSM in comparison to the complexities of FEM or FVM. Valous *et al.* (2002) studied the effect of moisture content on the steam pressure and speed of rotation. Increasing the steam pressure causes the moisture to decrease because of the increase in drum temperature. On the other hand, an increase in speed reduces the film thickness over the drum and results in higher heat fluxes. If the film dries beyond the free water, then the product layer is heated up and consequently decreases the heat and mass flux.

Supprung & Noomhorm (2003) varied the drum surface temperature, holding time and solid content of the feed to optimize the drying process. Holding time is dependent on the surface temperature of the drum and the solid content. An increase in surface temperature (115–135 °C) did have a significant effect on moisture content because in a single rotation on the drum dryer the increase in steam pressure above a certain point could not reduce the moisture content of the feed. Pua *et al.* (2010) varied the drum rotation speed (1–3 rpm) and the steam pressure (300–440 kPa) to investigate the effect on product quality, moisture content, water activity, solubility, Hunter *L*, *a*, *b* values, quantitative descriptive analysis (QDA) and the hedonic test. Moisture content and water activity decreased on increasing the drum temperature. Moisture content decreased rapidly at a lower steam pressure and the decrease gradually slowed at high steam pressure (Figure 8.7).

8.3.2 Spray drying

Spray drying is a well-established technique that has been used for more than a century, but it remains an active field of innovation, driven by the ever-increasing demand for more sophisticated particles (Vehring *et al.* 2007). Spray drying is defined as the process of transformation of a feed from a fluid state (solution, dispersion or paste) into a dried particulate form by spraying the feed into a hot drying medium (Masters 1985). It is a

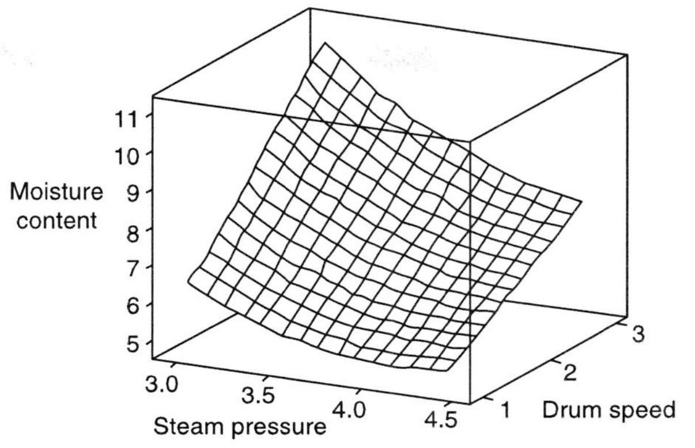

Figure 8.7 Optimization of drum-drying parameters (Pua *et al.* 2010).

continuous process involving several stages: (i) atomization, (ii) mixing of spray and air, (iii) evaporation and (iv) product separation. Important process parameters include:

1. drying air inlet temperature (T_{inlet})
2. drying air outlet temperature (T_{outlet})
3. drying air-flow rate (Q_{DA})
4. atomizing air-flow rate (Q_{AR})
5. residence time.

Particle formation in spray drying: The final particle morphology is governed by the process parameters listed abover along with feed properties such as solute diffusion coefficient and solvent latent heat. Further drying kinetics play a key role in determining the particle morphology (Walton & Mumford 1999; Vehring *et al.* 2007). Particle formation in the spray dryer is governed by two dimensionless parameters, the Peclet number and the evaporation rate (Tsapis *et al.* 2002; Vehring 2008). The Peclet number (Pe_i) is a dimensionless mass transport number that characterizes the relative importance of diffusion and convection:

$$Pe_i = \frac{\kappa}{8D_i}$$

where κ is the evaporation rate and D is the diffusion rate. If the Peclet number is equal to 1, then the evaporation rate of the solvent is of the same magnitude as the diffusion rate of the solute.

For $Pe \leq 1$, the diffusional velocity of the solute is faster than the recession rate of the surface, which results in spherical particles with little or no void space (Vehring *et al.* 2007). Similarly, for $Pe > 1$, the recession rate is faster than the diffusional velocity and the concentration of the solute increases sharply at the surface during evaporation. High surface concentration increases the viscosity, with subsequent skin or shell formation. As the drying proceeds, the skin becomes thick and the remaining water diffuses through the shell or evaporates through the openings in the shell. Finally, the hollow particle might collapse or wrinkle depending on the thickness and mechanical properties of the skin.

8.3.2.1 Exergy efficiency

In the dairy industry energy consumption is the most important process parameter for optimization (IDF 2005). Exergy analysis is an important tool for analysing and improving the energy efficiency of the spray dryer. Exergy (ε) is defined as the maximum amount of work and is a measure of the potential of a stream to cause change (Dincer & Sahin 2004). Theoretically it is defined as the ratio of total exergy out (Ex_{out}) to total exergy in (Ex_{in}) where 'out' refers to net output or product or desired value, and 'in' refers to given or used or fuel (Dincer & Sahin 2004; Erbay *et al.* 2015).

$$\varepsilon = \frac{Ex_{out}}{Ex_{in}} \times 100 \tag{8.11}$$

For spray drying, the exergetic product is the evaporation rate (ε), which can be calculated by (Erbay & Koca 2012):

$$\varepsilon = \frac{Ex_{vap}}{W} \times 100 \tag{8.12}$$

RSM has been used in most exergy efficiency studies. Erbay *et al.* (2015) made an extensive study of the exergy efficiency of spray drying for cheese powders. Three process parameters, inlet temperature, outlet temperature and atomization pressure, were used to optimize the process. The exergy efficiency for the spray-drying process was improved by increasing the inlet temperature and decreasing the outlet temperature and atomization pressure. Similarly, Al-Mansour *et al.* (2011) showed that increasing the temperature difference across the dryer and increasing the feed flow rate caused a decrease in energy consumption values.

8.3.2.2 Atomization

The atomization stage during spray drying is very important since it affects the final particle size. Many researchers have studied optimizing the atomization process using various nozzles, such as a pressure nozzle atomizer and a rotary disc atomizer. Huang *et al.* (2006) developed a three-dimensional computational fluid dynamic model for pressure nozzle and rotating disc spray dryers. High velocity variation is found at the centre of the spray-drying chamber for the pressure nozzle, whereas a disc rotation nozzle induces high air swirl. Furthermore, a rotating disk atomizer produces smaller droplets with a narrower size distribution than those produced by a pressure nozzle (Huang *et al.* 2006).

Initial droplet size: The initial droplet size emerging from the spray nozzle has a direct impact on the spray drying. Mono-dispersed droplets are capable of producing uniformly sized particles. Droplet size has a direct impact on the trajectory of the droplets in the spray dryer tower. Smaller particles tend to disperse more widely and stay longer in the tower (Yang *et al.* 2014), reaching the drying temperature rapidly, whereas larger droplets first approach the wet-bulb temperature, which is lower than the outlet air temperature, and then the droplet temperature slowly increases (see Figure 8.8). Truong *et al.* (2005) observed that droplets that were less than or equal to 100 μm reached the air temperature (145 °C) 0.2 m from the atomizer. Droplets below 40 μm reached a temperature above 100 °C in a short time and then decreased in accordance with the air outlet temperature. The Scott formula helps to predict the Sauter mean of droplets produced

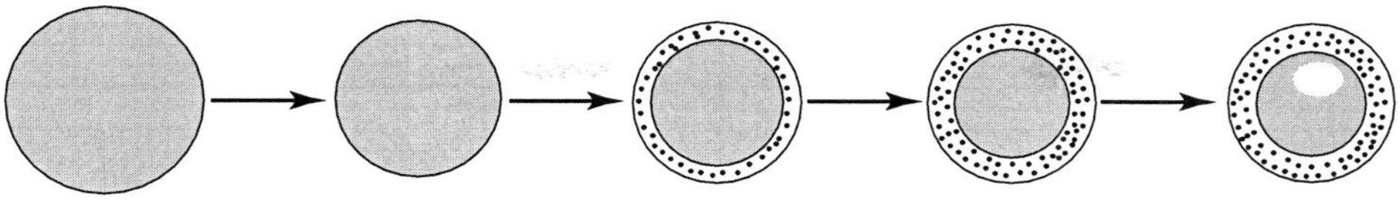

Figure 8.8 Drying of a liquid droplet containing solid (Anandharamakrishnan & Ishwarya 2015).

from the atomizer (Truong *et al.* 2005):

$$D_{vs} = 5904.7(M_p)^{0.171}(\pi d_a N_a)^{-0.537}(\mu_F)^{-0.017} \quad (8.13)$$

where M_p is the mass flow rate per total periphery of the rotary atomizer (kg/ms), d_a is the diameter of the atomizer disc (m), N_a is the speed of rotation of the atomizer (rpm) and μ_F is the viscosity of the feed liquid (Pas). Yang *et al.* (2014) investigated a spray dryer with three different initial droplet sizes (190, 200 and 210 μm). Smaller droplets (190 μm) reach the maximum temperature compared to larger droplets (210 μm) and finally moisture reabsorption due to rewetting is observed. In case of dairy products, the proteins in milk will denature if the maximum temperature is higher (more than 70 °C) and moisture reabsorption affects the product characteristics. Hence, optimizing the initial droplet size for the specific product is necessary for both product recovery and energy efficiency.

Islam *et al.* (2013) optimized the process parameters to crystallize lactose in a spray-drying process. Lactose has a high glass transition temperature and therefore is difficult to crystallize in a one-step spray-drying process. The rate of crystallization is explained by the Williams–Landel–Ferry (WLF) equation. According to the WLF equation, the rate of crystallization depends on the difference between the temperature of the material and the glass transition temperature ($T - T_g$). T_g depends on the moisture content, which is affected by the relative humidity in the spray dryer. Many researchers have reported that the interplay between process temperature and glass transition temperature affects the crystallization rate during spray drying (Langrish 2007). Hence, high humidity inside the spray dryer ensures solid-phase crystallization of lactose. Another difficulty with the solid-phase crystallization process is the stickiness of the particles on to the dryer wall. The authors used whey protein isolate to increase the process yield of crystalline lactose particles. Whey protein isolate forms a glassy film during drying and thereby reduces particle–particle and particle–wall stickiness (Anandharamakrishnan *et al.* 2007). An increase in whey protein isolate (WPI) in the lactose solution increases recovery because of the formation of a thin protein-rich film at the droplet–air interface.

8.3.3 Freeze drying

Freeze drying is a well-known dehydration process in the food and dairy industries. Freeze drying not only preserves heat-sensitive dairy products like yoghurts, but also helps to maintain the quantity of probiotics (Capela *et al.* 2006). If the freeze-drying process is not optimized, it may take weeks to complete or result in product collapse. The stability of the product during freeze drying and the duration of the freeze-drying cycle are the two major considerations in the optimization of the freeze-drying process (Tang *et al.* 2005).

Lyophilization is a three stage process: (i) cooling the product to freezing point (chilling stage or precooling stage), (ii) primary drying, removing the latent heat of crystallization (phase transition stage) and (iii) secondary drying, cooling the product to the final storage temperature (tempering stage) (Kiani & Sun 2011). Of the three stages, the primary drying stage consumes considerable freeze-drying cycle time, therefore optimizing this stage has a huge economic impact (Pikal 1990). The primary drying time can be reduced by increasing the product temperature (T_P), which increases the sublimation rate. The main problem associated with increasing the product temperature is product collapse. To avoid the collapse of freeze-dried product, T_P should be below the collapse temperature (Pikal 1990; Tang *et al.* 2005). However, the product temperature depends on many process variables, for example shelf temperature (T_s), chamber pressure (P_c), the heat-transfer coefficient of the container (K_v), product resistance to vapour flow (R_p) (Milton *et al.* 1997; Tang *et al.* 2005). In the early 1990s, Sharma & Arora (1995) investigated the effect of chamber pressure on the product temperature profile for freeze-dried yoghurt. Reduction in chamber pressure from 0.50 to 0.01 mmHg increased the drying time, and the maximum production rate was obtained for a chamber pressure of 0.01 mmHg and 6.2 mm product thickness.

Capela *et al.* (2006) studied the effect of cryoprotectants on the viability of *Lactobacillus acidophilus, Lactobacillus casei, Lactobacillus rhamnosus* and *Bifidobacterium longum*. Cryoprotectants like Unipectine^{TM} RS 150 and Satialgine ^{TM} GCF 639 were used for the study. Cryoprotectants improve cell viablity by binding with water and inhibiting the intracellular ice formation. The degree of protection offered by cryoprotectants varies between each probiotic organism due to the difference in cell membrane composition. For instance, Unipectine^{TM} RS 150 showed an 80% improvement in the survival of *B. longum*, but for *Lactobacillus* spp. survival was increased by only 30%.

8.3.3.1 Temperature measurement

Measuring the parameters of interest without interfering with the process dynamics presents a challenge. The development of non-invasive techniques is a valuable alternative for thermocouples to monitor temperature dependent processes (Barresi *et al.* 2009). Up to now, the non-invasive techniques proposed have been based on manometric temperature measurement (Milton *et al.* 1997), dynamic pressure rise (Liapis & Sadikoglu 1998) and pressure rise analysis (Chouvenc *et al.* 2004).

Manometric temperature measurement Manometric temperature measurement (MTM) is a technique that helps to determine the product temperature during primary drying at the sublimation interface by quickly isolating the chamber (in less than a second) from the condenser for a short period of time. The resulting pressure rise is analysed with the algorithm, which finally leads to direct measurement of both the pressure at the sublimation interface (P_{ice}) and the mass-transfer resistance (R_p) (Figure 8.9). The other parameters, such as the product temperature (T_P), heat transfer into the product (dQ/dt), actual ice thickness (L_{ice}), temperature at the bottom (T_b), heat-transfer coefficient of the container (K_v) and sublimation rate (dm/dt), can be calculated based on the MTM data.

Dynamic pressure rise Liapis & Sadikoglu (1998) developed a complex algorithm for the dynamic pressure rise (DPR) to determine the temperature profile of the dried and

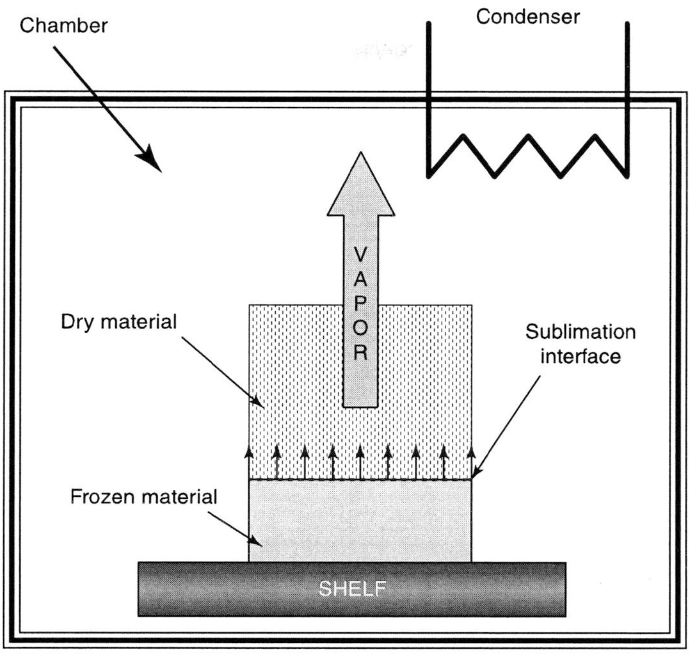

Figure 8.9 The freeze-drying process (Lopez-Quiroga *et al.* 2012).

frozen layers of the product from the pressure rise test. The algorithm is based on the unsteady-state mathematical model developed by Sadikoglu & Liapis (1997). In this method, a freeze dryer was designed with rapid closing (5–15 s) of the valve between the chamber and the condenser. This increases the chamber pressure rapidly at first, after which it approaches the vapour pressure of the sublimated ice. During this period, the increase in the chamber pressure with time is monitored and analysed by a computer program. The algorithm relates the pressure rise data with the heat- and mass-transfer expressions, developed by Sadikoglu & Liapis (1997), to calculate the product temperature at the moving interface, the top of the dried layer and the bottom of the frozen layer (Fissore *et al.* 2010). However, the DPR method requires many parameters, such as diffusivity, porous layer permeability, heat-transfer coefficient and temperature and partial pressure, to perform the analysis, which makes it more complex.

Pressure rise analysis Chouvenc *et al.* (2004) developed the pressure rise analysis based on the macroscopic heat balance, accounting for the desorption of bound water and the thermal capacity of the portion of the vial glass in contact with the frozen product. The PRA algorithm assumes that the rate of water vapour pressure increases during the pressure rise test (Fissore *et al.* 2010):

$$\frac{\mathrm{d}p_{\mathrm{w,c}}}{\mathrm{d}t} = \frac{N_{\mathrm{v}}A_{\mathrm{p}}RT_{\mathrm{c}}}{M_{\mathrm{w}}V_{\mathrm{c}}R_{\mathrm{p}}}(p_{\mathrm{w,i}} - p_{\mathrm{w,c}}) + D_{\mathrm{des}} \tag{8.14}$$

This equation is combined with the macroscopic heat balance for the frozen layer, assuming that the temperature increase at the interface, $\mathrm{d}T_{\mathrm{i}}/\mathrm{d}t$, is the same as the mean product temperature rise.

8.3.3.2 Computational modelling

Optimization of freeze-drying process parameters has been usually carried out using a time-consuming trial-and-error experimental approach. Theoretical modelling helps to predict the position of the ice sublimation interface, moisture distribution and the glass transition temperature in the dried layer, which is very difficult to calculate experimentally. With these theoretical models, the effect of various process parameters, such as shelf temperature, ramp rate, chamber pressure, sublimation rate and drying time, on critical product quality attributes (product temperature, residual water) can be predicted. The theoretical approach to the optimization of the freeze-drying process was started by the Liapis research group. They developed equations based on classical heat and mass transfer (Millman *et al.* 1985; Liapis & Bruttini 1995; Sadikoglu & Liapis 1997). Many mathematical models have been developed for pharmaceutical freeze drying using the heat and mass balance equations. However, pharmaceutical solutions are usually dried from a glass vial and sublimation occurs from the top by receiving conductive heat from the vial bottom. Earlier models were developed assuming one-dimensional movement of the sublimation interface. This was governed by the mass balance equations and heat balance at the sublimation interface. Later, the primary and secondary freeze-drying stages were simulated using multidimensional models (Sheehan & Liapis 1998).

Moving boundary approach The moving boundary approach is a successful technique for tracking the moving interface due to phase change (see Figure 8.10). Moving boundary analysis can be classified into two types: (i) solid–liquid phase change (e.g. freezing and thawing) and (ii) liquid–vapour (sublimation, e.g. frying, spray drying) or solid–vapour (sublimation, e.g. freeze drying) phase change. A numerical solution to the moving boundary problems can be found by moving and fixed grid methods, which have been developed to solve free surface problems (Floryan & Rasmussen 1989; Scardovelli & Zaleski 1999). The moving grid method is referred to as the Lagrangian method, and the fixed grid method is the Eulerian method. The moving grid method is the most suitable technique for solving the boundary conditions at the moving interface. On the other hand, the fixed grid approach is combined with momentum and energy equations at the interface boundary. The major advantage of the fixed grid approach over the moving grid method is that it can handle the complex geometry of free and moving surfaces (Li 2006).

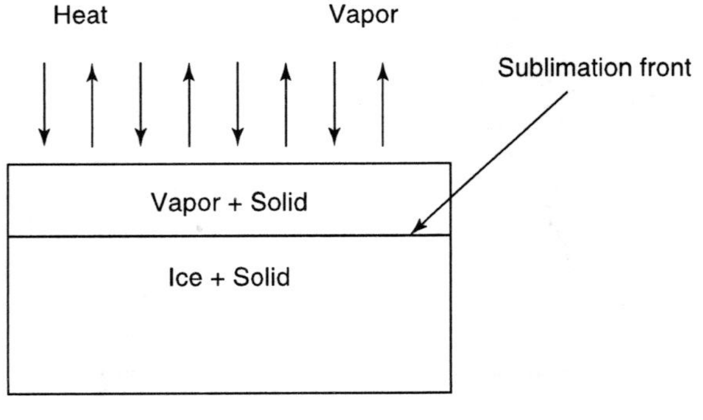

Figure 8.10 Moving boundary approach for solving freeze drying (Farid 2002).

Finite volume method Optimization of the freeze-drying process has been carried out using the FVM, which is conservative and facilitates easy understanding of the discretization procedure. Along with the FVM, Song *et al.* (2002) adopted the moving grid method to handle irregular and continuously changing domains during the primary drying stage. Freeze drying was carried out for skim milk, loaded in trays and vials. At the beginning of the primary drying stage, a whole vial consists of the frozen region because free water in skim milk is frozen completely during the prefreezing stage. During the primary drying stage, frozen free water starts sublimating and the vapour is expelled through the sublimation interface. Heat transfer from the side wall is not negligible for the vial and the sublimation interface is in a curved configuration, whereas for tray sublimation the interface is in a one-dimensional planar configuration. It is difficult to calculate the temperature at the sublimation interface because it is not possible to locate the interface as it changes continuously. Song & Kim (2003) calculated the temperature profile at the sublimation interface by collecting the minimum temperature of all measured temperature histories at different positions. At the end of primary drying, there was a rapid increase in temperature because of the lack of ice for sublimation. The moving grid approach is considered to be the most suitable method for freeze drying. Nam & Song (2005) compared the fixed grid and moving grid approaches for the freeze-drying problems. The fixed grid method requires about 70% more computation time than the moving grid method for two-dimensional freeze-drying problems.

The heat- and mass-transfer rates are the limiting factors in the freeze-drying process, and result in a long drying time and low process productivity. Continuous effort has been devoted to improve the freeze-drying process in addition to shelf heating, microwave heating (Wang *et al.* 2005) and infrared heating (Chakraborty *et al.* 2011). Infrared radiation penetrates the material and transfers thermal energy to a depth that depends on the absorbance of the material. Bae *et al.* (2010) developed a numerical model for the freeze-drying process accompanying infrared radiation heating.

Another important variable during the freeze-drying process is dryer load. Full dryer load defines the material occupying the available shelf surface area. If there is insufficient raw material, or in order to meet immediate production needs, freeze drying can be carried out under partial load conditions. Under partial load, the freeze-drying process operates optimally and results in higher shelf surface temperature than under full load conditions (Patel & Pikal 2011). Searles (2004) described the overload phenomenon as 'choked flow or condenser overload' in a high sublimation rate process. During high sublimation flux, the chamber pressure rises above the set point, choked flow occurs in the duct connecting the chamber and the condenser, and the process runs out of control (Fissore *et al.* 2010). Choked flow arises when the flow velocity in the duct approaches the speed of sound. Patel *et al.* (2010) examined flow of water vapour in the duct connecting the chamber and the condenser using the FVM.

Patel *et al.* (2010) simulated the freeze-drying process for choked flow conditions to predict gas temperature, pressure, velocity and Mach number in the duct. The velocity profile across the diameter and length of the duct is shown in Figure 8.11. At the duct entrance a sharp increase in velocity is observed due to sudden area reduction and then the velocity increases gradually. Velocity is zero near the wall and reaches its maximum

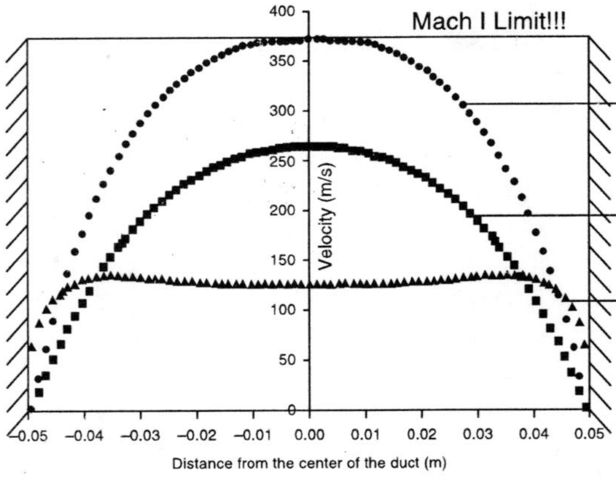

Figure 8.11 Velocity profile across the diameter and length of the duct (Patel *et al.* 2010).

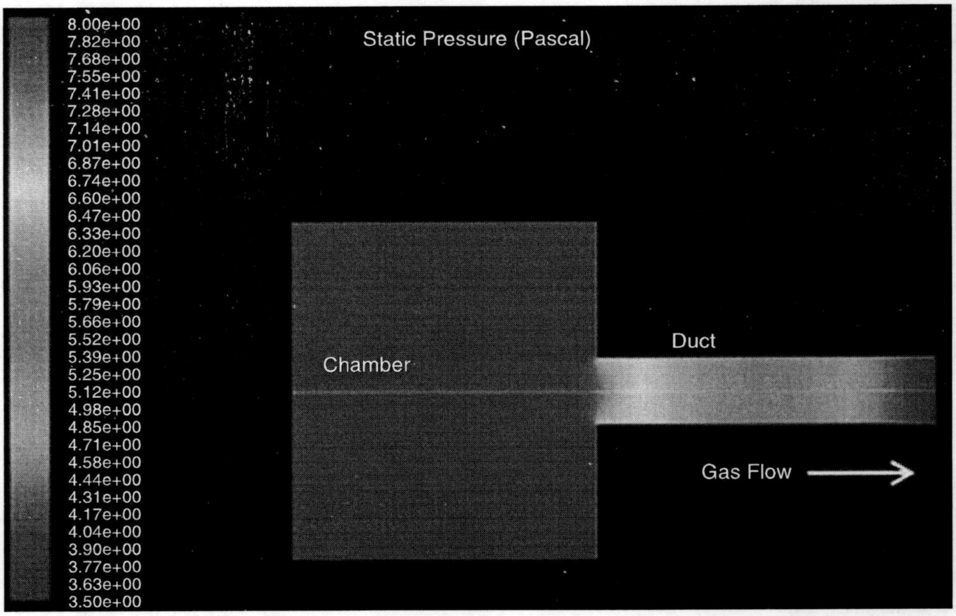

Figure 8.12 Simulation results of static pressure contour during choked flow (Patel *et al.* 2010).

velocity at the centre of the duct. At the duct exit, the gas-flow velocity reaches the speed of sound (Mach I). Under choked conditions there was a sudden increase in chamber pressure about the set point (Figure 8.12), which subsequently resulted in increased product temperature (Figure 8.13).

Finite element method The FEM came into practice in the 1990s to solve the governing equations of freeze-drying operations. Several commercial FEM software programs are available, for example Adina, Ansys and Comsol (formerly FEMLAB). Many researchers

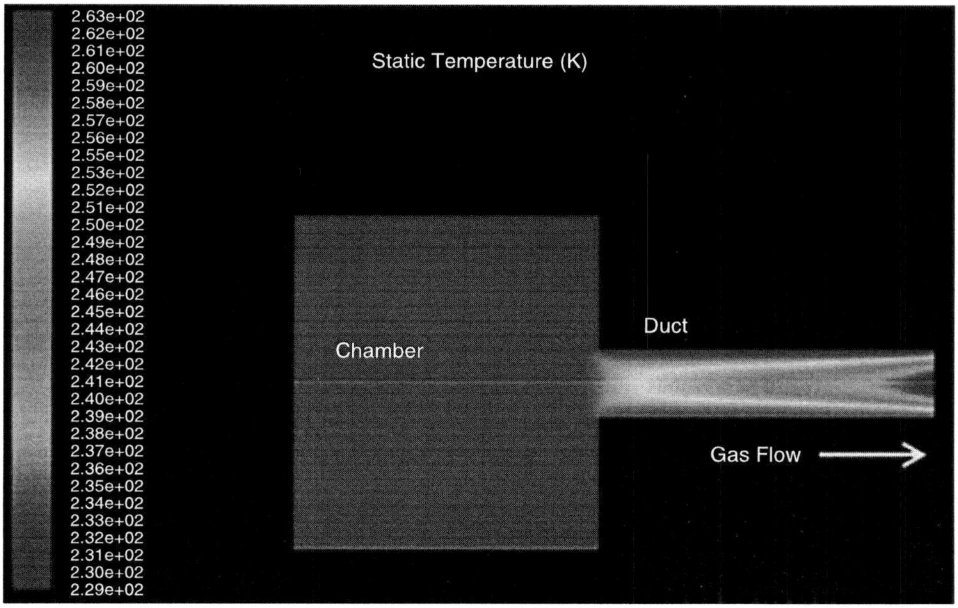

Figure 8.13 Simulation results of static temperature contour during choked flow (Patel *et al.* 2010).

have highlighted the combination of computational and experimental approaches to predict process parameters for freeze drying. Heat- and mass-transfer coefficients vary at different lyophilization scales, and different approaches have used a variety of factors that are considered appropriate for scaling-up from lab scale to commercial scale. Chen *et al.* (2015) determined heat- and mass-transfer coefficients through experimental gravimetric analysis.

The ice sublimation rate can be determined by:

$$\frac{\mathrm{d}\,m_{\mathrm{ice}}}{\mathrm{d}\,t} = A_{\mathrm{p}}\frac{P_{\mathrm{ice}} - P_{\mathrm{c}}}{R_{\mathrm{ps}}} \tag{8.15}$$

where $\mathrm{d}m/\mathrm{d}t$ is the ice sublimation rate, A_{p} is the internal cross-section area, P_{ice} is the vapour pressure of ice (Torr) at the temperature of the sublimation interface, determined through MTM, and R_{ps} is the sum of the area normalized dry layer and stopper resistance (cm^2/Torr.h/g).

The heat transfer into the product can be determined by:

$$\frac{\mathrm{d}\,Q}{\mathrm{d}\,t} = \Delta H_{\mathrm{s}}\frac{\mathrm{d}\,m_{\mathrm{ice}}}{\mathrm{d}\,t} \tag{8.16}$$

where ΔH_{s} is the heat of ice sublimation (cal/g). The heat transfer coefficient can be expressed as the sum of three terms:

$$K_{\mathrm{v}} = K_{\mathrm{c}} + K_{\mathrm{r}} + K_{\mathrm{g}} \tag{8.17}$$

where K_{c} is the contribution from direct conduction from the shelf to the product at the points of contact, K_{r} is the contribution from radioactive heat transfer and K_{g} is the contribution from conduction through the gas between the shelf and the product bottom (Pikal *et al.* 1984).

Tang *et al.* (2005) calculated the frozen layer thickness from the cumulative amount of sublimed ice:

$$\mathrm{I}(t) = \frac{m_0 - m(t)}{\rho_i A_p \varepsilon} \tag{8.18}$$

where $l(t)$ is the ice thickness at time t (cm), m_0 is the initial mass of water, $m(t)$ is the mass of ice sublimed, which can be calculated by numerical integration of dm/dt over the primary drying time, t, ρ_i is density of the ice (g/cm^3) and ε is the volume fraction of the ice.

Chen *et al.* (2015) used an FEM based on software called PASSAGE to compute primary drying parameters such as drying time and temperature. These authors determined the heat- and mass-transfer coefficients through gravimetric analysis using vial sublimation experiments. With the developed computational model, Chen *et al.* were able to predict primary drying time and product temperature, along with the interface temperature and ice thickness (Figure 8.14a). The dark region in Figure 8.14a represents the frozen part, whereas the gray region represents the dried material. Figure 8.14b shows snapshots of product temperature at the sublimation interface during lyophilization. The computational model is able to predict the product temperature and track the sublimation front interface at any time step, which is difficult to monitor experimentally during lyophilization.

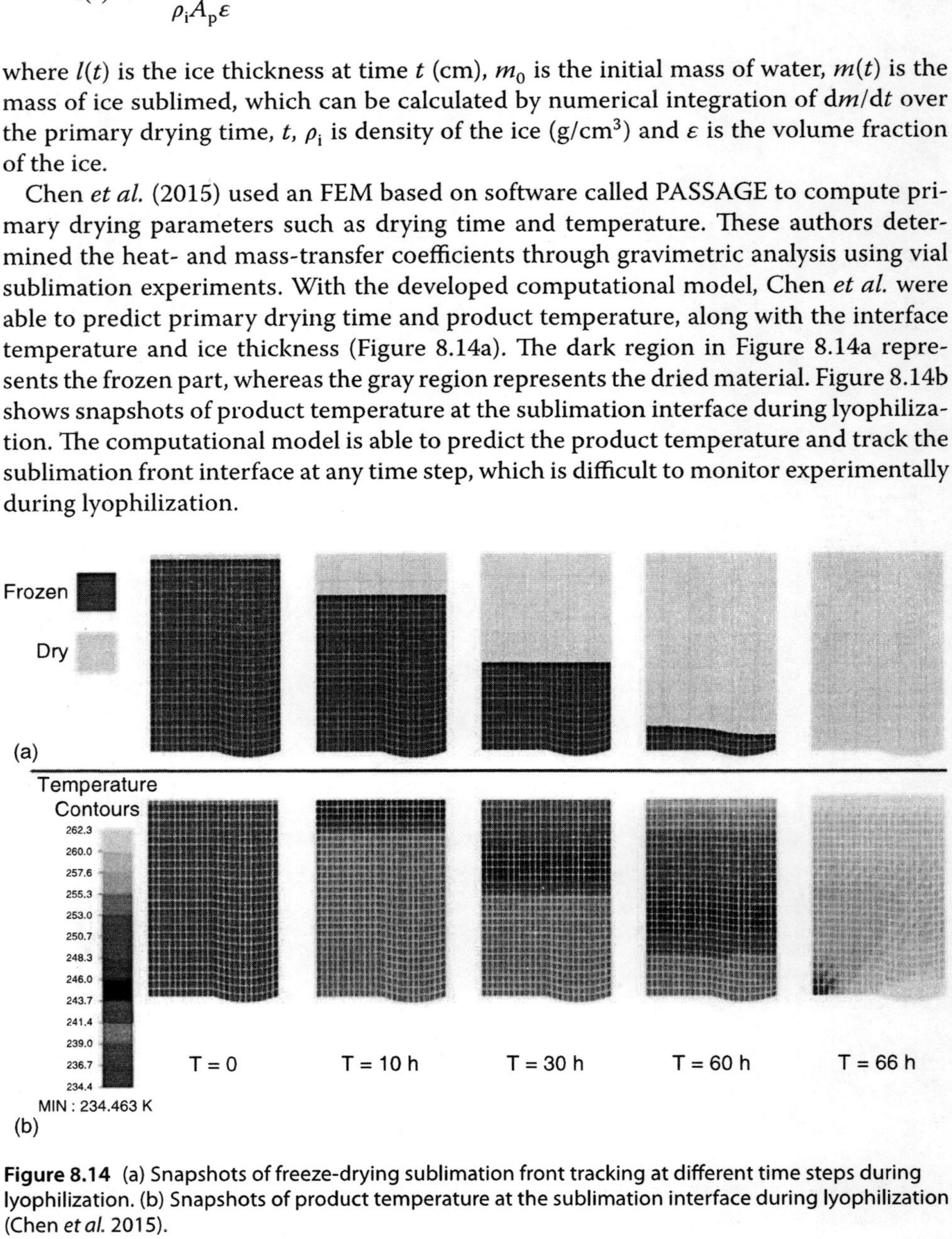

Figure 8.14 (a) Snapshots of freeze-drying sublimation front tracking at different time steps during lyophilization. (b) Snapshots of product temperature at the sublimation interface during lyophilization (Chen *et al.* 2015).

8.3.4 Spray freeze drying

The production of dairy products by drying can be divided into destructive and constructive particle formation methods. The destructive method involves milling of a product, for example freeze drying of an aqueous protein solution. Milling is required after the freeze-drying process, which may cause difficulty in controlling the particle size distribution. Constructive methods include spray-drying and spray freeze drying (SFD) techniques (Sonner *et al.* 2002). Freeze drying is a popular method of producing shelf-stable particulate products, and is most suitable for drying thermally sensitive biological-based materials. Freeze-dried products are porous in nature due to formation of ice crystals in the freezing stage, which subsequently sublime in the drying stage. Their porous structure enables good rehydration behaviour in the powdered product. However, freeze drying involves high capital and operating cost due to high-vacuum, low-temperature operation and longer residence time (Anandharamakrishnan *et al.* 2010). Lack of control of particle size distribution is the major limitation in freeze drying (Ezhilarasi *et al.* 2013; Rajam & Anandharamakrishnan 2015).

Spray drying is a suspended particle processing technique. Liquid feed is atomized in a hot gas to instantaneously produce powder. Air is mostly used as the gas, although nitrogen, an inert gas, is also sometimes used. The liquid feed can be a solution, emulsion or suspension, depending on the operating conditions and feed properties. Spray dryers can produce particulate powders consisting of particles of 10–50 μm to 2–3 mm (Gharsallaoui *et al.* 2007). The main drawback with spray dryers is the high temperature operation, which restricts their application for heat labile biological products (Dolly *et al.* 2011; Ezhilarasi *et al.* 2014). One solution to the problems of high temperature (spray drying) and long residence time (freeze drying) is SFD.

SFD is a three-stage process: (i) atomization, (ii) freezing and (iii) freeze drying of the frozen particles in a freeze dryer (Anandharamakrishnan *et al.* 2010; Ishwarya *et al.* 2015). The geometry of the spray-freezing chamber is shown in Figure 8.15. SFD has successfully been applied to drying/encapsulating probiotics (Dolly *et al.* 2011; Rajam & Anandharamakrishnan 2015), DHA oil (Karthik & Anandharamakrishnan 2013), vanillin (Hundre *et al.* 2015) and coffee (Ishwarya & Anandharamakrishnan 2015). SFD is a complex process which involves number of mechanisms: (i) the fluid mechanics of the spray determine the formation and motion of individual droplets in the spray freezing chamber, (ii) heat transfer between the gas and the droplet depends on the gas temperature, droplet temperature and droplet-gas slip velocity, and (iii) freezing and ice crystallization within the drops (Anandharamakrishnan *et al.* 2010).

Atomization and rapid freezing of droplets will impose stress on the biomaterials in the SFD process. Costantino *et al.* (2000) studied the effect of atomization conditions on particle size and stability of spray-freeze-dried protein. The authors used various atomization conditions (atomization N_2 pressure 2–130 psi, atomization N_2 flow 20–330 l/min, liquid feed pressure 3–120 psi, feed liquid flow rate 33–850 l/min and mass flow ratio 0.08–3.6) to spray freeze dry the model compound protein. The effect of the atomization conditions was correlated with the effect of the mass flow ratio. The mass flow ratio is defined as the ratio of atomization gas to liquid feed:

$$\frac{Q_{\text{atomN}_2}}{Q_{\text{liquid}}} = \frac{M_{\text{atomN}_2} \cdot \rho_{\text{atomN}_2}}{M_{\text{liquid}} \cdot \rho_{\text{liquid}}} \tag{8.19}$$

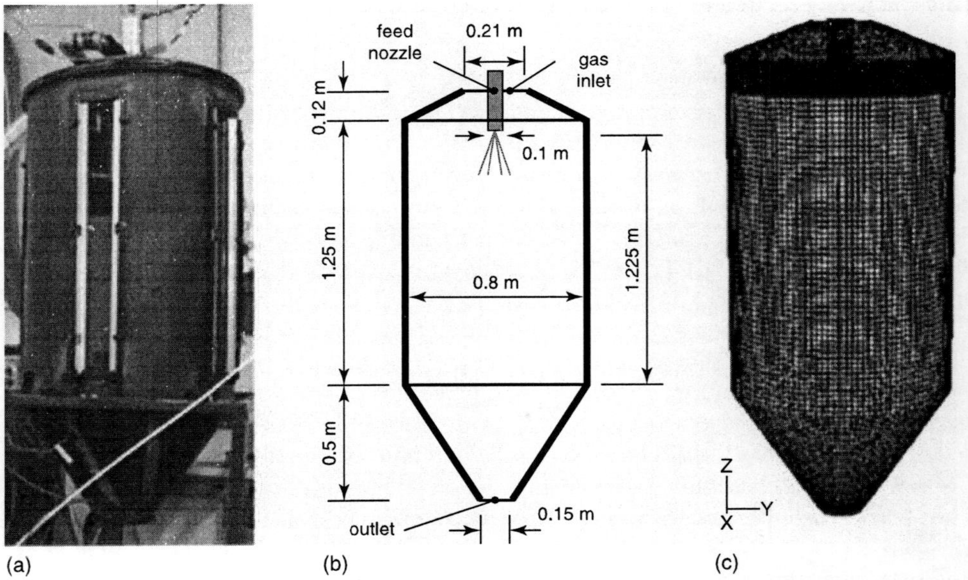

Figure 8.15 CFD simulation of the spray-freezing chamber: (a) photograph, (b) dimensions and (c) computational domain and meshing (Anandharamakrishnan *et al.* 2009).

where Q is the mass flow rate, M is the volumetric flow rate and ρ is the density. The subscripts atomN_2 and liquid refer to the atomization nitrogen gas and liquid feed streams, respectively. Decreasing the size of the atomized droplet results in more rapid freezing, which aids in the formation of smaller ice crystals, leading to production of porous microstructures. The formulation parameters also have an impact on the particle size and stability during SFD operation. Among many excipients studied, such as trehalose, mannitol, Tween 20 and ammonium sulphate, trehalose was found to be a suitable wall material for effective shielding of protein after SFD (Costantino *et al.* 2002).

Atomized droplet size plays an important role in determining the morphology of the dried powder. Al-Hakim (2004) investigated droplet size and axial velocity during spray freezing using phase Doppler anemometry (PDA). PDA was able to measure the droplet size at a maximum axial distance of 0.2 m below the nozzle. At greater distances authors observed difficulties due to (i) reduction in refractive scatter due to droplet freezing, (ii) few droplets entering the measurement volume due to spreading of the spray and (iii) greater interference of droplets not in the measurement volume due to a generally greater degree of 'fog' further away from the nozzle. To overcome these practical difficulties in measuring in-line process parameters, Anandharamakrishnan *et al.* (2015) simulated a spray-freezing operation with computational fluid dynamics (CFD) to predict gas and particle temperature, velocity and residence time.

8.4 Conclusion

Optimization is finding the best solution when there are several conflicting demands. Aris (1964) defines the optimization as 'getting the best you can out of a given situation'.

This chapter discussed the different types of optimization procedures used for drying dairy products. Optimization tools like RSM, ANNs, the FEM and the FVM were discussed in detail. Optimization studies using CFD are still limited in the food processing sector when compared to other disciplines, however advanced simulation techniques such as the FEM and the lattice Boltzmann method will play an increasingly important role in optimizing foods systems. The major problem in using advanced simulation techniques such as CFD is the computing time required. Computing time is an important issue in optimization studies involving heat transfer and fluid flow. A computational tool with minimal computing time is required to optimize complex food-processing operations.

References

Al-Hakim, K. (2004). *An investigation of spray-freezing and spray-freeze-dryings*, PhD thesis, Loughborough University.

Al-Mansour, H.E., Al-Busairi, B.H. & Baker, C.G.J. (2011). Energy consumption of a pilot-scale spray dryer. *Drying Technology*, **29(16)**, 1901–1910.

Anandharamakrishnan, C. & Ishwarya, S.P. (2015). *Spray drying techniques for food ingredient encapsulation*, John Wiley & Sons, Ltd, Chichester.

Anandharamakrishnan, C., Rielly, C.D. & Stapley, A.G.F. (2007). Effects of process variables on the denaturation of whey proteins during spray drying. *Drying Technology*, **25(5)**, 799–807.

Anandharamakrishnan, C., Gimbun, J., Stapley, A.G.F. & Rielly, C.D. (2009). Application of computational fluid dynamics (CFD) simulations to spray-freezing operations. *Drying Technology*, **28(1)**, 94–102.

Anandharamakrishnan, C., Rielly, C.D. & Stapley, A.G.F. (2010). Spray-freeze-drying of whey proteins at sub-atmospheric pressures. *Dairy Science & Technology*, **90(2–3)**, 321–334.

Aprajeeta, J., Gopirajah, R. & Anandharamakrishnan, C. (2015). Shrinkage and porosity effects on heat and mass transfer during potato drying. *Journal of Food Engineering*, **144**, 119–128.

Aris, R. (1964). *Discrete Dynamic Programming*, pp. 10–39, Blaisdell Publishing Co., New York.

Atalar, I. & Dervisoglu, M. (2015). Optimization of spray drying process parameters for kefir powder using response surface methodology. *LWT – Food Science and Technology*, **60(2)**, 751–757.

Bae, S.H., Nam, J.H., Song, C.S. & Kim, C.J. (2010). A numerical model for freeze drying processes with infrared radiation heating. *Numerical Heat Transfer, Part A: Applications: An International Journal of Computation and Methodology*, **58(5)**, 333–355.

Barresi, A.A., Pisano, R., Fissore, D., Rasetto, V., Velardi, S.A., Vallan, A., Parvis, M. & Galan, M. (2009). Monitoring of the primary drying of a lyophilization process in vials. *Chemical Engineering and Processing: Process Intensification*, **48(1)**, 408–423.

Baş, D. & Boyacı, İ. (2007a). Modeling and optimization I: Usability of response surface methodology. *Journal of Food Engineering*, **78(3)**, 836–845.

Baş, D. & Boyacı, İ. (2007b). Modeling and optimization II: Comparison of estimation capabilities of response surface methodology with artificial neural networks in a biochemical reaction. *Journal of Food Engineering*, **78**(**3**), 846–854.

Bates, P.D., Lane, S.N. & Ferguson, R.I. (2005). *Computational Fluid Dynamics: Applications in Environmental Hydraulics*, John Wiley & Sons.

Brenner, S. & Scott, R. (2008) *The Mathematical Theory of Finite Element Methods*, 3rd edn, Springer, New York.

Bustamante, M., Villarroel, M., Rubilar, M. & Shene, C. (2015). *Lactobacillus acidophilus* La-05 encapsulated by spray drying: Effect of mucilage and protein from flaxseed (*Linum usitatissimum* L.). *LWT – Food Science and Technology*, **62**(**2**), 1162–1168.

Capela, P., Hay, T.K.C. & Shah, N.P. (2006). Effect of cryoprotectants, prebiotics and microencapsulation on survival of probiotic organisms in yoghurt and freeze-dried yoghurt. *Food Research International*, **39**(**2**), 203–211.

Carlyle, W., Montgomery, D.C. & Runger, G.C. (2000). Optimization problems and methods in quality control and improvement. *Journal of Quality Technology*, **32**(**1**), 1–17.

Chakraborty, R., Mukhopadhyay, P., Bera, M. & Suman, S. (2011). Infrared-assisted freeze drying of tiger prawn: Parameter optimization and quality assessment. *Drying Technology*, **29**(**5**), 508–519.

Chegini, G.R., Khazaei, J., Ghobadian, B. & Goudarzi, A. M. (2008). Prediction of process and product parameters in an orange juice spray dryer using artificial neural networks. *Journal of Food Engineering*, **84**(**4**), 534–543.

Chen, X., Sadineni, V., Maity, M., Quan, Y., Enterline, M. & Mantri, R. V. (2015). Finite element method (FEM) modeling of freeze-drying: monitoring pharmaceutical product robustness during lyophilization. *AAPS PharmSciTech*, **16**(**6**), 1317–1326.

Chong, E.K.P. & Zak, S.H. (2001) *An Introduction to Optimization*, Wiley-Interscience Publications, New York.

Chouvenc, P., Vessot, S., Andrieu, J. & Vacus, P. (2004). Optimization of the freeze-drying cycle: A new model for pressure rise analysis. *Drying Technology*, **22**(7), 1577–1601.

Costantino, H.R., Firouzabadian, L., Hogeland, K., Wu, C., Beganski, C., Carrasquillo, K.G., Córdova, M., Griebenow, K., Zale, S.E. & Tracy, M.A. (2000). Protein spray-freeze drying. Effect of atomization conditions on particle size and stability. *Pharmaceutical Research*, **17**(**11**), 1374–1382.

Costantino, H.R., Firouzabadian, L., Wu, C., Carrasquillo, K.G., Griebenow, K., Zale, S.E., & Tracy, M.A. (2002) Protein spray freeze drying. 2. Effect of formulation variables on particle size and stability. *Journal of Pharmaceutical Sciences*, **91**, 388–395.

De Jong, P. (1996). *Modelling an optimisation of thermal treatments in the dairy industry*, PhD thesis, University of Technology, Delft.

Dincer, I. & Sahin, A.Z. (2004). A new model for thermodynamic analysis of a drying process. *International Journal of Heat and Mass Transfer*, **47**(**4**), 645–652.

Dolly, P., Anishaparvin, A., Joseph, G.S. & Anandharamakrishnan, C. (2011). Microencapsulation of *Lactobacillus plantarum* (mtcc 5422) by spray-freeze-drying method and evaluation of survival in simulated gastrointestinal conditions. *Journal of Microencapsulation*, **28**(**6**), 568–574.

Erbay, Z. & Koca, N. (2012). Energetic, exergetic, and exergoeconomic analyses of spray-drying process during white cheese powder production. *Drying Technology*, **30**(**4**), 435–444.

Erbay, Z., Koca, N., Kaymak-Ertekin, F. & Ucuncu, M. (2015). Optimization of spray drying process in cheese powder production. *Food and Bioproducts Processing*, **93**, 156–165.

Erenturk, S. & Erenturk, K. (2007). Comparison of genetic algorithm and neural network approaches for the drying process of carrot. *Journal of Food Engineering*, **78**(**3**), 905–912.

Ezhilarasi, P.N., Indrani, D., Jena, B.S. & Anandharamakrishnan, C. (2013). Freeze drying technique for microencapsulation of Garcinia fruit extract and its effect on bread quality. *Journal of Food Engineering*, **117**(**4**), 513–520.

Ezhilarasi, P.N., Indrani, D., Jena, B.S. & Anandharamakrishnan, C. (2014). Microencapsulation of Garcinia fruit extract by spray drying and its effect on bread quality. *Journal of the Science of Food and Agriculture*, **94**(**6**), 1116–1123.

Farid, M. (2002). The moving boundary problems from melting and freezing to drying and frying of food. *Chemical Engineering and Processing: Process Intensification*, **41**(**1**), 1–10.

Fausett, L.V. (1994). *Fundamentals of Neural Networks: Architectures, Algorithms, and Applications*, Prentice-Hall, New Jersey.

Fissore, D., Pisano, R. & Barresi, A.A. (2010). On the methods based on the pressure rise test for monitoring a freeze-drying process. *Drying Technology*, **29**(**1**), 73–90.

Floryan, J.M. & Rasmussen, H. (1989). Numerical methods for viscous flow with moving boundaries. *Applied Mechanics Reviews*, **42**(**12**), 323–341.

Gavrielidou, M.A., Valous, N. A., Karapantsios, T. D. & Raphaelides, S. N. (2002). Heat transport to a starch slurry gelatinizing between the drums of a double drum dryer. *Journal of Food Engineering*, **54**(**1**), 45–58.

Gharsallaoui, A., Roudaut, G., Chambin, O., Voilley, A. & Saurel, R. (2007). Applications of spray-drying in microencapsulation of food ingredients: An overview. *Food Research International*, **40**(**9**), 1107–1121.

Hall, G.M. & Iglesias, O. (1997) Functional properties of dried milk whey. *Food Science and Technology International*, **3**(**5**), 381–383.

Haykin, S. (1998). *Neural networks: A Comprehensive formulation*, 2nd edn, Macmillan, New York.

Haykin, S. (2005). Cognitive radio: brain-empowered wireless communications. *Selected Areas in Communications, IEEE Journal*, **23**(**2**), 201–220.

Henika, R.G. (1972). Simple and effective system for use with response surface methodology. *Cereal Science Today*, **17**(**10**), 309–314.

Henselman, M.R., Donatoni, S.M. & Henika, R.G. (1974). Use of response surface methodology in the development of acceptable high protein bread. *Journal of Food Science*, **39**(**5**), 943–946.

Huang, L.X., Kumar, K. & Mujumdar, A.S. (2006). A comparative study of a spray dryer with rotary disc atomizer and pressure nozzle using computational fluid dynamic simulations. *Chemical Engineering and Processing: Process Intensification*, **45**(**6**), 461–470.

Huang, Y., Kangas, L.J. & Rasco, B.A. (2007). Applications of artificial neural networks (ANNs) in food science. *Critical Reviews in Food Science and Nutrition*, **47**(**2**), 113–126.

Hundre, S.Y., Karthik, P. & Anandharamakrishnan, C. (2015). Effect of whey protein isolate and β-cyclodextrin wall systems on stability of microencapsulated vanillin by spray–freeze drying method. *Food Chemistry*, **174**, 16–24.

IDF (2005). Energy use in dairy processing. In: *Bulletin of the International Dairy Federation*, p. 401, International Dairy Federation.

Ishwarya, S.P. & Anandharamakrishnan, C. (2015). Spray-freeze-drying approach for soluble coffee processing and its effect on quality characteristics. *Journal of Food Engineering*, **149**, 171–180.

Ishwarya, S.P., Anandharamakrishnan, C. & Stapley, A.G.F. (2015). Spray-freeze-drying: A novel process for the drying of foods and bioproducts. *Trends in Food Science & Technology*, **41**(**2**), 161–181.

Islam, M.I.-U., Edrisi, M. & Langrish, T. (2013). Improving process yield by adding WPI to lactose during crystallization and spray drying under high-humidity conditions. *Drying Technology*, **31**(**4**), 393–404.

Karthik, P. & Anandharamakrishnan, C. (2013). Microencapsulation of docosahexaenoic acid by spray-freeze-drying method and comparison of its stability with spray-drying and freeze-drying methods. *Food and Bioprocess Technology*, **6**(**10**), 2780–2790.

Khuri, A.I. & Mukhopadhyay, S. (2010). Response surface methodology. *Wiley Interdisciplinary Reviews: Computational Statistics*, **2**(**2**), 128–149.

Kiani, H. & Sun, D.W. (2011). Water crystallization and its importance to freezing of foods: A review. *Trends in Food Science and Technology*, **22**(**8**), 407–426.

Langrish, T.A.G. (2007). New engineered particles from spray dryers: research needs in spray drying. *Drying Technology*, **25**(**6**), 971–983.

Lertworasirikul, S. & Tipsuwan, Y. (2008). Moisture content and water activity prediction of semi-finished cassava crackers from drying process with artificial neural network. *Journal of Food Engineering*, **84**(**1**), 65–74.

LeVeque, R.J. (2002) *Finite Volume Methods for Hyperbolic Problems*, Cambridge University Press, Cambridge.

Li (2006). *Discontinuous finite elements in fluid dynamics and heat transfer*, Springer-Verlag, London.

Liapis, A.I. & Bruttini, R. (1995). Freeze-drying of pharmaceutical crystalline and amorphous solutes in vials: dynamic multidimensional models of the primary and secondary drying stages and qualitative features of the moving interface. *Drying Technology*, **13**(**1–2**), 43–72.

Liapis, A.I. & Sadikoglu, H. (1998). Dynamic pressure rise in the drying chamber as a remote sensing method for monitoring the temperature of the product during the primary drying stage of freeze drying. *Drying Technology*, **16**(**6**), 1153–1171.

Lopez-Quiroga, E., Antelo, L.T. & Alonso, A.A. (2012). Time-scale modeling and optimal control of freeze-drying. *Journal of Food Engineering*, **111**(**4**), 655–666.

Manzini, G. & Putti, M. (2007). Mesh locking effects in the finite volume solution of 2-D anisotropic diffusion equations. *Journal of Computational Physics*, **220**(**2**), 751–771.

Masters, K. (1985) *Spray Drying Handbook*, John Wiley & Sons, New York.

Millman, M.J., Liapis, A.I. & Marchello, J.M. (1985). An analysis of the lyophilization process using a sorption-sublimation model and various operational policies. *AIChE Journal*, **31**(**10**), 1594–1604.

Milton, N., Pikal, M.J., Roy, M.L. & Nail, S.L. (1997). Evaluation of manometric temperature measurement as a method of monitoring product temperature during lyophilization. *PDA Journal of Pharmaceutical Science and Technology*, **51**(**1**), 7–16.

Moore, J.G. (1995) Drum dryers. In: *Handbook of Industrial Drying*, Vol. **1**, 2nd edn (ed. Mujumdar, A.S.), pp. 249–262. Marcel Dekker, New York.

Murphy, M., O'Mahony, M. & Shalloo, L. (2014). Comparison of modelling techniques for milk-production forecasting. *Journal of Dairy*, **97(6)**, 3352–3363.

Nam, J.H. & Song, C.S. (2005). An efficient calculation of multidimensional freeze-drying problems using fixed grid method. *Drying Technology*, **23(12)**, 2491–2511.

Nindo, C.I. & Tang, J. (2007). Refractance window dehydration technology: a novel contact drying method. *Drying Technology*, **25(1)**, 37–48.

Okos, M.R., Narsimhan, G., Singh, R.K. & Weinauer, A.C. (1992). Food dehydration. In: *Handbook of Food Engineering* (eds D.R. Heldman & D.B. Lund), Taylor and Fransis, New York.

Ozdemir, M. & Floros, J.D. (2008). Optimization of edible whey protein films containing preservatives for water vapor permeability, water solubility and sensory characteristics. *Journal of Food Engineering*, **86(2)**, 215–224.

Palzer, S., Dubois, C. & Gianfrancesco, A. (2012). Generation of product structures during drying of food products. *Drying Technology*, **30(1)**, 97–105.

Patel, S.M., Chaudhuri, S. & Pikal, M.J. (2010). Choked flow and importance of Mach I in freeze-drying process design. *Chemical Engineering Science*, **65(21)**, 5716–5727.

Patel, S.M. & Pikal, M.J. (2011). Emerging freeze-drying process development and scale-up issues. *AAPS PharmSciTech*, **12(1)**, 372–378.

Pikal, M.J. (1990). Freeze-drying of proteins Part 1: Process design, *BioPharm*, **3(8)**, 18–27.

Pikal, M.J., Roy, M.L. & Shah, S. (1984). Mass and heat transfer in vial freeze-drying of pharmaceuticals: role of the vial. *Journal of Pharmaceutical Sciences*, **73(9)**, 1224–1237.

Pua, C.K., Hamid, N.S.A., Tan, C.P., Mirhosseini, H., Rahman, R.B.A. & Rusul, G. (2010). Optimization of drum drying processing parameters for production of jackfruit (*Artocarpus heterophyllus*) powder using response surface methodology. *LWT – Food Science and Technology*, **43(2)**, 343–349.

Rajam, R. & Anandharamakrishnan, C. (2015). Spray freeze drying method for microencapsulation of *Lactobacillus plantarum*. *Journal of Food Engineering*, **166**, 95–103.

Ranade, V.V. (2002). *Computational Flow Modeling for Chemical Reactor Engineering*, Academic Press, London.

Rao, S.S. (2009). *Engineering Optimization: Theory and Practice*, John Wiley & Sons, New Jersey.

Rao, R. (2010). *Advanced modeling and optimization of manufacturing processes: International research and development*, Springer-Verlag, London.

Rebouças, M.C., Rodrigues, M.D.C.P. & Afonso, M.R.A. (2014). Optimization of the acceptance of prebiotic beverage made from cashew nut kernels and passion fruit juice. *Journal of Food Science*, **79**(7), S1393–S1398.

Sadikoglu, H. & Liapis, A.I. (1997). Mathematical modelling of the primary and secondary drying stages of bulk solution freeze-drying in trays: Parameter estimation and model discrimination by comparison of theoretical results with experimental data. *Drying Technology*, **15(3–4)**, 791–810.

Scardovelli, R. & Zaleski, S. (1999). Direct numerical simulation of free-surface and interfacial flow. *Annual Review of Fluid Mechanics*, **31(1)**, 567–603.

Searles J. (2004). Observation and implications of sonic water vapour flow during freeze-drying. *American Pharmaceutical Review*, **7**, 58–69.

Sharma, N. & Arora, C. (1995). Influence of product thickness, chamber pressure and heating conditions on production rate of freeze-dried yoghurt. *International Journal of Refrigeration*, **18(95)**, 297–307.

Sheehan, P. & Liapis, A.I. (1998). Modeling of the primary and secondary drying stages of the freeze drying of pharmaceutical products in vials: Numerical results obtained from the solution of a dynamic and spatially multi-dimensional lyophilization model for different operational policies. *Biotechnology and Bioengineering*, **60(6)**, 712–728.

Singh, A.K., Tiwari, S., Singh, R.R.B., Tyagi, R.K. & Arora, S. (2008). Optimization of ingredient levels for manufacturing malted milk beverage using response surface methodology. *International Journal of Dairy Technology*, **61(2)**, 192–198.

Song, C.S. & Kim, C.J. (2003). Freeze-drying of skim milk in a cylindrical container. *Drying Technology*, **21(9)**, 1811–1838.

Song, C.S., Nam, J.H., Kim, C.J. & Ro, S.T. (2002). A finite volume analysis of vacuum freeze drying processes of skim milk solution in trays and vials. *Drying Technology*, **20(2)**, 283–305.

Sonner, C., Maa, Y.F. & Lee, G. (2002). Spray-freeze-drying for protein powder preparation: Particle characterization and a case study with trypsinogen stability. *Journal of Pharmaceutical Sciences*, **91(10)**, 2122–2139.

Straatsma, J., Van Houwelingen, G., Steenbergen, A. E. & De Jong, P. (1999). Spray drying of food products: 1. Simulation model. *Journal of Food Engineering*, **42(2)**, 67–72.

Supprung, P. & Noomhorm, A. (2003). Optimization of drum drying parameters for low amylose rice (kdml105) starch and flour. *Drying Technology*, **21(9)**, 1781–1795.

Tang, J., Feng, H. & Shen, G.-Q. (2003). Drum drying. In: *Encyclopedia of Agricultural, Food, and Biological Engineering* (ed. D.R. Heldman), Marcel Dekker, New York.

Tang, X., Nail, S.L. & Pikal, M.J. (2005). Freeze-drying process design by manometric temperature measurement: Design of a smart freeze-dryer. *Pharmaceutical Research*, **22(4)**, 685–700.

Thomas, M.E.C., Scher, J., Desobry-Banon, S. & Desobry, S. (2004). Milk powders ageing: effect on physical and functional properties. *Critical Reviews in Food Science and Nutrition*, **44(5)**, 297–322.

Tijskens, L.M.M., Konopacki, P. & Simcic, M. (2001). Biological variance in agricultural products. Experimental examples. In: *Proceedings of the 8th International Controlled Atmosphere Research Conference*, 8–13 July 2001, Rotterdam.

Truong, V., Bhandari, B.R. & Howes, T. (2005). Optimization of cocurrent spray drying process for sugar-rich foods. Part II – Optimization of spray drying process based on glass transition concept. *Journal of Food Engineering*, **71(1)**, 66–72.

Tsapis, N., Bennett, D., Jackson, B., Weitz, D.A. & Edwards, D.A. (2002). Trojan particles: large porous carriers of nanoparticles for drug delivery. *Proceedings of the National Academy of Sciences of the United States of America*, **99(19)**, 12001–12005.

Valous, N.A., Gavrielidou, M.A., Karapantsios, T.D. & Kostoglou, M. (2002). Performance of a double drum dryer for producing pregelatinized maize starches. *Journal of Food Engineering*, **51(3)**, 171–183.

Vehring, R. (2008). Pharmaceutical particle engineering via spray drying. *Pharmaceutical Research*, **25(5)**, 999–1022.

Vehring, R., Foss, W.R. & Lechuga-Ballesteros, D. (2007). Particle formation in spray drying. *Journal of Aerosol Science*, **38(7)**, 728–746.

Versteeg, H.K. & Malalasekera, W. (1995). *An Introduction to Computational Fluid Dynamics: The Finite Volume Method*, Pearson Education.

Walton, D.E. & Mumford, C.J. (1999). The morphology of spray-dried particles. *Chemical Engineering Research and Design*, **77**(**5**), 442–460.

Wang, W., Chen, G. & Gao, F. (2005). Effect of dielectric material on microwave freeze drying of skim milk. *Drying Technology*, **23**(**1–2**), 317–340.

Yang, X., Xiao, J., Woo M.-W. & Chen, X.D. (2014). Three-dimensional numerical investigation of a mono-disperse droplet spray dryer: validation aspects and multi-physics exploration. *Drying Technology*, **33**(**6**), 742–756.

Zhang, Z. & Friedrich, K. (2003). Artificial neural networks applied to polymer composites: A review. *Composites Science and Technology*, **63**(**14**), 2029–2044.

Zurada, J.M. (1992). *Introduction to Artificial Neural Systems*, PWS Publishing Company, Boston.

9

Computational Fluid Dynamics Modelling of the Dairy Drying Processes

J. Gimbun[1], *W.P. Law*[1] *and C. Anandharamakrishnan*[2]

[1] *Centre of Excellence for Advanced Research in Fluid Flow, Universiti Malaysia Pahang 26300, Gambang, Pahang, Malaysia*
[2] *Indian Institute of Crop Processing Technology, Ministry of Food Processing Industries, Government of India, Thanjavur 613 005, Tamil Nadu*

9.1 Introduction

Computer simulations of the drying of dairy products are now widely used in food engineering research. They enable the underlying mechanism and performance of the drying process to be predicted, which is useful for design retrofits to improve performance or to optimize the operation of existing equipment. Mathematical equations describing momentum, heat and mass transfer coupled with equilibrium and kinetic equations, which usually form a model for a drying operation, are often solved together. In this chapter, a brief description of the principles of computational fluid dynamics (CFD) for the solution of fluid flows with heat and mass transfer is outlined in the case of spray drying, spray freeze drying and freeze drying. CFD solutions are routinely used to optimize and develop drying equipment as evident in the significant increase in the number of peer-reviewed papers related to CFD applications to drying processes over the years (Figure 9.1). Figure 9.1 shows that the number of papers that specifically analyse freeze and spray-freezing operations using CFD is still relatively small. Current progress and best modelling practices are also reviewed. Finally, future direction and challenges in the modelling of dairy drying are presented.

9.2 Spray drying

9.2.1 Spray-drying process

Spray drying is a well-established method for producing dry powder from liquid feed materials with the aid of hot air. It is widely applied to produce powder-based versions of food, healthcare and pharmaceutical products. Products in powder form have advantages such as longer shelf life and easy of transportation. Thus having an optimized spray-drying process is an important step to ensure good product quality. In most cases, a spray dryer is installed on the endpoint of the processing line. The main advantages of spray drying are rapid drying rates, a wide range of operating temperatures and

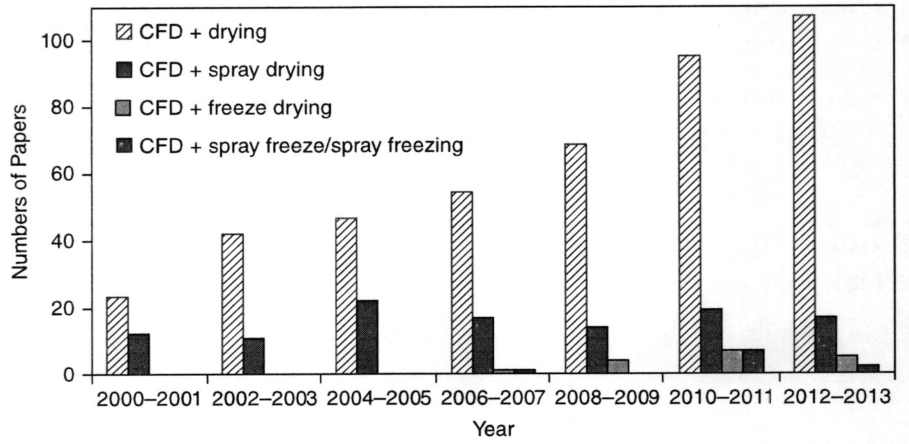

Figure 9.1 Number of papers published on CFD and drying from 2000 to 2013 (www.scopus.com).

short residence times. Simulation of the spray-drying process is vital to understand the underlying mechanism and improve performance, which will result in improvement in the final product quality. A combination of advanced measurement techniques such as particle image velocimetry (PIV), which provides detailed visualization on the flow pattern, temperature and humidity sensors may provide detailed information inside the drying chamber. However, experimental measurement using PIV has an inherent limitation, namely, it is not applicable to an opaque wall. Furthermore, it is not convenient and is potentially dangerous to employ PIV measurement for a large industrial-scale spray dryer operating at a temperature above 400 K. Alternatively, a CFD technique can provide comprehensive information on the fluid flow, heat and mass transfer in a spray dryer. CFD simulation is gaining popularity for modelling spray drying operations because experimental measurements of air flow, temperature, particle size and humidity within the drying chamber are very difficult and expensive to obtain for large-scale dryers (Kuriakose & Anandharamakrishnan 2010).

CFD simulation of spray dryers comprises many submodels, all of which are built following the fluid flow model. Among these submodels are the particle tracking model (which includes particle interaction and deposition), the droplet drying model (which is accounted for by the heat- and mass-transfer model) and the particle quality model. A detailed review of the current numerical practice for CFD modelling of spray drying is presented in this section.

9.2.2 Flow field simulation

All components of the CFD simulation of spray drying are dependent on the air-flow field model. The air-flow pattern in the chamber affects the movement of the particles, which subsequently effects the residence time of the particles, drying rates (i.e. heat and mass transfer) and whether the particles are deposited on the chamber wall or escape through the outlet pipe. Hence, getting an accurate prediction of the flow field in the drying chamber is a prerequisite for the subsequent modelling effort. Most spray dryer CFD simulations are performed using commercial codes such as FLUENT, CFX, STAR-CD or PHOENICS, as well as in-house codes or user defined functions (UDF). The basic CFD equations are well established and will not be repeated here. Details of the CFD model can be found in the textbook *An introduction to computational fluid dynamics: the finite volume method* (Versteeg & Malalasekera 2007) meanwhile details of the model used for spray-drying simulation can be found elsewhere (Birchal *et al.* 2006). Discussion in this chapter focuses on the specific CFD issues when modelling the flow field in the spray-dryer chamber.

9.2.2.1 Steady or unsteady approach

Most of the CFD studies on the spray-drying process (Kieviet 1997; Harvie *et al.* 2001; Huang *et al.* 2004; Li & Zbicinski 2005; Anandharamakrishnan *et al.* 2010) have been performed using a steady state solver. This implies that there are no long-term scale changes in the flow inside the chamber despite the presence of turbulence fluctuations. Kieviet (1997) showed that the steady simulation can predict the averaged velocity, temperature and humidity satisfactorily, but an accurate prediction is not achieved. Furthermore, the spray-dryer geometry studied by Kieviet (1997) is fairly simple and hence the flow field can be assumed to be pseudo-steady over a long time period. This simplistic steady simulation may not be applicable for a complex dryer such as the counter-current spray dryer, which features a turbulent swirling flow. A contrasting effect of swirls in stabilizing or destabilizing the flow field was presented by Harvie *et al.* (2002) and Langrish *et al.* (2004), which may demand an unsteady approach to obtaining an accurate prediction. The transient CFD simulation by Gabites *et al.* (2006) also shows that the air-flow jet does not follow any distinct plane, but flaps through the central axis of the dryer. Despite the previous success of steady simulation in predicting the flow field in a spray dryer, it is theoretically flawed from the fluid dynamics point of view. However, the unsteady solver represents experimental measurements better and should be able to produce a more accurate result. Experimental measurements are often taken as time-averaged quantities and the unsteady solver mimics this situation much better. Moreover, the transient flow provides richer information that may be useful to understand the drying mechanism better.

Experimental work by Southwell & Langrish (2000, 2001) suggests that the internal air-flow pattern exhibits significant transient behaviour. The central core flow tends to fluctuate sidewards, with transient eddies being formed near the wall. This phenomenon usually occurs when the outlet pipe is located inside the drying chamber, whereby the fluid flow is obstructed by the protruding pipe in one side of the dryer whereas the other side is relatively free from any obstruction. The sideward movement of the core flow tends to be in the obstruction-free region, like the one presented by Lebarbier *et al.* (2001). A numerical study by Jin & Chen (2009a), Mezhericher *et al.* (2009), Jongsma *et al.* (2013) and Gimbun *et al.* (2015) also suggests that the transient solver may improve the prediction accuracy of the flow field in the spray-dryer chamber. According to Gimbun *et al.* (2015), the turbulent flow inside the spray dryer can be best resolved by using an unsteady solver rather than a steady solver due to the inherent nature of turbulence, although it should be noted that the numerical requirement of transient simulation is significantly higher than that of the steady solver. This is mainly due to a longer simulation required to attain the pseudo-steady condition, which can be performed by monitoring changes in the velocity, temperature or turbulence stresses over computation time until the changes are no longer significant. Certainly, more CFD and fluid dynamics experimental work is needed on the effect of unsteady simulation to predict the accuracy of various spray-dryer configurations and understand the significance of performing an unsteady simulation.

9.2.2.2 Turbulence modelling

The selection of a turbulence model for spray-drying simulation is very important. Each model has its advantages and disadvantages, mainly in terms of prediction accuracy and computational requirement. It is therefore important to select an optimum turbulence model in simulating the fluid flow in a spray dryer.

The standard $k-\varepsilon$ (where k is the turbulence kinetic energy and ε is the turbulence dissipation rate) model is the simplest turbulence model and the one most commonly used in engineering practice due to its robustness, low computational requirement and reasonable accuracy. It can produce a faster convergence than the Reynolds stress model (RSM) (Kieviet 1997). The standard $k-\varepsilon$ model performs well in simple flow cases but fails to model more complex flow patterns, such as highly swirling flows, which often occur in counter-current spray dryers. This is because the standard $k-\varepsilon$ model does not consider the swirling effects on the flows. Bayly *et al.* (2004) reported that the RSM can give better results for turbulent swirling flows in a counter-current spray dryer. The shortcoming of the standard $k-\varepsilon$ model is attributed to its assumption of isotropic turbulence. Other $k-\varepsilon$ related models, such as the renormalized $k-\varepsilon$ (RNG) and realizable $k-\varepsilon$ (RKE), accommodate the swirling flow model and mathematical function to ensure positivity in turbulence stresses. However, they still use the assumption of isotropic turbulence and hence may not be able to provide a complete solution. Huang *et al.* (2004) tested the performance of various turbulence models, namely, RNG, RSM and RKE, on highly swirling flow induced by a rotating atomizer. Their findings must be treated with caution since no validation of the modelling result was presented with respect to swirling flow. In addition, the RSM is not a fully pledged anisotropic turbulence model since the turbulence dissipation rate is modelled according to an isotropic turbulence analogy. Langrish and co-workers (Langrish *et al.* 2004; Kota & Langrish 2007) studied the shear stress transport (SST) model, which is a hybrid of $k-\varepsilon$ and $k-\omega$ closures, and concluded that the model has the ability to capture fluctuation frequencies relatively well.

The direct numerical simulation (DNS) model is also used for CFD simulation, but it is not convenient for use on industrial flows due to its intensive computational demand. The computational cost of DNS is proportional to the Reynolds number of the flow, thus it is relatively expensive to solve in high Reynolds number flows. The large eddy simulation (LES) model resolves large eddies of the flow field and models the small eddies. LES can predict unsteady turbulent flow in semi-industrial scale spray dryers well (Jongsma *et al.* 2013). However, it requires relatively high computational resources to solve high Reynolds number industrial flows. In addition, it requires large memory and computational time to obtain a stable statistic of flow during simulation (Turnbull & Thompson 2005). The detached eddy simulation (DES) model belongs to a hybrid model which combines the best aspects of Reynolds-averaged Navier–Stokes (RANS) near the wall and LES away from the boundary layer. This model was introduced by Spalart *et al.* (1997) in an effort to reduce the overall computational effort of LES modelling by allowing a coarser grid within the boundary layers. The DES approach was shown to give an excellent prediction of flow field in co-current spray drying by Gimbun *et al.* (2015),

Aspari & Gimbun (2010) and Muhammad & Gimbun (2011). Scale adaptive simulation (SAS) is a relatively new development in turbulence modelling and has the potential for better prediction because it is less sensitive to the computational mesh compared to either LES or DES. A study by Fletcher & Langrish (2009) showed that the SAS technique can capture some of the transient eddies commonly observed in pilot-scale spray-dryer experiments, although no comparison with experimental measurement was presented. Certainly, RANS is not the way forward for simulation of spray dryers, and advanced turbulence models such as the SAS, DES, LES and DNS should be considered to improve prediction accuracy.

9.2.3 Discrete phase modelling

The Euler–Lagrange method is the preferred method for multiphase modelling of a dilute dispersed phase like that in a spray dryer. In fact, the Euler–Lagrange method is suitable for multiphase flows up to 10% of particle loading. The particle source in the cell (PSI-cell) model was first proposed by Crowe *et al.* (1977) and become the basis of the Euler–Lagrange method or discrete phase model (DPM). The flow field is divided into a grid defining computational cells around each grid point. Each computational cell is treated as a control volume for the continuous phase (gas phase) whereas the droplets are treated as source of mass, momentum and energy inside each control volume. The background continuous fluid flow is obtained by solving the single-phase (Eulerian) equations assuming no droplets present. The droplet trajectories, together with size and temperature histories along the trajectories, are calculated using the continuous phase flow field. The source terms for mass, momentum and energy in each cell of the domain are then determined. The source terms are obtained by the Lagrangian equation and are integrated over the time needed to cross the length of the trajectory inside each control volume. The results are scaled by the number of flow rate of droplets associated with this trajectory (Crowe *et al.* 1977). These source terms are included in the subsequent solution of the gas flow field. This approach provides the two-way coupling solution between the fluid flow and the particle since the air-flow affects the momentum and heat transfer of the particle and vice versa. The method continues iteratively to obtain the gas and particle velocity fields. The atomizer produces a range of droplet sizes represented by a number of discrete droplet sizes. Each initial droplet size is associated with one trajectory where the number of droplets is constant, assuming that no coalescence or breakup occurs. The transport equations for the droplets of each size are integrated over time to the gas-flow field to obtain the droplet trajectories, velocities, sizes and temperatures. The calculations continue until the volatile fractions, such as water in the droplets, evaporate completely or until the droplets exit the domain (Papadakis & King 1988). The droplets are often assumed to leave the domain through the exit pipe or if the particle impacts the chamber wall.

In the CFD codes, a combined Eulerian and Lagrangian model is used to obtain particle trajectories by solving the force balance equation as follows:

$$\frac{du_p}{dt} = \frac{18\mu}{\rho_p d_p^2}\frac{C_D\mathrm{Re}}{24}(v - u_p) + g\left[\frac{\rho_p - \rho_g}{\rho_p}\right] \tag{9.1}$$

where v is the fluid phase velocity, u_p is the particle velocity, ρ_g is the density of the fluid, g is gravity and ρ_p is the density of the particle. Re is the particle's slip Reynolds number and C_D is a drag coefficient obtained by assuming a spherical droplet.

The stochastic effects of particle interactions with eddies can also be included. Turbulent particle dispersion is considered in the DPM model as a discrete eddy concept (Langrish & Zbicinski 1994). The turbulent air-flow pattern is assumed to be made up of a collection of randomly directed eddies, each with its own lifetime and size. Particles are injected into the flow domain at the nozzle point, then successively enter and leave these random eddies until they impact the wall or leave the flow domain through the outlet pipe.

The heat- and mass-transfer values between the dispersed and continuous phases are obtained in the same way as interphase momentum exchange. The heat and mass transfer between the particles and the hot gas is calculated as follows:

$$m_p c_p \frac{dT_p}{dt} = hA_p(T_g - T_p) + \frac{dm_p}{dt} h_{fg} \tag{9.2}$$

where m_p is the mass of the particle, c_p is the particle's specific heat, T_p is the particle temperature, h_{fg} is the latent heat, A_p is the surface area of the particle and h is the heat-transfer coefficient.

The heat-transfer coefficient (h) is obtained from the Ranz–Marshall equation as a function of the particle Reynolds number (Re) and the Prandtl number (Pr):

$$Nu = \frac{hd_p}{k_{ta}} = 2 + 0.6(\mathrm{Re}_d)^{1/2}(\mathrm{Pr})^{1/3} \tag{9.3}$$

where Nu is the Nusselt number for the transfer particle, d_p is the particle diameter and k_{ta} is the thermal conductivity of the fluid. The evaporative mass-transfer rate between the gas and the particles is calculated from the following equation:

$$\frac{dm_p}{dt} = -k_c A_p (Y_s^* - Y_g) \tag{9.4}$$

where Y^*_s is the saturation humidity, Y_g is the gas humidity and k_c is the mass-transfer co-efficient.

The boiling rate model is applied when the temperature of the droplet reaches the boiling point and the mass of the droplet exceeds the non-volatile fraction (Kuo 1986):

$$\frac{d(d_p)}{dt} = \frac{4k_{ta}}{\rho_p c_g d_p}(1 + 0.23\sqrt{\mathrm{Re}}) \ln \left[1 + \frac{c_g(T_g - T_p)}{h_{fg}}\right] \tag{9.5}$$

where k_{ta} is the thermal conductivity of the gas and c_g is the heat capacity of the gas (Kuriakose & Anandharamakrishnan 2010).

Transient particle tracking has also been the subject of study in recent research. Particles were injected and tracked until they reached the boundary of the simulation domain in the steady method (Kota & Langrish 2007; Woo *et al.* 2009). Meanwhile, particles were typically tracked and penetrated the flow field following the simulation time step and duration in the transient approach (Jin & Chen 2009b; Mezhericher *et al.* 2009). Transient particle tracking requires a longer simulation time to ensure all tracked particles penetrate the simulation domain. In addition, a long simulation time is needed to develop satisfactory statistics for the particle movement. This method also needs the particle to be injected at each time step, which increases the memory requirement significantly (Woo *et al.* 2010). The is no apparent improvement of the CFD prediction obtained from transient particle tracking from the work by Jin & Chen (2009b) and

Mezhericher *et al.* (2009) despite of enormous computational effort. Hence, transient particle tracking is not a smart computational choice for spray dryers since the improvement gain on accuracy is marginal.

Various types of atomizer, such as pressure nozzle, ultrasonic nozzle, two-fluid nozzle and rotary disk, can be found in spray drying. The choice of atomizer type affects the initial droplet distribution, velocity and spray angle. In most cases, commercial nozzles and rotary disks come with specifications of droplet size distribution and cone angle. However, if the nozzle specification is not available an advanced technique such as particle image velocimetry or phase Doppler anemometry (PDA) may be used (Al-Hakim *et al.* 2006). The initial droplet distribution from the nozzle may also be affected by the viscosity or concentration of the sample. For instance, the viscosities of maltodextrin and whey protein isolate increase proportionally with solid concentration (Pang *et al.* 2014a), therefore experimental data on the effect of the liquid feed viscosity on the initial droplet size are very important. However, correlations for the initial droplet size distribution and the initial velocity for several types of atomizer can be found from Masters (1979), which is useful in the absence of experimental data. CFD simulation cannot be performed without data on the nozzle specification.

The nozzle-type atomizer is modelled by adding an injection point for particles into the domain at a position analogous to the location of the nozzle. In the case of a rotary atomizer, a physical disc with a suitable boundary for the rotational momentum source term must be prepared. The rotating boundary may be prepared using a conformal or non-conformal mesh depending on the type of model chosen, which can be either a frozen rotor or sliding mesh.

Droplets are treated as species in CFD simulation of spray drying (Anandharamakrishnan *et al.* 2010; Gimbun *et al.* 2015). The evaporative species, which often consists of water, is assumed to evaporate in a saturated manner without any falling rate period. If the boiling point is reached, the moisture is assumed to boil off from the particle or droplet leaving only the non-evaporative component. This simple method can produce a good prediction of the velocity, temperature and humidity profile in the spray dryer (Huang *et al.* 2006; Anandharamakrishnan *et al.* 2010; Gimbun *et al.* 2015). Other methods are also used, including the diffusion model (Lo 2005), characteristic drying curve (CDC) (Langrish & Kockel 2001; Woo *et al.* 2008) and reaction engineering approach (REA) (Chen & Xie 1997; Jin & Chen 2009b; Woo *et al.* 2008). According to Woo *et al.* (2010) a good comparison was obtained between CDC and REA prediction with experimental data, although a detailed comparison of CFD prediction of temperature, humidity and velocity profile in conjunction with CDC and REA was not given. Jin & Chen (2009a) only presented a comparison with a point measured temperature at the chamber wall, while Woo *et al.* (2008) presented a limited comparison on outlet temperature and bulk moisture. In contrast, there are many detailed validations of temperature, humidity and gas velocity profile in the literature for the simplified evaporation approach (Huang *et al.* 2006, Anandharamakrishnan *et al.* 2010; Gimbun *et al.* 2015). From the perspective of a CFD simulation, this limited comparison does not warrant full applicability. Moreover, the residence time of a spray dryer is very short and hence the simplistic method of evaporation in a saturated manner is thought to be sufficient.

At present the droplet drying model does not include information about product quality. Dried powder often contains thermally sensitive material such as whey protein, phenolic compound, vitamins or probiotic organisms. Thermal degradation kinetics of the thermoabile material should be implemented in the particle tracking model to ensure prediction of product quality.

9.2.4 Wall deposition and the particle build-up model

In most cases, particles are assumed to stick on contact (trap boundary) at the drying chamber wall (Huang *et al.* 2006; Anandharamakrishnan *et al.* 2010; Gimbun *et al.* 2015). While this assumption may be true for sticky particles (above glass state, hygroscopic material or wet particles), it is not true for all, and especially not for dried particles. Dried particles do not always stick to the chamber wall and a force balance model like the one proposed by Shi & Bayless (2007) may be useful to determine the particle fate. A good wall deposition model should include hygroscopicity, glass transition temperature, particle moisture content and the force balance model. At present, no comprehensive treatment is available in the commercial CFD code to model particle deposition apart from either a simplified trap or an escape boundary. Nevertheless, it is relatively easy to predict the particle impact positions inside the drying chamber using the method outlined by Anandharamakrishnan *et al.* (2010).

Another aspect to consider is the particle build-up on the chamber wall. At present, it is not convenient to model the particle build-up using DPM, but it is possible to use other modelling techniques such as discrete element modelling (DEM). Li & Marshall (2007) successfully demonstrated prediction of particle deposition on fibre using DEM (Figure 9.2). However, the computational effort required to track all particles in a spray dryer via DEM it is still too expensive for a fast practical solution. The particle build-up model should be combined with the particle deposition and particle tracking models to ensure all physics are considered correctly in the computation.

9.2.5 Particle interaction

The droplet–droplet interaction during spray drying has become a subject of interest in recent years. The most commonly used approach to predict particle interaction via two-way coupling only permits momentum exchange between droplets and air flow, in the way that both phases affect each other. Ideally, spray dryer simulation should consider four-way coupling, which considers all the two-way coupling, droplet–droplet interaction and the effect of the local droplet size on the two-phase flow field, as shown in Figure 9.3.

Most authors use the Lagrangian-based droplet collision model (Nijdam *et al.* 2004; Mezhericher *et al.* 2008), although the population balance model (PBM) is also used (Nijdam *et al.* 2004). All work so far has been limited to the aggregation kernel, with the exception of that by Seydel *et al.* (2006), who considered nucleation and growth. Although nucleation and growth are observed in the crystallization process, the formation of amorphous crystalline solid might occur under certain conditions in spray drying (Chiou *et al.* 2007). At present, validation of the model prediction that takes into account droplet–droplet interactions is not available. Previous study has been limited to discussion of the result obtained from the CFD prediction without comparing it to experimental measurement. According to Mezhericher *et al.* (2008), droplet collisions may affect temperature and humidity patterns, but no significant effect on the velocity profile was observed. The humidity is affected by the water evaporation rates from the droplet, which are determined by temperature and droplet size. The surface area

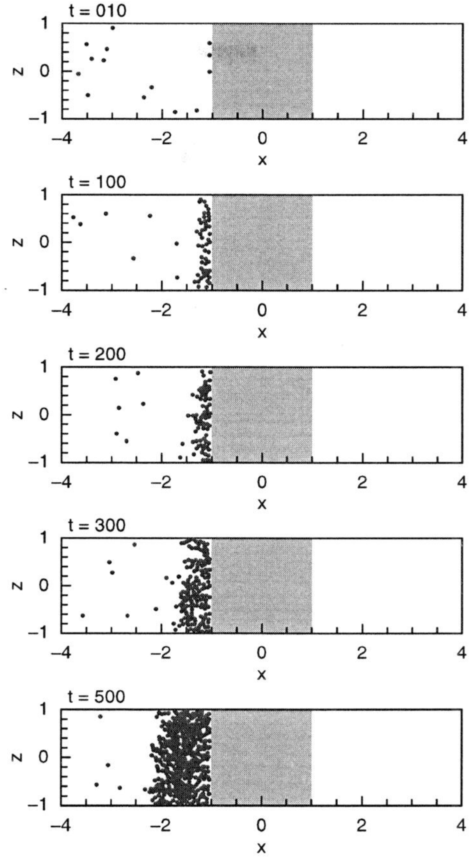

Figure 9.2 Particle deposition on fibre from $t = 10$ s to $t = 500$ s obtained using DEM (Li & Marshall 2007).

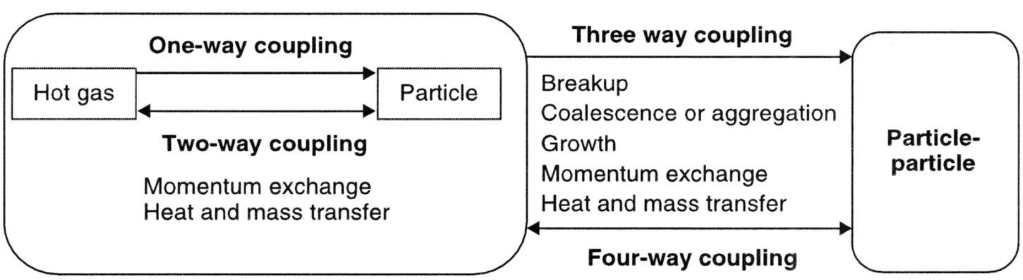

Figure 9.3 Illustration of phase coupling in spray-dryer modelling.

exposed to heat and mass transfer is proportional to the square of the droplet diameter, assuming a spherical shape. Hence, it is not surprising that heat and mass transfer are significantly affected by the inclusion of the droplet interaction model. Meanwhile, the effect of droplet momentum on the gas flow is limited because the droplet dispersion often very dilutes. Moreover, the initial size distribution of droplet is already introduced in most CFD simulations as part of the injection setting of the atomizer. Thus, the effect gained from considering a more complicated droplet interaction model is less significant. However, it will be useful if a full PBM is implemented that can predict the droplet interaction as well as the initial droplet distribution. In this case, the CFD model becomes fully predictive and is less reliant on the availability of experimental data.

An extensive particle interaction model coupled with CFD simulations was employed to predict the final particle size distribution in the EDECAD project (Verdurmen *et al.* 2004). All droplet–droplet, droplet–particle and particle–particle interactions were considered separately. A reasonably good prediction of the particle size distribution in comparison with measured pilot scale experiment was presented. Nijdam *et al.* (2004) modelled the droplet coalescence in the spray chamber using a population balance model solved by the method of classes in a Eulerian framework and a simplified droplet collision model (Lagrangian-based approach). Nijdam *et al.* compared their prediction using PDA measurement and concluded that the Lagrangian approach is preferred for modelling droplet coalescence due to ease of implementation. Nevertheless, the formulation of the coalescence model in the Lagrangian approach is doubtful because the probability of collision for the whole range of particle size distribution is not considered in as rigorous a manner as those in the population balance model (Gimbun *et al.* 2009a). Moreover, all aspects of droplet breakup and coalescence or aggregation should be considered, coupled with the fluid dynamics simulation, to perform a four-way interaction between the fluid flow and droplets.

The dynamic population balance equation for a closed homogeneous system can be written with the diameter as the internal coordinate as follows (Rod & Misek 1982):

$$\frac{\partial n(L)}{\partial t} = \underbrace{\int_L^\infty b(L,\lambda)a(\lambda)n(\lambda)\,\mathrm{d}\lambda}_{\text{birth due to breakage}} + \underbrace{\frac{L^2}{2}\int_0^L \frac{\beta((L^3-\lambda^3)^{1/3},\lambda)\,n((L^3+\lambda^3)^{1/3})\,n(\lambda)}{(L^3-\lambda^3)^{2/3}}\,\mathrm{d}\lambda}_{\text{birth due to coalescence or aggregation}}$$
$$-\underbrace{a(L)n(L)}_{\text{death due to breakage}} - \underbrace{n(L)\int_0^\infty \beta(L,\lambda)n(\lambda)\,\mathrm{d}\lambda}_{\text{death due to coalescence or aggregation}} + \underbrace{\frac{\partial(G(L)n(L))}{\partial L}}_{\text{growth}} + \underbrace{\delta(L_0,L)B}_{\text{nucleation}} \tag{9.6}$$

where β, a, G, B, b, L_0 and δ are the aggregation kernel, breakage kernel, growth rate, nucleation rate, daughter particle size distribution, size of the nuclei and Dirac delta function, respectively, whereas both L and λ are the particle characteristic length. The solution of the population balance equation (Eq. 9.6) may be obtained using the simultaneous quadrature method of moment, quadrature method of moment or method of classes (Gimbun *et al.* 2009b; Gimbun 2009). In the case of spray dryer, the aggregation or coalescence, growth and breakage kernels should be considered. Aggregation is important because droplets may coalesce upon collision. Meanwhile, growth occurs due to droplet expansion during the drying process to form a larger hollow particle. Particle expansion is often followed by shrinkage towards the end of the drying stage (Pang *et al.* 2014a), and the shrinking mechanism should be included in the growth kernel. Particle breakage may occur in some cases, as observed in Figure 9.4, when the material is not elastic enough to expand during drying (Pang *et al.* 2014b), although particle-shattering collision is not common. It can be seen from Figure 9.4 that phenolic compounds are susceptible to breakage without the presence of a carrier such as whey protein isolate, which act as a plasticizer. All the three mechanisms must be included in the PBM to

(a) (b)

Figure 9.4 Surface morphology of a spray-dried particle containing phenolic compounds from *Orthosiphon stamineus*: (a) with the presence of whey protein isolate as carrier and (b) without the presence of carrier (Pang *et al.* 2014b).

obtain a physically correct and descriptive model. At present, very limited data are available for validation of the particle interaction model and hence experimental work using an advanced method such as PDA or PIV is necessary in this area.

9.2.6 Validation and issues of CFD simulation

A relatively new development in CFD modelling must be validated thoroughly before it can routinely use for spray-dryer simulation. However, local measurement of the flow field and particle information in a spray dryer is challenging because it operates at high temperature (often above 400 K) and high humidity. Moreover, spray dryers are often made of an opaque material such as stainless steel, rendering use of laser-based techniques impossible. For this reason, research-dedicated spray dryers in research institutes or universities often featured several glass windows to enable PDA and laser Doppler anemometer (LDA) measurement. High-quality data on spray dryers suitable for CFD validation are few and far between, for example the measurements by Kieviet (1997), Bayly *et al.* (2004) and Southwell & Langrish (2001). Care should be taken when analysing results from CFD simulation to avoid mistakes that may occur as a result of inexperienced users, suitability of the grid used, discretization and convergence. It is also important to make sure that all the correct physics has been considered.

9.3 Freeze drying

Freeze drying is an important drying technology for highly heat-sensitive foods, pharmaceuticals and biochemicals because it can preserve most of biological activity, flavour, aroma and quality. However, freeze drying is an order of magnitude more expensive than spray drying because it needs refrigeration, vacuum and long running times (Liapis & Bruttini 1995). Freeze drying involves freezing the solvent in a material at low temperatures followed by ice sublimation from the solid state directly into the vapour state under vacuum. Freeze drying of an aqueous solution involves three steps: freezing (solidification), ice sublimation (primary drying) and removal of unfrozen solvent vapour (secondary drying).

The freeze-drying process has been widely studied experimentally and numerically. As discussed in the spray-drying section, an advanced technique such as PIV has its limitations as it is not applicable for an opaque system as well as being costly to set up. Furthermore, measurement by inserting the thermal sensor into the freeze dryer could affect nucleation and ice crystal growth (Fissore *et al.* 2009). Hence, a numerical method is useful to study the mechanism and performance of the freeze-drying process. Currently, numerical studies on freeze drying are performed using various numerical tools, for example FEMLAB, COMSOL, FLUENT and MATLAB (Hottot *et al.* 2007; Trelea *et al.* 2009; Bogdani *et al.* 2013; Ganguly *et al.* 2013). CFD is gaining popularity in predicting the freeze-drying process (Figure 9.1), although the number of papers specifically addressing freeze drying are relatively small. Freeze drying is a complex process that includes a solidification, sublimation and condensation cycle. It is important to understand the freeze-drying cycle since it may significantly affect the quality of the final product. Hence, this section discussed the progress in modelling of heat and mass transfer, as well as primary and secondary drying.

9.3.1 Modelling of freeze drying

9.3.1.1 Mass and heat-transfer modelling

The performance of a freeze dryer and the properties of the freeze-dried product depend on the heat- and mass-transfer process. A poor transfer of mass and heat may cause non-optimal freezing of liquid and sublimation of ice, thus producing an unsatisfactory quality of freeze-dried product. For instance, a large heat and mass transfer between the frozen sample and the drying medium increases the retention of volatile aromatic compounds in food (Kudra & Mujumdar 2009). Hence, a good understanding of mass and heat transfer in freeze dryers is a prerequisite for a subsequent modelling effort.

Mass- and heat-transfer mechanisms in the freeze-drying process are influenced by several factors, for example freeze-dryer configuration, operating conditions and the inlet air condition (Abdelwahed *et al.* 2006; Cheng *et al.* 2014). Recently, CFD simulation has been employed to model the heat and mass transfer in a freeze dryer (Li *et al.* 2007; Ganguly *et al.* 2013; Cheng *et al.* 2014). Recent work by Ganguly *et al.* (2013) indicated that the air-flow rate is significantly affected by the valve-baffle design in the freeze dryer. They reported that the heat-transfer rate increased proportionally with the inlet gas-flow rate. They suggested that a higher temperature gradient between the shelf and frozen materials can be obtained by increasing the shelf temperature and reducing the total gas pressure, which accelerates the heat- and mass-transfer rate. George & Datta (2002) reported that the freeze-drying process is mainly controlled by the mass-transfer mechanism as the resistance of mass transfer is much larger than the resistance of heat transfer. In their work George & Datta experimentally observed that a drying time of 240 min is needed to obtain 83% moisture removal compared to just 2.7 min when the process is assumed to solely depend on heat transfer. A better prediction was obtained when the mass transfer was considered, but this model is not fully predictive since the value for mass diffusivity changed over time and can only be obtained from analysis of experimental data (George & Datta 2002). Cheng *et al.* (2014) employed a laminar viscous model to evaluate the effect of freeze-dryer design on heat-transfer performance.

They found that the heat-transfer rate and distribution vary with different inlet and outlet duct locations. However, no validation is presented in their work. Li *et al.* (2007) performed a CFD simulation on the freeze drying of apple, but no result specific to heat and mass transfer was presented.

9.3.1.2 Primary drying modelling

The primary drying or sublimation process is the second stage after the freezing of material in a freeze-drying operation. This process occurs as a result of heat conducted to the sublimation interface through the dried and frozen layers. The sublimation process involves phase change from a solid state directly into a vapour state. The sublimed water vapour is then transported by convection and diffusion through the porous dried layer to the vacuum chamber and finally collected in the condenser.

The sublimation model was developed in order to predict the drying in this stage using the following assumptions (McCulloch & Sunderland 1970; Hill & Sunderland 1971; Liapis & Bruttini 2014):

i. One-dimensional heat and mass flows perpendicular to the surfaces and sublimation interface.
ii. Mass and heat transfer are solved by pseudo steady-state equations.
iii. There is no heat loss because of the insulated freeze-dryer wall.
iv. Sublimation of ice at the sublimation interface is parallel to the shelf.
v. The concentration of water vapour is in equilibrium with the ice at the sublimation interface.
vi. The solid matrix and gas are in thermal equilibrium at the porous region.
vii. There is uniform distribution of frozen and unfrozen water over the product thickness.

Liapis & Bruttini (2014) provided a detail description of the freeze-drying model for both the primary and secondary drying. The model considers the energy balance in the dried and frozen layers, as well as the physical properties of both gas and solid. The dried layer is distinguished by assuming a porous structure modelled according to Darcy's law. Radiation heat transfer at the upper dried surface is also considered. The model can predict the concentration of bound water with time as well as the material balance at interface that takes into account the moving boundary due to the sublimation process. Most of the previous modelling attempt on freeze drying was performed using a general tool such as Mathematica, Matlab, the finite element method (COMSOL, FEMLAB) or the finite volume method (FLUENT). A CFD simulation of the sublimation process had been studied by Hottot *et al.* (2007) and Bogdani *et al.* (2013) to evaluate the dependency of the sublimation process on the operating parameters. They reported an excellent prediction of the sublimation rate. They found that the sublimation rate is greatly enhanced when the shelf temperature and total gas pressure increase. However, a mild shelf temperature is preferred to prevent thermal degradation of the product. A well-controlled shelf temperature and total gas pressure may improve the sublimation rate. Borgognoni *et al.* (2012) modelled the freeze drying of bovine pericardium, focusing on the effect of chamber temperature and pressure. They showed that the temperature profile decreased slightly as the chamber pressure increased, resulting in a longer drying time and a lower sublimation rate. They reported that good agreement was obtained between the simulation and experimentally measured temperature and drying time. Li *et al.* (2007) modelled the freeze drying of apple using the CFD approach. They used the uniformly retreating ice front (URIF) model developed by Sheehan *et al.* (1998) to simulate the ice–vapour interface during the drying stage and reported a satisfactory prediction of drying rate compared to experimental measurement. Schoen *et al.* (1995) investigated the primary drying time and temperature of the product vials during the primary drying phase. The prediction on drying time and temperature profile showed good agreement with the experimental measurement. Trelea *et al.* (2009) presented modelling of the freeze drying of pellets and reported a good prediction of temperature and partial vapour pressure. They showed that the change in internal structure of the freeze-dried product may affect the primary drying rate. During the sublimation of ice, the sublimation interface moves toward the pellet core (i.e. the frozen region), the frozen region area decreases and the heat- and mass-transfer resistance increases. Hence, a dramatic drop in the initial drying rate was observed as the sublimation process continued. During the drying of the frozen product, conduction is the dominant of total heat transfer, followed by the radiation (Hottot *et al.* 2006; Bogdani *et al.* 2013). The study by Hottot *et al.* (2007) also stated that the sublimation process is controlled by both conductive and radiative heat-transfer mechanisms. However, the primary drying modelling by Borgognoni *et al.* (2012) only considers the radiation mechanism due to the relatively low pressure in the freeze dryer.

At present, the sublimation model assumes a uniformly retreating ice front and in most cases the non-uniform surface present after solvent sublimation is not considered. This non-uniform surface affects the amount of heat- and mass-transfer area available, which is an important aspect of the model. A good drying model should consider precisely the surface of the porous solid and the available area for heat and mass transfer. This can be realized by performing a three-dimensional CFD simulation with a moving boundary that represents the solvent sublimation. The formation of the solid surface should also include the concentration of non-evaporating species in the product material.

9.3.1.3 Secondary drying modelling

The secondary drying state involves removal of the residual water vapour in the system after the sublimation process. The residual water vapour may affect the properties of the final product, therefore further drying is necessary once the primary drying stage has been completed. For instance, Chang & Patro (2004) found that the residual moisture content is the major factor that causes protein denaturation during the secondary drying process. Typically, a low moisture content (1% or less) is preferred to prolong the product shelf-life (Tang & Pikal 2004).

The mathematical model for the secondary drying process is described in detail by Liapis & Bruttini (2014). The model assumes no frozen layer and the dried layer is treated as porous material, modelled according to Darcy's law. The model consists of energy and material balance for water vapour, inert gas and dried material. Radiation heat transfer at the upper dried surface is also considered. Boss *et al.* (2004) performed a secondary drying model for skim milk and coffee to predict the residual moisture content profile. The predicted moisture content as a function of drying time was in good agreement with the experimental measurement. Sadikoglu & Liapis (1997) modelled the secondary drying process of skim milk. They considered three mass-transfer mechanisms, that is, residual water removal, bulk diffusion and Knudsen diffusion, and reported excellent agreement between the experimental data and the predicted residual moisture content.

The mathematical model for freeze drying is well established, but it has still not been implemented in a commercial CFD code such as FLUENT. The present model is simplified by assuming a porous dried material is characterized by its porosity. A fully predictive model should consider a three-dimensional simulation with a moving boundary modelled using a deforming mesh to account for the process of water sublimation. In this case, the surface area available for heat and mass transfer can be deduced accurately, which may lead to a more accurate prediction of the freeze drying process. Other than commercial CFD tools such as COMSOL and FLUENT, there is also a dedicated freeze drying software such as the PASSAGE from Technalysis Inc., which can perform a primary and secondary drying simulation.

9.4 Spray freeze drying

Spray freeze drying, as illustrated in Figure 9.5, combines two different drying processes, spray drying and freeze drying. It is an emerging technology for drying high-value products, especially for pharmaceutical and biochemical uses. The first step in spray freeze drying is to prepare frozen droplets, which can be achieved by atomizing the material in a cryogenic medium such as liquid nitrogen or a very cold gas stream. Two-fluid nozzles or pneumatic nozzles, similar to the nozzles used in spray drying, are used in spray freeze drying for atomization. The frozen droplet is taken out into the vacuum fluidized-bed dryer or ordinary vacuum freeze dryer for further drying. The free solvent (water) content of the material is removed by sublimation and bound water content is removed by desorption during the primary and secondary drying stages, in a similar manner to that of conventional freeze drying. The resultant products from spray freeze drying are very high-quality spherical porous particles in a wide range of sizes. Spray freeze drying is often used to produce relatively large porous micro/nanoparticles, which are ideal for pulmonary drug delivery. Unlike spray drying, modelling of spray freeze drying is highly complicated, therefore only a handful of attempts have been made to

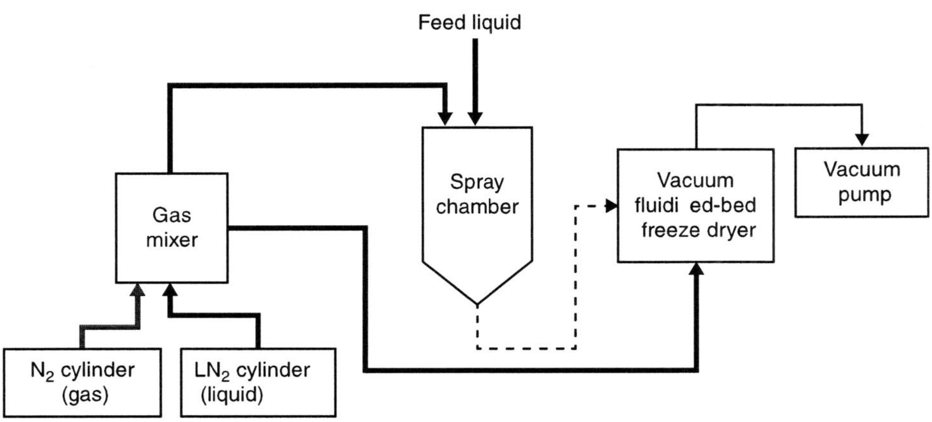

Figure 9.5 Block diagram of a spray freeze dryer (Anandharamakrishnan 2008).

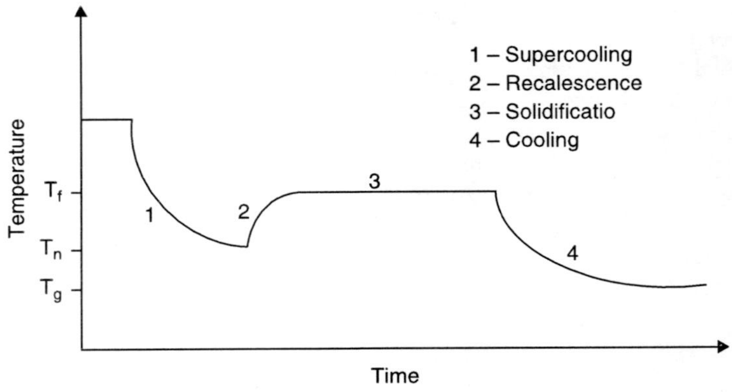

Figure 9.6 Temperature profile of the four-stage freezing process of a droplet.

simulate spray freeze dryers (Henczka *et al.* 2006; Song & Yeom 2009; Tanner *et al.* 2009; Anandharamakrishnan *et al.* 2010; Ivanov *et al.* 2011).

Anandharamakrishnan *et al.* (2010) employed a method similar to that applied for spray drying with turbulence and particle tracking except they omitted the evaporation model. The multistage droplet freezing process shown in Figure 9.6 also included using a piecewise linear function. According to Hindmarsh *et al.* (2003), water experiences supercooling before the solidification process starts. The freezing process of a droplet has four stages: supercooling, recalescence, solidification and cooling. The droplet cools from the initial temperature to the nucleation temperature (T_n), which lies below the equilibrium freezing temperature (T_f) during the supercooling stage. Rapid crystal growth releases latent heat during the recalescence stage, which results in a temperature increase until the droplet reached its equilibrium freezing temperature (T_f). The next process is a freezing stage in which crystal growth takes place at the constant equilibrium freezing temperature (T_f) until the droplet is completely solidified. Finally, during the cooling stage, the solidified droplet cools down to the temperature of the ambient gas (T_g). Hindmarsh *et al.* (2003) reported a satisfactory prediction of the temperature profile in the chamber and particle velocity. In the case of spray freeze drying, the frozen droplet often sticks to the chamber wall whenever a contact is made, based on our experimental observation while operating a pilot-scale spray freeze drying at Loughborough University (Gimbun 2009). Hence the assumption of stick-upon-contact (trap boundary) for the particle deposition model is valid. The CFD prediction of the particle impact positions is also in good agreement with the photographic evidence of the real SFD operation (Figure 9.7).

Tanner *et al.* (2009) also performed a CFD simulation of spray freezing using a KIVA-3 code. They implemented the three-stage model (supercooling, solidification and cooling) and reported good prediction of the average drop size. However, they did not present a comparison on particle velocity and temperature profile. Similarly, Henczka *et al.* (2006) modelled the spray-freezing process and reported satisfactory prediction of mean particle size, but no other validation was presented. Song & Yeom (2009) presented a one-dimensional model for spray freeze drying and predicted ice saturation and mass flux of water vapour without comparison to experimental measurement.

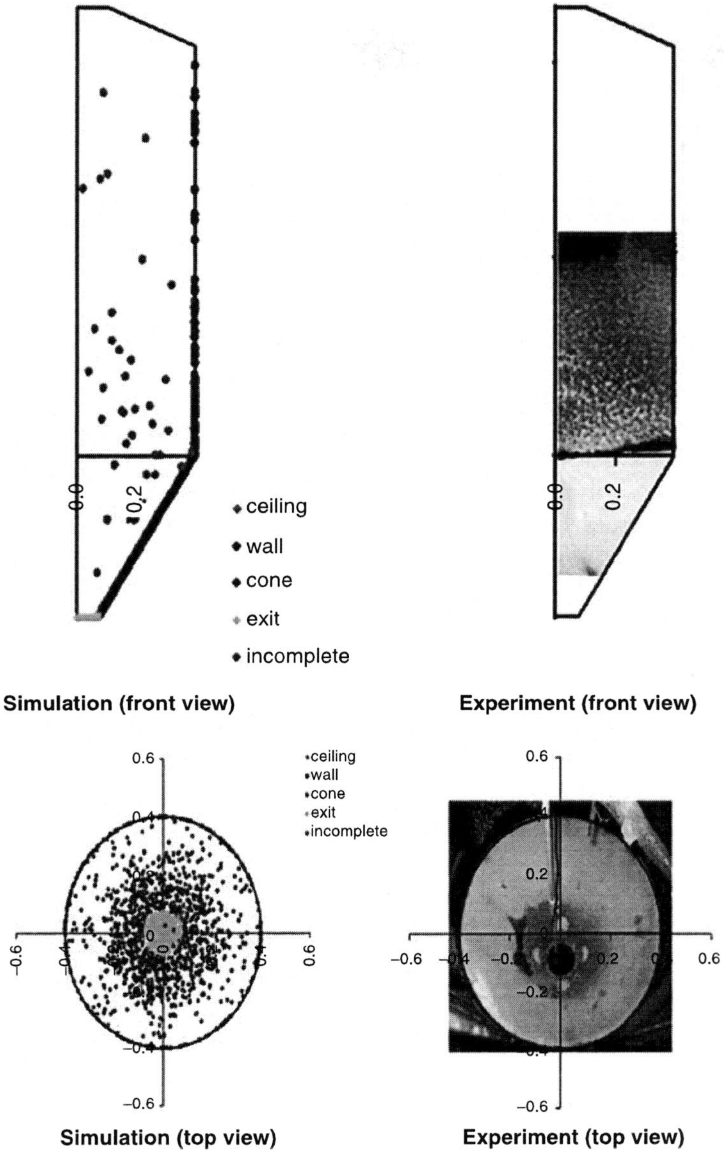

Figure 9.7 CFD simulation and experimental observation of the particle impact position on SFD walls (Anandharamakrishnan 2008).

At present no complete CFD modelling of spray freeze drying is available in the literature, but the drying of the frozen droplet is essentially similar to that presented for freeze drying in section 9.3. Among other issues for spray freeze drying modelling are the absence of a model for particle interaction and particle deposition. A fully predictive model should incorporate DEM, the population balance model and the droplet freezing model as well as the primary and secondary drying models of the frozen droplet.

9.5 Conclusions and future scope

Modelling and simulation of the drying process are challenging, although steady progress made in recent years. The CFD model for spray drying is relatively well developed and can be performed with reasonable accuracy using commercial CFD codes such as FLUENT and CFX. Based on the present capabilities of the CFD models, CFD can play an important role in addressing both design issues and troubleshooting in industrial spray dryers. However, further improvement can be made by including the particle interaction and particle deposition model. A fully predictive model should incorporate DEM, a population balance model and a product quality model. A relatively new turbulence model, such as DES or SAS, must be validated further before it can be routinely used with confidence. More work on flow stability or instability in the spray-dryer chamber of various configurations is also needed both experimentally and numerically. A suitable model to account for crystallization in spray dryers may be included within the population balance model. Limited work in the literature is currently available on CFD modelling of crystallization in a spray dryer.

The mathematical model for freeze drying is well developed, but it has still not been implemented in a commercial CFD code such as FLUENT. At present, most of the modelling work assumes a one-dimensional homogeneous system, which is not fully descriptive. The present model is simplified by simply assuming that a porous dried material is characterized by its porosity. A fully predictive model should consider a three-dimensional simulation with a moving boundary due to water sublimation modelled using a moving or deforming mesh. In this case, the surface area available for heat and mass transfer can be deduced precisely, which may lead to a more accurate prediction.

At present no complete three-dimensional CFD modelling of spray freeze drying is available. Among other issues for spray freeze drying modelling, the absence of a model for particle interaction and particle deposition is the major hurdle. A fully predictive model should incorporate DEM, the population balance model and the droplet freezing model as well as the primary and secondary drying models of the frozen droplet.

References

Abdelwahed, W., Degobert, G., Stainmesse, S. & Fessi, H. (2006). Freeze-drying of nanoparticles: formulation, process and storage considerations. *Advanced Drug Delivery Reviews*, **58**(**15**), 1688–1713.

Al-Hakim, K., Wigley, G. & Stapley. A.G.F. (2006). Phase Doppler anemometry studies of spray freezing. *Chemical Engineering Research and Design*, **84**(**12**), 1142–1151.

Anandharamakrishnan, C. (2008). *Experimental and computational fluid dynamics studies on spray-freeze-drying and spray-drying of proteins*, PhD thesis, Loughborough University.

Anandharamakrishnan, C., Gimbun, J., Stapley, A.G.F. & Rielly, C.D. (2010). A study of particle histories during spray drying using computational fluid dynamic simulations. *Drying Technology*, **28**(**5**), 566–576.

Aspari, S.F. & Gimbun, J. (2010). CFD simulation of hydrodynamics in counter-current spray dryer tower. In *ICPEAM 2010, Kuala Lumpur*.

Bayly, A.E., Jukes, P., Groombridge, M. & McNally, C. (2004). Airflow patterns in a counter-current spray drying tower-simulation and measurement. In: *Proceedings of the 14th International Drying Symposium*, pp. 775–781.

Birchal, V.S., Huang, L., Mujumdar, A.S. & Passos, M.L. (2006). Spray dryers: modeling and simulation. *Drying Technology*, **24(3)**, 359–371.

Bogdani, E., Vessot, S. & Andrieu, J. (2013). Experimental and modeling study during freeze drying of aqueous tert-butanol–based formulations in vials. *Drying Technology*, **31(15)**, 1772–1779.

Borgognoni, C.F., Bevilacqua, J.D.S. & Pitombo, R.N.D.M. (2012). Freeze-drying microscopy in mathematical modeling of a biomaterial freeze-drying. *Brazilian Journal of Pharmaceutical Sciences*, **48(2)**, 203–209.

Boss, E.A., Maciel Filho, R. & de Toledo, E.C.V. (2004). Freeze drying process: real time model and optimization. *Chemical Engineering and Processing: Process Intensification*, **43**(12), 1475–1485.

Chang, B.S. & Patro, S.Y. (2004). Freeze-drying process development for protein pharmaceuticals. *Lyophilization of Biopharmaceuticals*, **2**, 113.

Chen, X.D. & Xie, G.Z. (1997). Fingerprints of the drying behaviour of particulate or thin layer food materials established using a reaction engineering model. *Food and Bioproducts Processing*, **75(4)**, 213–222.

Cheng, H.P., Tsai, S.M. & Cheng, C.C. (2014). Analysis of heat transfer mechanism for shelf vacuum freeze-drying equipment. *Advances in Materials Science and Engineering*. doi:10.1155/2014/515180.

Chiou, D., Langrish, T.A.G. & Braham, R. (2007). Partial crystallization behavior during spray drying: simulations and experiments. *Drying Technology*, **26(1)**, 27–38.

Crowe, C.T., Sharma, M.P. & Stock, D.E. (1977). The particle-source-in cell (PSI-CELL) model for gas-droplet flows. *Journal of Fluids Engineering*, **99(2)**, 325–332.

Fissore, D., Pisano, R., Velardi, S., Barresi, A. & Galan, M. (2009). PAT tools for the optimization of the freeze-drying process. *Pharmaceutical Engineering*, **29(5)**, 58–70.

Fletcher, D.F. & Langrish, T.A.G. (2009). Scale-adaptive simulation (SAS) modelling of a pilot-scale spray dryer. *Chemical Engineering Research and Design*, **87(10)**, 1371–1378.

Gabites, J.R. Abrahamson, J. & Winchester, J.A. (2006). Air flow patterns in an industrial milk powder spray dryer. In: *Proceedings of the Fifth International Conference on CFD in the Process Industries*, Melbourne, Australia, pp. 13–15.

Ganguly, A., Alexeenko, A.A., Schultz, S.G. & Kim, S.G. (2013). Freeze-drying simulation framework coupling product attributes and equipment capability: Toward accelerating process by equipment modifications. *European Journal of Pharmaceutics and Biopharmaceutics*, **85(2)**, 223–235.

George, J.P. & Datta, A.K. (2002). Development and validation of heat and mass transfer models for freeze-drying of vegetable slices. *Journal of Food Engineering*, **52(1)**, 89–93.

Gimbun, J. (2009). *Scale-up of gas-liquid stirred tanks using coupled computational fluid dynamics and population balance modelling*, PhD thesis, Loughborough University.

Gimbun, J., Rielly, C.D. & Nagy, Z.K. (2009a). Modelling of mass transfer in gas–liquid stirred tanks agitated by Rushton turbine and CD-6 impeller: A scale-up study. *Chemical Engineering Research & Design*, **87(4)**, 437–451.

Gimbun, J., Rielly, C.D. & Nagy, Z.K. (2009b). Simultaneous quadrature method of moments for the solution of population balance equations, using a differential algebraic equation framework. *Industrial & Engineering Chemistry Research*, **48(16)**, 7798–7812.

Gimbun, J., Muhammad, N.I.S. & Law, W.P. (2015). Unsteady RANS and detached eddy simulation of the multiphase flow in a co-current spray drying. *Chinese Journal of Chemical Engineering*, **23**(**9**), 1421–1428.

Harvie, D.J.E., Langrish, T.A.G. & Fletcher, D.F. (2001). Numerical simulations of gas flow patterns within a tall-form spray dryer. *Chemical Engineering Research and Design*, **79**(**3**), 235–248.

Harvie, D.J.E., Langrish, T.A.G. & Fletcher, D.F. (2002). A computational fluid dynamics study of a tall-form spray dryer. *Food and Bioproducts Processing*, **80**(**3**), 163–175.

Henczka, M., Bałdyga, J. & Shekunov, B.Y. (2006). Modelling of spray-freezing with compressed carbon dioxide. *Chemical Engineering Science*, **61**(**9**), 2880–2887.

Hill, J.E. & Sunderland, J.E. (1971). Sublimation-dehydration in the continuum, transition and free-molecule flow regimes. *International Journal of Heat and Mass Transfer*, **14**(**4**), 625–638.

Hindmarsh, J.P., Russell, A.B. & Chen, X.D. (2003). Experimental and numerical analysis of the temperature transition of a suspended freezing water droplet. *International Journal of Heat and Mass Transfer*, **46**(**7**), 1199–1213.

Hottot, A., Peczalski, R., Vessot, S. & Andrieu, J. (2006). Freeze-drying of pharmaceutical proteins in vials: modeling of freezing and sublimation steps. *Drying Technology*, **24**(**5**), 561–570.

Hottot, A., Andrieu, J. & Vessot, S. (2007). Sublimation kinetics during freeze-drying of pharmaceutical protein formulation. *Drying Technology*, **25**(**5**), 753–758.

Huang, L., Kumar, K. & Mujumdar, A.S. (2004). Simulation of a spray dryer fitted with a rotary disk atomizer using a three-dimensional computional fluid dynamic model. *Drying Technology*, **22**(**6**), 1489–1515.

Huang, L.X., Kumar, K. & Mujumdar, A.S. (2006). A comparative study of a spray dryer with rotary disc atomizer and pressure nozzle using computational fluid dynamic simulations. *Chemical Engineering and Processing: Process Intensification*, **45**(**6**), 461–470.

Ivanov, S., Troyankin, A., Gurikov, P., Kolnoochenko, A. & Menshutinaa, N. (2011). 3D Cellular automata for modeling of spray freeze drying process. *Computer Aided Chemical Engineering*, **29**, 136–140.

Jin, Y. & Chen, X.D. (2009a). Numerical study of the drying process of different sized particles in an industrial-scale spray dryer. *Drying Technology*, **27**(**3**), 371–381.

Jin, Y. & Chen, X.D. (2009b). A three-dimensional numerical study of the gas/particle interactions in an industrial-scale spray dryer for milk powder production. *Drying Technology*, **27**(**10**), 1018–1027.

Jongsma, F.J., Innings, F., Olsson, M. & Carlsson, F. (2013). Large eddy simulation of unsteady turbulent flow in a semi-industrial size spray dryer. *Dairy Science & Technology*, **93**(**4–5**), 373–386.

Kieviet, F.G. (1997). *Modelling quality in spray drying*, Laboratory of Separation Processes and Transport Phenomena, Department of Chemical Engineering, Eindhoven University of Technology.

Kota, K. & Langrish, T. (2007). Prediction of deposition patterns in a pilot-scale spray dryer using computational fluid dynamics (CFD) simulations. *Chemical Product and Process Modeling*, **2**(**3**).

Kudra, T. & Mujumdar, A.S. (2009). *Advanced Drying Technologies*. CRC Press.

Kuo, K.K. (1986). *Principles of Combustion*, Wiley.

Kuriakose, R. & Anandharamakrishnan, C. (2010). Computational fluid dynamics (CFD) applications in spray drying of food products. *Trends in Food Science & Technology*, **21**(**8**), 383–398.

Langrish, T.A.G. & Kockel, T.K. (2001). The assessment of a characteristic drying curve for milk powder for use in computational fluid dynamics modelling. *Chemical Engineering Journal*, **84**(**1**), 69–74.

Langrish, T.A.G. & Zbicinski, I. (1994). The effects of air inlet geometry and spray cone angle on the wall deposition rate in spray dryers. *Chemical Engineering Research & Design*, **72**(**A3**), 420–430.

Langrish, T.A.G., Williams, J. & Fletcher, D.F. (2004). Simulation of the effects of inlet swirl on gas flow patterns in a pilot-scale spray dryer. *Chemical Engineering Research and Design*, **82**(**7**), 821–833.

Lebarbier, C., Kockel, T.K., Fletcher, D.F. & Langrish, T.A.G. (2001). Experimental measurement and numerical simulation of the effect of swirl on flow stability in spray dryers. *Chemical Engineering Research and Design*, **79**(**3**), 260–268.

Li, S.Q. & Marshall, J.S. (2007). Discrete element simulation of micro-particle deposition on a cylindrical fiber in an array. *Journal of Aerosol Science*, **38**(**10**), 1031–1046.

Li, X. & Zbicinski, I. (2005). A sensitivity study on CFD modeling of co-current spray drying process. *Drying Technology*, **23**(**8**), 1681–1691.

Li, S., Stawczyk, J. & Zbicinski, I. (2007). CFD model of apple atmospheric freeze drying at low temperature. *Drying Technology*, **25**(**7–8**), 1331–1339.

Liapis, A.I. & Bruttini, R. (1995). Freeze drying. In: *Handbook of Industrial Drying* (eds Mujumdar, A.S. & Menon, A.S.), pp. 309–344, CRC Press, New York.

Liapis, A.I. & Bruttini, R. (2014). Freeze drying. In: *Handbook of Industrial Drying* (eds Mujumdar, A. S.), 4th edn, pp. 259–282, CRC Press, New York.

Lo, S. (2005). Application of computational fluid dynamics to spray drying. *Le Lait*, **85**(**4–5**), 353–359.

Masters, K. (1979). *Spray Drying Handbook*, 3rd edn, George Godwin, UK.

McCulloch, J.O.H.N. & Sunderland, J.E. (1970). Integral techniques applied to sublimation drying with radiation boundary condition. *Journal of Food Science*, **35**(**6**), 834–838.

Mezhericher, M., Levy, A. & Borde, I. (2008). Droplet–droplet interactions in spray drying by using 2D computational fluid dynamics. *Drying Technology*, **26**(**3**), 265–282.

Mezhericher, M., Levy, A. & Borde, I. (2009). Modeling of droplet drying in spray chambers using 2D and 3D computational fluid dynamics. *Drying Technology*, **27**(**3**), 359–370.

Muhammad, N.I.S. & Gimbun, J. (2011). Detached eddy simulation (DES) of co-current spray dryer. In: *International Conference on Chemical Innovation 2011 (ICCI2011), Terengganu*.

Nijdam, J.J., Guo, B., Fletcher, D.F. & Langrish, T.A.G. (2004). Challenges of simulating droplet coalescence within a spray. *Drying Technology*, **22**(**6**), 1463–1488.

Pang, S.F., Yusoff, M.M. & Gimbun, J. (2014a). Assessment of phenolic compounds stability and retention during spray drying of Orthosiphon stamineus extracts. *Food Hydrocolloids*, **37**, 159–165.

Pang, S.F., Gimbun, J., Yusoff, M.M. & Chuah, T.G. (2014b). Microencapsulation of polyphenols from *Orthosiphon stamineus* leaves extracts using polysaccharide and protein. In: *Malaysia University Conference Engineering Technology*, Malacca.

Papadakis, S.E. & King, C.J. (1988). Air temperature and humidity profiles in spray drying. 1. Features predicted by the particle source in cell model. *Industrial & Engineering Chemistry Research*, **27**(**11**), 2111–2116.

Rod, V. & Misek, T. (1982). Stochastic modelling of dispersion formation in agitated liquid-liquid systems. *Transactions of the Institution of Chemical Engineers*, **60**(**1**), 48–53.

Sadikoglu, H. & Liapis, A.I. (1997). Mathematical modelling of the primary and secondary drying stages of bulk solution freeze-drying in trays: Parameter estimation and model discrimination by comparison of theoretical results with experimental data. *Drying Technology*, **15**(**3–4**), 791–810.

Schoen, M.P., Braxton, B.K., Gatlin, L.A. & Jefferis III, R.P. (1995). A simulation model for the primary drying phase of the freeze-drying cycle. *International Journal of Pharmaceutics*, **114**(**2**), 159–170.

Seydel, P., Blömer, J. & Bertling, J. (2006). Modeling particle formation at spray drying using population balances. *Drying Technology*, **24**(**2**), 137–146.

Sheehan, P., Sadikoglu, H. & Liapis, A. I. (1998). Dynamic behavior and process control of the primary and secondary drying stages of freeze drying of pharmaceuticals in vials. In: *Proceedings of the 11th International Drying Symposium (IDS'98)*, Vol. **100** (eds Mujumdar, C.A.S, Akritidis, C.B., Marinos-Kouris, D. & Saravacos, G.D.), pp. 1727–1740.

Shi, L. & Bayless, D.J. (2007). Comparison of boundary conditions for predicting the collection efficiency of cyclones. *Powder Technology*, **173**(**1**), 29–37.

Song, C.S. & Yeom, G.S. (2009). Experiment and numerical simulation of heat and mass transfer during a spray freeze-drying process of ovalbumin in a tray. *Heat and Mass Transfer*, **46**(**1**), 39–51.

Southwell, D.B. & Langrish, T.A.G. (2000). Observations of flow patterns in a spray dryer. *Drying Technology*, **18**(**3**), 661–685.

Southwell, D.B. & Langrish, T.A.G. (2001). The effect of swirl on flow stability in spray dryers. *Chemical Engineering Research and Design*, **79**(**3**), 222–234.

Spalart, P.R., Jou, W.H., Strelets, M. & Allmaras, S.R. (1997). Comments on the feasibility of LES for wings, and on a hybrid RANS/LES approach. *Advances in DNS/LES*, **1**, 4–8.

Tang, X.C. & Pikal, M.J. (2004). Design of freeze-drying processes for pharmaceuticals: practical advice. *Pharmaceutical Research*, **21**(**2**), 191–200.

Tanner, F.X., Srinivasan, S., Althaus, T.O., Feigl, K.A. & Windhab, E.J. (2009). Modeling and validation of a freezing model for food sprays. In: *Triennial International Annual Conference on Liquid Atomization and Spray Systems*, Colorado.

Trelea, I.C., Passot, S., Marin, M. & Fonseca, F. (2009). Model for heat and mass transfer in freeze-drying of pellets. *Journal of Biomechanical Engineering*, **131**(**7**), 074501.

Turnbull, J. & Thompson, C.P. (2005). Transient averaging to combine large eddy simulation with Reynolds averaged Navier–Stokes simulations. *Computers & Chemical Engineering*, **29**(**2**), 379–392.

Verdurmen, R.E.M., Menn, P., Ritzert, J., Blei, S., Nhumaio, G.C.S., Sonne Sørensen, T., Gunsing, M., Straatsma, J., Verschueren, M., Sibeijn, M., Schulte, G., Fritsching, U., Bauckhage, K., Tropea, C., Sommerfeld, M., Watkins, A.P., Yule, A.J. & Schønfeldt, H. (2004). Simulation of agglomeration in spray drying installations: the EDECAD project. *Drying Technology*, **22**(**6**), 1403–1461.

Versteeg, H.K. & Malalasekera, W. (2007). *An Introduction to Computational Fluid Dynamics: The finite volume method*, Pearson Education.

Woo, M.W., Daud, W.R.W., Mujumdar, A.S., Wu, Z., Meor Talib, M.Z. & Tasirin, S.M. (2008). CFD evaluation of droplet drying models in a spray dryer fitted with a rotary atomizer. *Drying Technology*, **26**(**10**), 1180–1198.

Woo, M.W., Daud, W.R.W., Mujumdar, A.S., Wu, Z., Talib, M.Z.M. & Tasirin, S.M. (2009). Non-swirling steady and transient flow simulations in short-form spray dryers. *Chemical Product and Process Modeling*, **4**(**1**), 1934–2659.

Woo, M.W., Huang, L.X., Mujumdar, A.S. & Daud, W.R.W. (2010). CFD simulation of spray dryers. In: *Spray Drying Technology* (eds Woo, M.W., Mujumdar, A.S. & Daud, W.R.W.), pp. 1–36, University Kebangsaan, Singapore.

10

Physicochemical and Sensory Properties of Dried Dairy Products

P.H. Amaladhas and F. Magdaline Eljeeva Emerald

Engineering Section, National Dairy Research Institute, Southern Campus, Bangalore 560030, India

10.1 Introduction

Drying is a technique that is widely used to deal with surplus milk during the flush season. With modern drying techniques, it is possible to process milk into powder without any significant loss in nutritive value. In Codex standard 207-1999, milk powders are defined as 'milk products that can be obtained by the partial removal of water from milk', with the moisture content generally reduced to less than 4% (Codex Alimentarius 1999). Cream is the only other ingredient permitted in the manufacture of milk powders. The raw milk used for powder production must be of the highest quality and comes under regulatory standards in many countries. The freshness of milk is significant because the shelf-life stability of the dried powder is dependent on the quality of raw milk used for its manufacture.

10.2 Milk Powder Manufacture

Manufacturing of milk powders involves primarily preheating the milk to 60–70 °C, homogenization (in the case of whole milk powders to protect the fat from coalescence during drying), concentration and drying.

Preheating is the most severe heat treatment step in the manufacture of milk powders. Preheat treatment of milk concentrate to a temperature of 60–80 °C reduces the viscosity of the feed prior to drying. This is particularly important in the case of spray drying, where the feed has to be atomized. Most of the physicochemical changes in whey proteins occur during the preheat treatment process. Preheating also improves the microbiological quality of the final product.

After preheating, milk is concentrated. Concentration is normally done in multiple-effect falling-film evaporators with mechanical or thermal vapour recompression for greater economy. These evaporators operate under vacuum with low boiling temperatures. The residence time in each stage of the evaporator is rather short. Most of the water in milk is removed by evaporation because this is a much more efficient way of water removal than drying. Only 10–20% is removed in the drying stage. The degree to which milk can be concentrated depends on the viscosity of the concentrate. Concentration of milk prior to drying has a significant effect on powder quality. A higher total solids (TS) in milk results in larger powder particles with higher density and less occluded air, thus resulting in better storage stability. However, a highly concentrated feed has several drawbacks: (i) higher TS feed results in the formation of larger spray particles (if spray dried), which because of their low surface area to volume ratio require a longer drying time, (ii) too much concentration also causes a reduction in the functionality of the spray-dried product, particularly the solubility of the powder (Baldwin *et al.* 1980) and (iii) intense heat treatment increases the viscosity of the concentrate, making evaporation of moisture from the feed difficult.

Homogenization, although not mandatory, is done to reduce the surface free-fat in the whole milk powder or full-cream powder (Free-fat is the fat that can be extracted using non-polar solvents such as hexane). Reduction of free-fat is desirable because a high level of it reduces the flowability, wettability and storage stability of the powder.

Dryers are used to remove water from the feed concentrate and produce a dry and shelf-stable product. Among all the processes in milk powder production, drying is most vital because it defines the properties of the final product. The two main types of milk powders produced directly from milk or skim milk are whole milk powder (WMP) and skim milk powder (SMP or non-fat dry milk). WMP is manufactured as instant or non-instant. The instantized WMP is mostly reconstituted, and it serves as an alternative to liquid milk. SMP is again classified as a low, medium or high-heat powder based on the level of denatured whey proteins, which in turn is dependent on the heat treatment given to milk/concentrate during the manufacture of the powder (Table 10.1). The changes to proteins during processing can have marked effects on the properties of dairy powders. The protein content should not be less than 34% on a milk solids–non-fat (SNF) basis for powder production (Singh 2007).

Low-heat SMP is typically preheated at about 72 °C for 2 min, medium-heat powder at 72–79.5 °C for 20 min, while for high-heat SMP the process temperature is about 90 °C for 30 min. Thus, the milk for low-heat SMP undergoes relatively mild heat treatment, with little protein denaturation. The properties of these different grades of SMP are therefore quite different. Most of the SMP made commercially is of the medium-heat category because reconstituted and sweetened condensed milk is made from this category of SMP. The level of protein denaturation during preheating helps to control viscosity development in these products. The medium-heat SMP is also widely used in confectionery, desserts, beverages, etc. For WMP manufacture, the milk is essentially heated to about 90–95 °C for 15–30 min to ensure inactivation of lipase. WMP has a minimum fat content of 26%. Low-heat WMPs are not recommended and manufactured

Table 10.1 Heat classification of skim milk powder (Augustin & Clarke 2008)

Milk powder class	Whey protein nitrogen index	Cysteine number	Thiol number	Heat number
Extra low heat		24–31		
Low heat	≥6	31–38	<7.5	≤80
Medium heat	1.51–5.99	38–62	7.5–13.3	80.1–88
High heat	≤1.5	>62	>13.3	≥88.1

because the lipases will not be inactivated completely and lipid oxidation will occur, leading to spoilage of the product and reduced shelf life. Moreover, the natural antioxidants produced during heating from amino acids such as cysteine, methionine, etc. may offer some protection against lipid oxidation during storage of WMP. The heat treatment used in the powder manufacture is quantified traditionally by the whey protein nitrogen index (WPNI). The bakery and confectionery industry prefers high-heat powder which has low WPNI.

The production of WMP differs from that of SMP in terms of the intensity of heat treatment and the absence of the homogenization step. The fat content is also standardized before the first heat treatment. The method of drying for WMP is almost the same as that for SMP. The drying process may cause a range of structural and physicochemical modifications, which in turn influence the properties of dairy powders (Westergaard 1994), therefore achieving desirable powder properties by controlling the drying operation becomes important. The final moisture content to which the powder is dried is critical. Typically, powders are dried to 2–4% moisture content (dry basis). This moisture content has a considerable influence on the keeping quality of the powder. Moisture determines the rate of deteriorative reactions such as lipid oxidation, non-enzymatic browning, protein denaturation and lactose crystallization during storage (Skanderby *et al.* 2009). Thus, depending on the raw milk composition and drying technique, a wide range of dairy powders with specific functional properties can be manufactured.

10.2.1 Roller drying

Roller drying was introduced in 1902 and it soon became a popular method of producing fruit powders and dried dairy products, particularly infant formulae. The roller dryer is ideal for drying viscous, concentrated solutions and slurries or pastes. It consists of hollow metal cylinder(s) that rotate on horizontal axes while being heated internally by steam or hot water. The feed is spread onto the drum(s) as a thin sheet that adheres to the hot surface. In the double roller dryer, the counter-rotation of the drums draws the liquid pool into the nip and spreads it into a thin sheet. Before drying, the TS level of milk is raised to 33–35%. Drying takes place by conduction, using the heat flux from the medium inside the drums. Mostly, dry steam at temperatures as high as 150 °C is used for heating. On contact with the drum, the slurry temperature is raised near to the boiling temperature, causing irreversible physicochemical transformations in the product. According to Caric (1993), the direct contact of concentrated milk on hot roller surfaces causes adverse changes to the constituents of milk (particularly the milk proteins) and also to the flavour and quality of the powder. Therefore, for heat-sensitive materials, drying in vacuum at a lower temperature is recommended. The operational variables are feed concentration, steam pressure, rotational speed and nip width (Gavrielidou *et al.* 2002; Valous *et al.* 2002; Kostoglou & Karapantsios 2003). The sheet of powder produced is ground to a predetermined size, which yields irregularly shaped powder particles with superior wetting properties to spray-dried powders (Kelly *et al.* 2003). The main advantages of roller drying are high drying rates and heat-transfer efficiency (Vega-Mercado *et al.* 2001).

The TS of the feed concentrate used is much less than that used in spray drying. This is because highly concentrated feed may form a thick layer on the rollers or may not form well, thereby slowing down the drying rate and causing irreversible damage to the milk proteins (Caric 1993). Also, the powders will have higher free-fat content and wettability. According to Keogh *et al.* (2003), WMP produced by roller drying had a free-fat content of 85–95% (of total fat) as compared to 2–3% for spray-dried powders. The removal of the thin film of product from the drum by the shearing and scraping action of blades/knives apparently contributes to the free-fat content, therefore roller-dried WMP is ideal for use in milk chocolate and the confectionery industry. If vacuum roller drying is done at temperatures below 100 °C it eliminates the oxygen effect, resulting in better powder characteristics (Caric 1993). Roller drying also produces milk powders with a certain amount of porous structure because of the rapid boiling off of water.

10.2.2 Spray drying

In the last century, many technological innovations occurred in the dehydration of foods, particularly the production of milk powders. Undoubtedly the most important one in the manufacture of dried milk is spray drying. Also, spray dryers have evolved in design and applications over the last few decades, and now a range of dryers to suit the wide range of dairy products is available. Milk is concentrated to about 42–48% and 45–50% TS, respectively, for the production of SMP and WMP before drying (Kyle 1993). The concentrated feed is atomized into a fine spray and exposed to the stream of hot air in the drying chamber. Atomization is the most important step in spray drying. The particle size and, more importantly, the distribution of particle sizes play a significant role in determining the properties of the milk powder. The droplet size is directly controlled by the feed liquid viscosity (Masters 1991). The two most common atomizing devices are the centrifugal disc and the pressure nozzle. The atomizer determines the size and shape of the droplets, their trajectory and speed, drying rate and consequently the particle size of the powder. Irrespective of the atomizer type, the powder particles gain a spherical shape during drying (because of surface tension) with trapped air, thus achieving a low bulk density (Caric 1993). The atomization parameters therefore influence the moisture content, particle size distribution, occluded air content and bulk density of the dried product (Caric *et al.* 2009).

The spray particles encounter the hot air at 150–250 °C. The drying air may flow in co-current, counter-current and mixed flow paths in relation to the product. Co-current air flow is mostly used for heat-sensitive materials (Early 1992). Drying is very rapid and the drying time is very short. The spray dryer operates at temperatures of 180–200 °C for WMP and slightly higher (180–230 °C) for the production of SMP (Bloore & O'Callaghan 2009). Despite using very hot air, the temperature of the milk droplets does not exceed 70 °C until they have lost almost all their water. When the drying air temperature is high, the initial evaporation rate of the wet particles is high enough to keep the product temperature low. Thus, most of the drying takes place at wet-bulb temperature conditions. Furthermore, as the diffusion rate decreases, the surface of the particle forms a dried shell and the falling rate period commences. The air temperature decreases to near wet-bulb temperature and the particle temperature approaches the temperature of the outlet air (70–90 °C). The drying rate should be controlled such that the falling rate drying period matches the end of drying (Kelly *et al.* 2003).

There are three classes of spray dryers based on the number of separate drying stages. In single-stage dryers, drying is completed within the spray chamber. In two-stage and three-stage dryers, the main spray chamber is followed sequentially by fluidized-bed and plug-flow type fluidized-bed dryers, respectively. The powder properties will depend on the type of spray-drying system used. The single-stage process necessitates the use of a higher air temperature, which results in a poor quality dusty milk powder. Westergaard (1994) noted remarkable weight loss, particle shrinkage and reduction in particle diameter during single-stage drying. The weight of spray droplets was found to decrease by about 50%, the volume by 60% and the diameter by 25%. The spray-dried powders have almost spherical particles with a diameter of 10–250 μm. The energy consumption is quite high and the thermal efficiency is quite low in these dryers. Ferguson (1989) recorded a specific steam consumption of 2.2–2.5 kg steam/kg of water evaporation in spray dryers.

In two-stage or multistage dryers, the final drying takes place outside the spray chamber on a vibrated fluidized bed (second stage). The powder is dried to 6–9% moisture in the main spray chamber and is further dried in the vibrated fluid bed to less than 4%. Two-stage and three-stage drying allow a lower outlet temperature in the main spray chamber. The vibrated fluid bed second stage is also ideal for the production of fat-rich powders. Also, single-stage dryers cannot easily handle the feed if the fat content is above 35%.The other advantage of two- or three-stage drying is that feed concentrates with high TS can be dried without much deterioration of the functional properties of the powder. Furthermore, by separating the drying process into two or three stages, improved process control is possible. The reduced outlet air temperature and longer drying time limits case hardening, thereby improving powder quality. The powder qualities, in terms of wettability, dispersibility, sinkability and solubility, will be improved to a great extent. Bulk density is also improved as levels of occluded air and free-fat are reduced. The higher drying temperatures practiced in single-stage dryers increase the formation of capillaries and vacuoles, which make fat unprotected (Farkye 2006), thereby resulting in higher free-fat content. The two-stage and three-stage dryers are also preferred because of their better thermal efficiency and reduced operating cost. The outlet air leaves at a much lower temperature, thereby improving heat utilization. For these reasons, single-stage dryers are now becoming obsolete.

Removal of water from milk by spray drying alone would be very expensive. In addition, spray drying of low-TS feed yields milk powders with significantly lower bulk density and increased powder losses. The degree to which milk can be concentrated depends on the type of powder (WMP or SMP), viscosity of the concentrate and the ability of the spray-drying plant to handle and atomize the feed, therefore the spray-dried product could be similar in quality to that obtained by freeze drying (Bloore & O'Callaghan 2009). Quality degradation in the spray dryer depends mostly on both moisture content and temperature.

The heat exposure of milk during spray drying varies considerably depending on the dryer design, operating conditions and drying time before cooling. In particular, the inlet and outlet temperatures of the drying air are critical in determining the final product quality. If the droplets are exposed to a high air temperature at the end of drying, the air within the powder particles may expand, forming large vacuoles, fracturing the particle and resulting in increased proportions of fines, therefore blowing up of the particles in the final phase of drying should be avoided (Bloore & O'Callaghan 2009). Pisecky (1997) recommended a maximum temperature of 180 °C to avoid thermal damage in spray-dried milk powders. If spray drying is done properly, the native properties of the milk components are essentially unmodified. Very little denaturation of whey proteins occurs during evaporation and drying because the final temperature of the dried product is quite low (Singh & Creamer 1991; Kelly *et al.* 2003). The size distribution of casein micelles, their heat stabilities and renneting characteristics are substantially recovered on reconstitution of spray-dried milk (Singh 2007).

The dryer type also plays a major role in determining the powder properties. Although roller and spray dryers were used in the past for milk powder manufacture, spray drying gives a better product. Thus, most milk powders today are made by spray drying, except for WMP destined for use in milk chocolate and confectionery. As discussed before, the high free-fat and low solubility of drum-dried milk powders are ideal for the manufacture of chocolate.

10.2.3 Freeze drying

Spray- and drum-dried WMP has some limitations for use in beverages. Normal WMP, even when fresh, has a marginally cooked and somewhat astringent taste. Freeze drying, also known as lyophilization, is one of the most sophisticated drying techniques in which the food is first frozen and water is sublimated by supplying heat through conduction. It is generally regarded as the best method for the production of heat-sensitive and high-quality dried products. The size and shape of the frozen milk particles do not change during freeze drying and as a result the density of the powder varies directly with the TS content of the milk used. Freeze drying provides products of porous structure with little or no shrinkage, superior taste and aroma retention. However, there will be no difference in the colour of spray- or freeze-dried WMP (Nickerson *et al.* 1952; Fyfe *et al.* 2011).

10.2.4 Spray freeze drying

Another innovative and recent dehydration technique involves spraying a solution into a refrigerant medium, followed by freeze drying the resultant frozen particles at atmospheric pressure (Anandharamakrishnan *et al.* 2010). The freeze drying is achieved conventionally, with the latent heat of sublimation supplied by conductive or radiative heating. This method of drying is quite fast. The major problem with atmospheric spray freeze drying is that very large quantities of cold dry gas have to be circulated through the bed. This technique has possible applications in dairy and coffee whiteners, where wettability and dispersibility are major desirable attributes.

10.3 Properties of dairy powders as influenced by drying method

Milk powders for each specific use require certain desirable physical and functional properties. They are characterized by their physical (particle size, bulk and particle densities, occluded air, interstitial air, flowability, hygroscopicity and heat stability), functional (insolubility index, dispersibility index, wettability index, sinkability index, free-fat, flowability, stickiness and caking), morphological, biochemical (non-enzymatic browning, oxidative changes), microbiological and sensory qualities. Furthermore, milk proteins have emulsification, foaming and gelling properties in foods. Many of these properties are significantly interrelated and affect the final quality of the powder (Caric & Milanovic 2002). The functionality and properties of milk powder depend on the feed composition, such as proteins, lactose and lipids, physicochemical characteristics, viscosity, TS content and feed rate, during manufacture and on dryer parameters (for spray dryers), such as dryer type, nozzle type, air pressure (for pressure nozzles), and agglomeration, and operating parameters, such as air temperature, etc.

Tuohy (1989) reported that the fat content, type of fat and drying process affected the physical properties of milk powder. The free-fat content has a greater influence on the physical properties of the powder than surface fat. Singh (2007) reported that pre-heat treatment of milk prior to evaporation and drying has been the most widely used method to produce milk powders with different functional properties. Thus, by simply changing the drying rate and process conditions, powders with varying characteristics can be obtained.

The viscosity of feed is influenced by milk composition, the effect of heat treatment on whey protein denaturation and the TS content to which it is concentrated (Early 1992). Whole milk concentrate at the same TS will therefore be less viscous than skim milk concentrate due to differences in fat content. If the initial viscosity is increased by preheat treatment or by increasing the TS, the milk powder produced will have larger particle size, increased densities and moisture content, and decreased porosity. Similarly, increasing the feed flow rate increases the density, particle size, moisture content and wettability of the powder and decreases its insolubility. The influences of process conditions and drying on powder quality are discussed in detail in the following sections.

10.4 Physical properties

10.4.1 Morphology, particle size, shape and distribution

The shape of milk powders varies from spherical to irregular. The particles may differ in their microstructural characteristics due to the differences in the drying conditions to which individual particles are exposed (Caric & Milanovic 2002). Powders produced by roller drying have a compact structure and irregular shape with no or much less occluded air. They also have sharp edges and are compact in size (Caric & Kalab 1987). Spray-dried powders, on the other hand, are spherical, containing occluded air or vacuoles, which are distributed through the interior of particles. The spray-dried powders tend to have a solid and relatively dense surface layer. Of all major drying techniques, spray drying produces the least porous particles, while freeze drying produces relatively porous particles. The surface of spray-dried powders is smooth except in cases where very high air temperatures are used. Wrinkles can occur at high inlet air temperature or at a large temperature differential between hot air and the product. Greenwald & King (1981) investigated the effects of design and operating conditions on the morphology of spray-dried foods. Spray freeze-dried powders are also found to be highly porous, with uniform pores throughout the particles (Rogers *et al.* 2008).

The particle size of milk powders ranges from 10 to 250 μm in diameter. The particle size and its distribution have a major influence on the powder properties and the products derived out of them (Barbosa-Canovas *et al.* 2005). The particle size of spray-dried powder is in turn dependent on the process conditions such as atomizer type, atomization conditions, viscosity of the feed and the lactose content (Elversson *et al.* 2003).

Higher spraying pressure, with nozzle size being constant, decreases droplet size and consequently particle size. In contrast, particle size increases with increasing TS level (Enríquez-Fernández *et al.* 2013) of the feed. However, if the droplet size becomes too large, case hardening occurs owing to the formation of a shell that prevents further evaporation, leading to cracking or bursting of the spray particles (Early 1992). The spread of particle size distribution is also important because most of the physical properties of milk powders are related to it (Baldwin & Pearce 2005). The feed viscosity influences various parameters of the drying process and the physicochemical properties of the powders as a result of influencing droplet size distribution.

10.4.2 Density

The bulk density of powders is important for economical and functional reasons. Powders with low bulk densities are not ideal because their higher amounts of occluded air favour oxidation and other deteriorative reactions. Bulk density is dependent on the shape of the particles, the surface characteristics, the width of the particle size distribution and fat content. The bulk density of spray-dried milk powders ranges from 500 to 700 kg/m^3 while roller-dried powders are about 300–500 kg/m^3 (Caric 1993). If the density is measured after tapping, the action should be mechanized and the number of taps (500 or 1000) should be specified. The low bulk density of roller-dried powders is attributed to its irregular structure. This is because the irregular particles lock together and cannot pack together closely. In contrast, the uniform and smooth-surfaced spray-dried powders generally give higher bulk density (Kelly *et al.* 2002).

Powders with a range of particle sizes give a higher bulk density than a powder with a narrow particle size distribution. In such cases, smaller particles fill up the space between larger and medium-sized particles. Duffie & Marshall (1953) observed that the bulk density was dependent on the particle size distribution even though the mean particle size remains unchanged. These authors also reported that smaller particles may be inherently denser than larger particles when dried under the same conditions. Hayashi *et al.* (1970) studied the relationship between powder particle size and bulk density, and stated that the bulk density decreased with increasing particle size. They attributed the lower density to the larger interstices between the larger powder particles. The amounts of occluded and interstitial air are very important. A higher amount of air in the feed produces milk powders of higher occluded air content. Occluded air in the droplets lowers the bulk density of the powder (Caric *et al.* 2009). The amount of interstitial air between powder particles also depends on the particle size, shape and size distribution. The interstitial air content in milk powders may be as high as 127 ml/100 g of powder (Caric 2002). Agglomerated milk powders are therefore expected to have lower bulk densities than conventional counterparts.

The density of milk powders is also dependent on the milk composition and the density of the ingredients that make up the product. Thus, the bulk density of WMP is expected to be lower than that of SMP because of the lower density of milk fat relative to other constituents. The lactose content in milk also influences the bulk density. Powders from low-lactose milk have a more porous structure with a larger amount (larger vacuoles) of occluded air (Aguilar & Ziegler 1994). The same authors reported a gradual increase in both the bulk and true densities of WMP with an increase in lactose content in the feed.

The mean true densities of WMP and SMP are about 1300 and 1480 kg/m^3, respectively (Berlin & Pallansch 1963).

Processing factors that contribute to the density of powders are preheat treatment to milk, feed viscosity, atomization method in the case of spray drying, incorporation of air into the feed prior to drying and drying conditions. Hayashi *et al.* (1970) reported that the bulk density of milk powders increased with intensity of preheat treatment. The trend was more pronounced at higher TS content in the feed. The authors attributed the increase to the physicochemical changes in the milk proteins and to the removal of air from feed concentrate at higher preheat temperatures. Similarly, a high degree of whey protein denaturation gives a low occluded air content and *vice versa*. The viscosity of feed concentrate plays a major role since it influences various drying parameters. (Oldfield *et al.* 2000). Lower feed viscosities reduce particle size. Increasing the TS of the feed concentrate, in general, results in dense powders (Mettler 1980; Caric *et al.* 2009; Enríquez-Fernández *et al.* 2013). This is because porosity decreases with increasing TS in the feed. Also, the TS content is related to the difference between tapped and loose bulk densities, which parallels the effect of TS on the interstitial air between the particulates. However, too much increase in feed viscosity produces larger droplet size with reduced surface area in the drying medium (hot air), resulting in improper drying of the product. Berlin & Pallansch (1963) reported little difference in the true densities of WMP made by different drying techniques. The bulk density of SMP produced by various processes is presented in Table 10.2 and the effect of heat treatment of milk on the bulk density of SMP is given in Figure 10.1. The desired physical characteristics of milk powders are given in Table 10.3.

Hayashi *et al.* (1970) varied the spray-drying conditions for the manufacture of milk powder, and concluded that such variations in drying conditions produced a range of bulk and particle densities. The nozzle type, air temperature and atomization conditions affect the bulk density of the powder, mainly because of their effect on the particle size and occluded air content. Air entrained in the feed during handling and pumping leads to reduction in bulk density. This entrapped air in the feed will be carried to the dried product if it does not escape during spraying and drying. According to Palmer & Dahle (1922), the spray droplets from atomizers contain occluded air in the form of spherical cells. The type of nozzle used for atomizing the feed governs the amount of air inclusion. In general, centrifugal nozzles incorporate more air into the product than pressure nozzles (Twomey & Keough 1998; Kelly *et al.* 2003). Boersen (1990) observed that the bulk density of milk powder produced by a nozzle atomizer was higher than

Table 10.2 Bulk density of SMP produced by various processes (Carić & Kalab 1987, copyright material of DigitalCommons@USU © 2016)

Drying process	Bulk density (g/cm^3)
Spray drying	0.50–0.60
Roller drying	0.30–0.50
Foam spray drying	0.32
Spray drying (commercial)	0.26
Instant spray drying (commercial)	0.59

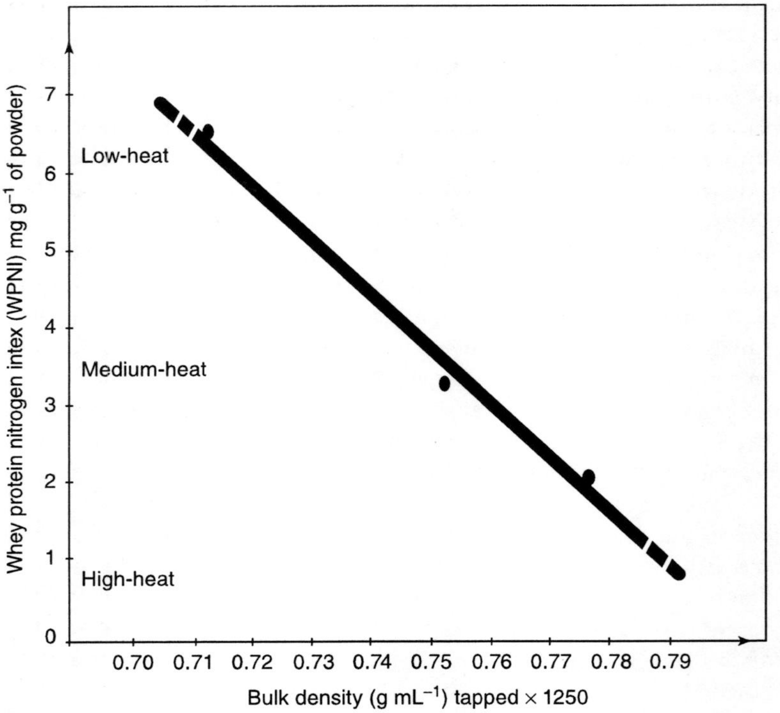

Figure 10.1 Bulk density of SMP as influenced by the severity of the heat treatment of milk (Skanderby *et al.* 2009).

Table 10.3 Desired physical properties of milk powders (Skanderby *et al.* 2009)

Parameter	Type of powder				
	SMP	Agglomerated SMP	WMP	Agglomerated WMP	Instant WMP
Physical properties					
Bulk density: tapped × 1250 (g/mL)	0.67–0.70	0.43–0.52	0.58–0.64	0.43–0.50	0.42–0.49
Solubility index (mL)	≤0.1	≤0.2	≤0.2	≤0.2	≤0.2
Free fat (g/100 g)	–	–	≤2.0	≤2.0	≤2.5
WPNI (mg/g)	1.5–6.0	2.5–4.5	2.5–3.5	2.5–3.5	3.5–4.5
Wettability (s)	–	≤30	–	–	≤15
Dispersibility (%)	–	>98	–	–	>95
Sludge (at 25 °C)	–	≤0.1	–	–	≤0.1

SMP, skim milk powder; WMP, whole milk powder; WPNI, whey protein nitrogen index.

that of the powder produced by centrifugal counterparts because of the compactness of the particulates. Centrifugal atomizers incorporated more air into the spray, producing powders with higher occluded air content. Despite its superior spray characteristics, pressure nozzles may not offer a significant advantage because the TS of feed handled is considerably lower than that of centrifugal nozzles. The centrifugal atomizers can handle higher concentrate feed viscosities. However, Verhey (1973) observed that particles atomized using a pressure nozzle contained occluded air at their centre, and thus had lower density. The differences in the findings of various authors could be due to the varied TS content of the feed. Thus, it could be stated that nozzle type may or may not have any significant influence on particle structure (Caric 1993). However, higher powder density can be obtained by using larger nozzle diameters (Caric 2002). Similarly, reducing the atomization pressure (Kelly *et al.* 2002) improves the bulk density (Caric 2002).

Higher inlet and outlet temperatures produce large-sized particles that have higher porosities. De Vilder *et al.* (1976) observed that increasing the outlet air temperature from 70 to 95 °C caused cracking of the powder particles, creating high vacuole volume and lower powder density. Kelly *et al.* (2003) added that increasing the inlet air temperature reduced bulk density and particle density due to an increase in occluded air and the rapid formation of a hard shell on the particles, which prevented expansion of the particle. Such occluded air is formed if the rate of evaporation exceeds the rate of diffusion of vapour through the spherical droplet (Hayashi *et al.* 1970). However, if case hardening is avoided, higher spray-drying temperatures may improve the powder density (Kelly *et al.* 2002). Thus, it can be stated that the variation in bulk density of milk powders can be explained by a complex series of interrelationships between milk composition, and the inlet and outlet temperatures of the dryer and process conditions (Oldfield *et al.* 2000).

10.4.3 Reconstitution properties

Reconstitution is the ability of the powder to become wet, sink, disperse and dissolve. It is a combination of several steps: wettability, immersibility, dispersibility and solubility. Wettability is the penetration of water into the particles while dispersibility is the ability of the powder to disperse and mix in water, forming a homogeneous liquid. Solubility is the final step in the reconstitution process. All these individual reconstitution features are interrelated, and they occur consecutively or simultaneously during reconstitution. The optimum temperature for particle dissolution is 50–60 °C. To have good reconstitution the powder should be wettable, the agglomeration optimal, and fine particles should be absent (Rollema & Muir 2009). A schematic representation of an agglomerated milk powder along with its scanning electron micrograph is given in Figure 10.2. The powder should segregate into agglomerates, which again should disintegrate into single primary particles. However, all the constituents of milk powder are not soluble in water. Lactose, undenaturated whey proteins and some of the salts are naturally soluble (Caric 2002). Normal spray-dried powders are very fine and therefore difficult to hydrate. They may form lumps when mixed with water, and require mechanical stirring to dissolve. The particles prefer to associate with themselves rather than be wetted by water, and thus, 'balling-up' occurs. This is because the first layer of powder particles that comes into contact with water dissolves, forming a thick outer layer which is resistant to water, thereby stopping any further wetting. The degree and rate of wetting is

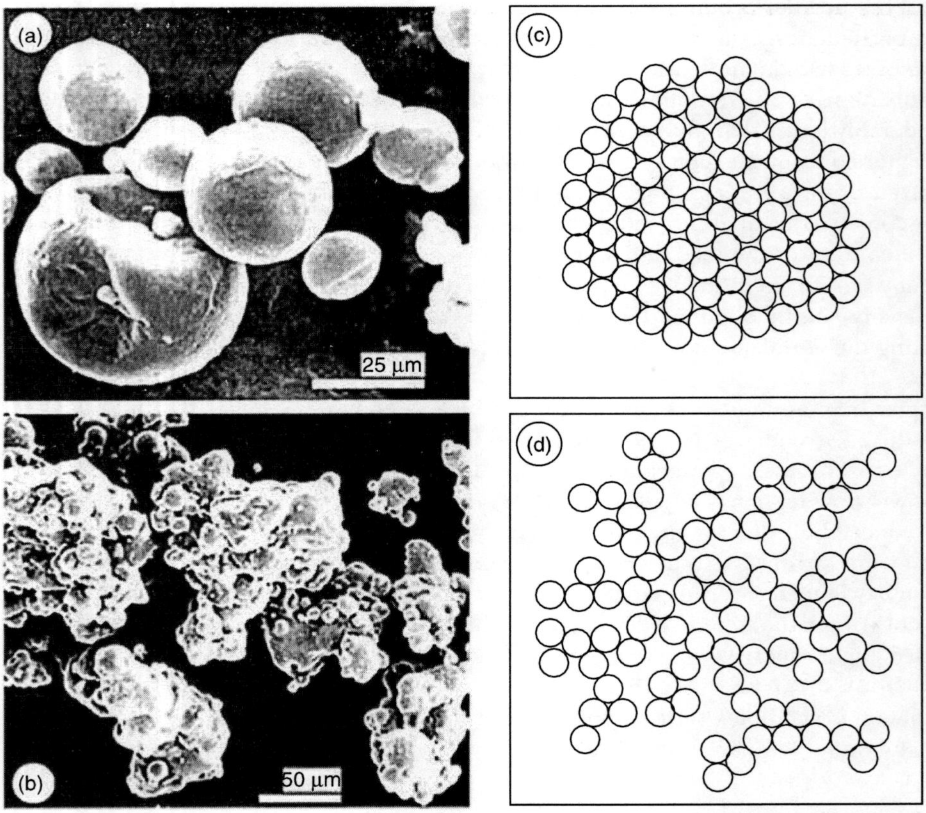

Figure 10.2 Scanning electron micrographs and schematic drawings of spray-dried milk powder: (a) one-stage dried (scanning electron micrograph), (b) agglomerated (scanning electron micrograph), (c) one-stage dried (schematic) and (d) agglomerated (schematic) (Caric 2004).

related to the interfacial tension between the particle surface and the water. In order to become wet, the powder must overcome the surface tension between itself and water (Caric 2002). Thus, individual particles will wet and dissolve more readily than clumps. The ideal particle size for rapid reconstitution is 150 ± 200 µm (Caric 2002). Powders consisting of large aggregates of 180–840 µm may have reduced dispersibility (Bockian *et al.* 1957). The solid and relatively dense surface of spray-dried milk powders may not be good for water penetration during reconstitution.

The reconstitution properties are also dependent on the moisture, fat and lactose contents, shape and size distribution of the particulates, bulk and particle density, drying conditions, etc. (Birchal *et al.* 2005). The optimum moisture content for rewetting is 11–12% (Neff & Morris 1968). SMP readily wets in water because it consists of mostly lactose and proteins, which are more hydrophilic. On the other hand, WMP particles are covered by a layer of fat, making them more hydrophobic and difficult to hydrate (Woodhams & Murray 1974). Litman & Ashworth (1957) reported that the inability of milk powder to wet completely was due to the formation of free-fat–protein complexes (Kelly *et al.* 2003). The amount of free-fat varies between 0.5 and 3 g/100 g of the powder

(Deeth & Hartanto 2009). As fat content in the feed concentrate increases, the powders produced will have a pasty surface with more deposits of fat that prevents diffusion of water. In this regard, nozzle atomizers are better than centrifugal discs in reducing the free-fat in the powder (Kelly *et al.* 2002). Wettability is also influenced by the state in which lactose is present (Augustin & Clarke 2008). If lactose crystallizes in the powder during storage, it reduces wettability by expelling the free-fat from the particles. This is the reason that addition of lecithin is recommended in the manufacture of WMP or high-fat milk powders. Also, the higher outlet air temperatures in a spray dryer tend to promote higher free-fat in the powder (Kelly *et al.* 2002) and thus reduced wettability can be expected.

Hall & Hedrick (1961) observed a strong correlation between porosity/bulk density and dispersibility of SMP. A higher value of bulk density and lower porosity will favour submergence into water and consequently the reconstitution properties (Birchal *et al.* 2005). Hahne & Passos (2002) recommended a tapped density range of 360–580 kg/m^3 for autodispersion of WMP. Wettability decreases when there is increased interstitial air between the powder particles. Similarly, occluded air within the particles impairs the ability to reconstitute in water. Other factors that affect reconstitution and solubility are the intensity of the preheat treatment to milk, type of dryer used, atomizer configuration, number of stages in drying and concentration (viscosity) of the feed. Spray-dried powders exhibit poor solubility because of their small particle size and symmetrical shape, which favour close packing of particles, thereby inhibiting penetration of water (Caric 2002). One another major reason for insolubility in milk powders is the formation of casein–whey protein–lactose complexes during processing. The denatured β-lactoglobulin is involved in the interaction with casein and lactose (Parris *et al.* 1990; Augustin & Clarke 2008). Anandharamakrishnan *et al.* (2010) observed significantly higher losses in solubility for β-lactoglobulin as compared to α-lactalbumin. Denaturation of milk proteins depends on the intensity of preheating and evaporation treatments. Intensive heat denatures the proteins irreversibly, thereby reducing their hydration potential (Anandharamakrishnan *et al.* 2008). Insolubility is also affected by crystallization of lactose.

Case hardening also may adversely affect the solubility of milk powders (Early 1992), hence milk powders from single-stage spray dryers are expected to have a higher insolubility than those from two- or multistage dryers. However, there is little loss of solubility of whey proteins during spray freeze drying (Anandharamakrishnan *et al.* 2010). The solubility of spray-freeze-dried milk powder was found to be three times more than conventional spray-dried milk powder (Rogers *et al.* 2008). Spray-dried powders from highly viscous feeds may have poor solubility and a high proportion of burnt particles. Chauhan & Kumar (2004) reported that the optimum operating parameters for obtaining the best-quality WMP in terms of moisture content and solubility index were an inlet air temperature of 190 °C, a feed concentration of 46% TS and a feed temperature of 70 °C. Birchal *et al.* (2005) proposed that the spray dryer should be operated at a range of 160–175 °C to achieve a Hausner ratio of 1.4, which favoured reconstitution. The authors also recommended a higher atomizer speed and lower hot air temperature during spray drying. Similarly, increasing the feed concentration at higher outlet temperatures caused a noticeable decrease in solubility. Instantization and agglomeration improve the wettability and dispersibility of milk powders. Neff & Morris (1968) reported that agglomeration improved the dispersibility of milk powder from 41% to

62%. The reconstitution properties are improved because water penetrates the pores of the agglomerates instead of penetrating the capillaries of individual particles.

10.4.4 Agglomeration and instantization

Single-stage dryers produce non-instant and non-agglomerated powders with rather high bulk density. These fine powders are very difficult to disperse in water, but the requirement today is for milk powders that reconstitute well when dispersed in cold water. Agglomeration is usually done for low- or medium-heat powders that are 25–50 μm in diameter (Sanderson 1978). Agglomeration is done by returning the powder particles to the dryer so that they aggregate to form agglomerates. Lecithin is normally added to facilitate agglomeration and instantization. If agglomeration is done in a two-stage spray dryer, the powder leaving the spray chamber should have a moisture content of 5–8%, bulk density 450–550 kg/m^3, wide particle size distribution and good cohesion between particles (Birchal *et al.* 2005).

Kelly *et al.* (2003) recommended the use of lecithin at 0.2% between the spray-drying stage and the final fluidized-bed drying for agglomeration. If lecithin is added to WMP, the mixture should be held at about 50 °C for a few minutes to ensure complete coating of the particles with the surface active agent. The external vibrated fluid bed is required to dry the agglomerated powder to the desired moisture content. Cooling and crystallization of the fat is ensured in vibrated fluid beds. Thus, the agglomerated powder from a two- or multistage dryer will be coarse and almost dust free. Agglomeration has a significant influence on the physical properties of milk powders. Larger agglomerated particles will be non-dusty and have lower bulk density, improved flowability, solubility and reconstitution properties. Agglomeration also results in porous clusters, ranging from 250 to 750 μm in diameter, with a high level of entrapped air (Neff & Morris 1968).

10.4.5 Flowability and stickiness

Flowability refers to the free-flow characteristics of a powder. It is important in powder handling and packaging because poor flowability can cause bridging inside hoppers, silos and storage vessels. Flowability depends on the cohesive forces that develop due to inter- and intraparticle attractions in the powder. The widely followed index of flowability is the angle of repose. The lower the angle of repose, the more flowable is the powder (Tuohy 1989). Flowability is influenced by the moisture and fat contents of the powder, as well as its particle size distribution and particle shape. The latter two properties are a function of the dryer used for the powder manufacture. Pisecky (1997) demonstrated that the flowability of agglomerated SMP was better than that of agglomerated WMP, which in turn was better than full-cream milk powder (Augustin & Clarke 2008). Woodhams & Murray (1974) reported that the flowability of milk powder decreased as the fat content increased to 20%. Uniformly sized and shaped particles have better flowability, and it is improved if the powder is agglomerated. When it comes to dryer design, nozzle atomizers produce powders with better flowability than wheel atomizers. This is because nozzles produce powders with lower free-fat content than with the wheel, mainly due to the homogenization effect of the nozzle. Flowability and bulk density are inversely related. A powder with higher bulk density will have poor flowability, and *vice versa*. The more compressible a powder is, the less flowable it is expected to be (Tuohy 1989).

Stickiness is a measure of the particle cohesion and adhesion to equipment surfaces. In spray drying, rapid drying of the liquid feed droplets at temperatures above the glass transition (T_g) point creates amorphous particles that can become very sticky. A powder that is cohesive will have reduced flowability. Temperature, moisture content, particle size and milk composition, which can influence the efficiency of drying, impact the cohesion of milk powder. The stickiness problems of milk powders have been investigated in detail (Jouppila & Roos 1994). Drying aids such as maltodextrins, starches, etc are added to the feed to increase the T_g and reduce stickiness. Hahne & Passos (2002) recommend a Hausner number (which quantifies the strength of the cohesive force between particles) of 1.4 for processing WMP in vibro-fluidized beds (Birchal *et al.* 2005). If the amorphous to crystalline transition can be promoted by altering the design or operation of the dryers, then the deposition of particles on dryer walls could be reduced.

Generally, spray-dried powders are amorphous in character and hence, under normal conditions of relative humidty (RH) and temperature, stickiness and caking are not a problem (Baldwin & Pearce 2005). However, stickiness and caking can occur in stored milk powders due to the presence of lactose and fat, and under the right conditions of temperature, ambient RH and water activity. WMP is therefore twice as cohesive as SMP because of the fat content. The cohesion increases as particle size decreases (Rennie *et al.* 1999). To avoid stickiness and caking during storage, milk powders should be dried to less than 4% moisture content (Early 1992). As they adsorb moisture from air, the lactose undergoes transition from amorphous form to crystalline state depending on the temperature. When the lactose is converted into α-lactose monohydrate, some of the moisture condenses within the powder particles, facilitating stickiness and caking (Early 1998). Ingress of moisture also accelerates the Maillard reaction, which in turn promotes physical changes. As the powders become cohesive and stick to process equipment, packaging materials, etc., they pose difficulties in handling and storage.

10.4.6 Heat and coffee stability

The heat stability of milk powder is important in recombined evaporated milk. The heat stability of milk powder is affected by the composition and quality of the fresh milk for powder production, processing conditions, additives and the food system in which the powder is used. A number of reviews have been published on the heat stability of milk (McCrae & Muir 1992; Singh *et al.* 1992). The maximum heat stability is usually obtained at around pH 6.6–6.7. Heat stability is also determined by the preheat treatment applied to milk before evaporation and drying (Sweetsur & Muir 1981; Chandan 2008). Low- and medium-heat SMPs exhibit poor heat stability because of the nature of whey protein interactions. Adjustment of preheat treatment to maximize whey protein–casein interactions results in improved heat stability (Early 1992). Chandan (2008) reported that heat treatment to milk with WPNI < 1.5 mg was adequate for heat stability in concentrated milks. However, high-heat pretreatment that is used to enhance the heat stability of SMP in concentrated products may detrimentally affect the heat stability in unconcentrated milk.

Coffee stability is a measure of a milk powder's resistance to provide instability, as exhibited by resistance to feathering (Early 1992). Powders with poor coffee stability are characterized by undissolved particles, typically described as sinkers and floaters (Oldfield *et al.* 2000). The main reason for sedimentation is the flocculation of whey proteins. Denaturation of whey proteins affects the coffee stability, while undenatured whey proteins exhibit the best stability. Early (1992) reported that the coffee stability of a coffee creamer could be more than 400 ppm $CaCO_3$ equivalent at about 80 °C as compared to less than 250 ppm for SMP at the same temperature. Oldfield *et al.* (2000) found that the preheat temperature of instant WMP manufactured by spray drying had a significant effect on the coffee stability while homogenization and mode of emulsion formation had only marginal effect. Addition of lecithin prior to evaporation and drying enhances both heat and coffee stabilities (Hardy *et al.* 1985; Singh *et al.* 1992). A possible explanation is that lecithin interacts with fat globule membrane components to increase surface charge (McCrae & Muir 1992).

10.5 Chemical and sensory properties

10.5.1 Protein quality

The application of heat during milk powder manufacture gives rise to reactions between the constituents. Casein micelles mostly retain their integrity during processing and thus provide vital functional properties such as heat stability and emulsification. In contrast, whey protein denaturation, Maillard reactions or lactose caramelization occur depending on the intensity of heat treatment during preprocessing or drying, and are irreversible. For example, in roller drying direct contact of concentrated milk with the hot surface adversely affects the components of milk, particularly the proteins and lactose. In contrast, under typical spray-drying conditions the whey proteins are largely not denatured.

The functionality provided by milk proteins in food applications is very important. Oldfield & Singh (2005) reported that the functional properties the proteins impart to milk powder depend on the extent of heat treatment applied during powder manufacture. Preheating denatures and unfolds the whey proteins, enabling them to interact with each other before forming a complex with κ-casein. The order of sensitivity of whey proteins to heat-induced denaturation is: immunoglobulins > serum albumin > β-lactoglobulin > α-lactoalbumin. The nature of the interaction depends on the severity of the heat treatment. The protein aggregation in medium-heat SMP is mainly due to whey–protein–whey protein interactions *via* sulphydryl groups. In high-heat milk powders, the protein interactions involve complexation between β-lactoglobulin (β-Lg) and κ-casein (Early 1992) through sulphydryl–disulphide linkage. Preconcentration induce changes in casein micelles predispose the milk system to enhance protein–protein interactions during spray drying, which subsequently interacted with κ-casein (Singh 2007). The two disulphide bridges and a free thiol group in the native structure of β-Lg are involved in this interaction. Oldfield *et al.* (1998) also reported that β-Lg also formed disulphide-linked aggregates with bovine serum albumin. However, some authors proposed that β-Lg formed intermediary complexes with α-lactalbumin (α-LA). Patel *et al.* (2006) reported that heat treatment caused β-Lg to interchange into non-native monomer and non-native disulphide-bonded dimmer, which subsequently interacted with α-LA or κ-casein.

In spray drying, protein denaturation and aggregation depend on the inlet temperature of the air into which the milk is sprayed, the degree of concentration and the temperature of the feed concentrate, the droplet size and the temperature of the outlet air. The protein–protein interaction affects the functionality of the milk powder. Other interactions, such as lactosylation, which is the conjugation of lactose to protein, can occur during drying and also influence the powder properties. This reaction is facilitated at high outlet air temperatures during drying (Guyomarc'h *et al.* 2000).

10.5.2 Non-enzymatic browning

Non-enzymatic browning (NEB) *via* the Maillard reaction can change the colour of milk powder. This reaction occurs between the proteins in milk and lactose-reducing sugar in thermally-processed foods, therefore dairy foods containing lactose are readily susceptible to NEB. Lactose-hydrolysed milk powders and infant formulae are highly susceptible to the Maillard reaction. Lysine is the main amino acid involved in the reaction in the early stages. Infant formulae have a high level of lysine. Freeze-dried milk powders have less loss of reactive and total lysine. Moisture and water activity (a_w) are also important factors in determining the rate of browning in milk powder. The optimum moisture and a_w values for browning in SMP are about 12–14% and 0.5–0.75, respectively. Acevedo *et al.* (2006) reported that NEB was maximum for milk powders heated at temperatures from 70 to 130 °C over a wide range of a_w (0.44–0.85). The influence of a_w on the kinetics of the Maillard reaction in milk powders can be described as a consequence of concentration and diffusion of reactants. The Maillard reaction leads to loss of nutritive value (from proteins and sugars), development of undesirable colour and off-flavour, decreased solubility and loss of vitamins (Labuza & Saltmarch 1982). Sometimes, even though browning is not significant, off-flavour formation renders milk powders unacceptable.

Scorched particles result from deposits of charred powder, discoloured by Maillard browning at low water activities. They are generally accepted to be a measure for any deposits in the drying chamber having been exposed to high temperatures. Since they are discoloured and insoluble, they are undesirable. Essentially, milk powders should be free of scorched particles and foreign matter. The dirt or sediment in raw milk may also contribute to scorched particles in the powder (Bloore & O'Callaghan 2009). The air flow in the spray dryer also must be managed effectively to avoid powder accumulation, which may be transformed into scorched particles. NEB is also reported in low-lactose products such as cheese powders during storage (Kilic *et al.* 1997).

10.5.3 Oxidation and chemical quality

Milk powders have a long shelf life of up to 12 months or more without considerable deterioration in organoleptic quality. However, during storage, the fat on the surface of milk powder particulates is prone to oxidation, leading to the development of rancidity. In the initial stages, primary products such as free radicals and hydroperoxides are formed, which further react to yield a complex mixture of secondary lipid oxidation products such as aldehydes and ketones. The amount of free-fat is an indicator of fat instability in milk powders. In general, heat treatment improves the oxidative stability of milk powders, and thus extends the shelf life of full-cream milk powder (Baldwin & Ackland 1991). This is because the heat treatment inactivates lipase present in milk and also develops the natural antioxidant activity of the milk components. It also causes exposure of the sulphydryl groups and enhances the Maillard reaction. The sulphydryl groups are antioxygenic and therefore reduce the oxidation of fat and the rate of flavour deterioration (Early 1992). Taylor & Richardson (1980) reported that the products of the Maillard reaction during preheating may be partly responsible for the increase in the oxidative stability of milk powders. Both these chemical events contribute to an increased oxidative stability (Taylor & Richardson 1980). However, the best way to improve the oxidative stability is to reduce the free-fat in WMP during manufacture. The amount of free and total fat in milk decreases as the lactose concentration increases. The oxidative stability also can be improved by the addition of certain antioxidants to milk powders during manufacture.

10.5.4 Sensory quality

Milk powders should have a slightly sweet taste and clean flavour, with no indication of rancidity (Early 1998). WMP has a rich and sweet flavour similar to whole milk, while SMP has the characteristics of fluid skim milk with stronger cooked flavours (Lloyd *et al.* 2009). In SMP the more intense cooked, sulphurous and caramelized flavours are associated with high heat treatment. The volatile flavours in milk powders originate from the breakdown of milk fat. Free fatty acids such as butanoic, pentanoic and hexanoic acids caused by fat hydrolysis are the major contributors to flavour. They impart a cheesy flavour to dried milk products, therefore low-fat dried milk products will have lower oxidative flavours when reconstituted after storage. Cooked flavour is due to the Maillard reaction and the production of free sulphydryl groups (Shiratsuchi *et al.* 1995). Lactones contribute to sweet and fatty flavours. Other aroma-active compounds that are found in milk powders are furaneol, maltol, sotolon and vanillin. These flavour compounds are found at higher levels in the high-heat treated powders since most of them are heat-generated. Thus, the preheat treatment of milk plays a major role in the development of specific flavours in milk powders. A wide variety of other compounds, such as aldehydes, alcohols esters and nitrogen-containing compounds, are associated with off-flavour compounds (Karagül-Yüceer *et al.* 2001). The sensory defects in milk powders are influenced by raw milk quality, processing, handling, drying method and inappropriate storage (Bodyfelt *et al.* 1988).

10.6 Properties of special powders

10.6.1 Whey powders

Whey powder is a good source of milk solids for many applications in the dairy and food industry. It may contain as high as 70% lactose and 12% protein. The whey proteins provide good functional and nutritional properties, which are useful to improve the quality of many food products. Both sweet and acid whey can be dehydrated into powder form and can be used (Písecký 2005). Compositionally, the acid and sweet whey powders (SWP) differ in protein, lactic acid, mineral elements and lecithin. Acid whey powders are generally high in minerals compared to SWP. They are also more stable during storage than SWP. However, the vast majority of the whey powder manufactured worldwide is made from sweet whey. SWP is a free-flowing powder with a slight yellowish colour. It contains a relatively high concentration of lactose and lysine.

Whey powder in its native form has limited applicability in food because of its hygroscopicity and salty taste. Hence, customized whey powders are made by process modifications so that they are more appropriate for the desired use. The whey is generally delactosed and demineralized before drying. For powder-making, the whey is evaporated to about 40–60% TS before spray drying. The lactose is in an amorphous state, and hence the powders produced are very hygroscopic (Kelly *et al.* 2003). It is very difficult to dry and handle such powder. Crystallization of lactose to α-lactose monohydrate is often desirable, therefore milk is concentrated and held under conditions that favour crystallization of lactose crystals. The type of spray dryer used for whey powder production also affects the product quality. As in the case of milk powders, single-stage dryers produce fine whey powders that are heavy, cohesive and dusty. Two- and multistage dryers, in contrast, produce agglomerated free-flowing and non-caking whey powders. Spray-dried whey powders, generally, are fine grained and resemble instant SMP. In contrast, freeze-dried whey powders may be flaky in texture. Yoghurt powder is almost spherical, with a porous surface (Koc *et al.* 2014). The bulk and true densities are about 530 and 1400 kg/m^3 (Barbosa-Canovas *et al.* 2005). The solubility of whey powders in water may be as high as 90–95% (Mavropoulou & Kosikowski 1973).

10.6.2 Whey protein concentrates

Another way to process whey is to remove the lactose and transform it into whey protein concentrate (WPC) (Písecký 2005). WPC is utilized in the food industry as a component of baby food and as a fortificant. WPC also has a high content of sulphur-containing amino acids, which support antioxidant functions. Such protein-enriched powders could be substituted for SMP as they possess good functional properties for a range of food applications. Because of the high protein content, WPC tends to be very light with a high content of occluded air, therefore atomization by pressure nozzles is usually preferred to improve the bulk density. Two-stage spray drying is common as the proteins are more likely to be protected against denaturation. Heino *et al.* (2007) observed no differences among drying methods on functional properties such as solubility and water-holding capacity but higher heat load decreased the solubility of WPC powder.

10.6.3 Cheese powder

One of the most important dehydrated cheese products is the cheese powder. It finds widespread use as a flavouring agent and/or nutritional supplement in a variety of foods, such as baked goods, snacks, potato chips, biscuits, savouries, etc. Preparation involves concentration of cheese slurry to 35–45% TS, followed by homogenization and spray drying. The cheese has to be melted and brought to a liquid form at a temperature of about 75 °C so that it is not too viscous for atomization. Co-current flow tall-form chambers with external fluidized beds and nozzle atomizers are preferred for drying. Because of its high fat content (18–25%), secondary air cooling is required to avoid the powder sticking to the dryer walls. The spray-drying conditions optimized by Erbay *et al.* (2015) for cheese powder production are an inlet air temperature of 174 °C, atomization pressure of 354 kPa and an outlet drying temperature of 68 °C. The bulk density, dispersibility, flowability and sinkability can be improved by increasing the TS content. Low drying temperatures produce dense powders and aid flavour retention. The temperature of the outlet air is the most significant process parameter affecting the browning and solubility indices of cheese powder. Masters (2002) reported that use of high-melting fat fractions resulted in low free-fat development and enhanced the flowability of resultant powders.

10.6.4 Yoghurt powder

Yoghurt powder, produced by drying fresh yoghurt, is widely used as an ingredient in confectioneries, bakery foods, beverages, instant drink mixes, soups, etc. The powder, besides its nutritive quality, also contains the same probiotic starter bacteria (*Lactobacillus bulgaricus* and *Streptococcus thermophilus*) as found in fresh yoghurt. The major problem in spray drying yoghurt is the preservation of the starter bacteria during and after drying. Because the probiotic bacteria are heat sensitive, the spray-drying conditions must be mild enough to avoid damage to these microbes but sufficient to yield a powder with moisture content less than 4%, which is required for stability during storage. Kirn & Bhowmik (1994) outlined the method for production of dried yoghurt and reported that preconcentration of stirred yoghurt under vacuum did not have any effect on the survivability of the probiotics. The survival of yoghurt bacteria during spray drying is affected by air temperature, type of atomization and direction of air flow in the spray dryer. Selvamuthukumaran & Shukla (2006) studied the production of spray-dried Bifidus milk powder and reported that both inlet air temperature and air pressure had a significant influence on the survival of bifidobacteria. The optimum survival of bacteria is at an outlet temperature of 65–75 °C. Other researchers have reported that protective agents such as hydrocolloids, starches, oligosaccharides and stabilizing gums (sodium alginate, gum acacia) influence the survivability of these starter bacteria (Crittenden *et al.* 2001). Also, the addition of thermoprotectants such as trehalose and milk proteins have been used in an attempt to improve the viability of probiotics during drying (Conrad *et al.* 2000). The survival rate of *Streptococcus thermophilus* is higher than *Lactobacillus bulgaricus* during spray drying. The colour and sensory properties of yoghurt powder are influenced by the outlet air temperature and drying conditions (Bielecka & Majkowska 2000). The oxidative stability of yoghurt powder is dependent on the heat treatment to milk, storage temperature and water activity. The bulk, tapped and true densities of yoghurt powder are 538, 746 and 1177 kg/m^3, respectively (Koc *et al.* 2014). *Dahi* powder can be made in a similar way to yoghurt powder.

10.6.5 Infant milk powders

Manufacturing of infant formulae involves blending water-soluble milk proteins, carbohydrates, vitamins and minerals with vegetable oils, followed by dehydrating the emulsion to achieve shelf-life stability. Thus, infant formulae are a diverse group of products that reconstitute well in warm water and are free of undissolved lumps. Most of the dried infant formulae sold today are produced by spray drying. The lysine amino acid content is important for the nutritional quality of infant formulae. The extent of lysine loss is influenced by outlet air temperature, drying kinetics and residence time inside the spray dryer (Schmitz-Schug *et al.* 2013). Infant milk powders are agglomerated for better reconstitution. Formulations with higher concentrations of carbohydrates are easier to agglomerate compared to protein-rich ones, therefore the degree of agglomeration and the compactness of the agglomerated particles are important and have to be controlled.

The typical density of a well-agglomerated infant formula is between 400 and 500 g/l. Chuy & Labuza (1994) investigated the effects of heating rate and storage temperature on the caking of infant milk powders. Powders with smaller particles are more cohesive and are easily flowable. Lactose content improves the flowability.

10.6.6 Dairy whiteners

Dairy whitener or creamer is a vital ingredient for use with coffee. Both SMP and WMP can be used as coffee whiteners. The rich taste and stable foam are desirable functionalities. The milk proteins in the whitener react with tannins in coffee and give the drink a pleasant and creamy taste. Besides coffee stability, the whitener should have whitening ability, provide body and mouthfeel, and impart a rich, creamy dairy flavour (Oldfield & Singh 2005). Drying can have a huge effect on the ability of these powders to act as whitening agents. High preheat treatment, long holding times and high TS concentration negatively affect the whitening potential of the powder. Since dairy whiteners are reconstituted, they also should have flowability, reconstitution properties and resistance to caking and rancidity. Most of these properties are affected by particle size, density, shape and composition (Kim *et al.* 2005), therefore dairy whiteners are agglomerated to a reasonable extent. Two-stage spray drying with integrated fluidized beds is used for manufacture. Surface free-fat also has a significant influence in determining the properties of dairy whitener.

10.7 Conclusion

The physical, chemical and sensory properties of dairy powders are largely influenced by the drying method. Spray drying, the most commonly used drying technique in the dairy industry, produces dried powders with desirable morphology, particle size, density and reconsitution properties. In spite of the high inlet drying temperatures involved, the denaturation of whey proteins is negligible as the temperature of the drying droplet is maintained below 70 °C. On the other hand, non-thermal methods, such as freeze drying and spray freeze drying, produce dried products with high porosity, which aids in instant solubility, and low degradation of micronutrients. Hence, careful selection of the drying technique is essential to obtain suitable dried dairy products for the end applications.

References

Acevedo, N., Schebor, C. & Buera, M.P. (2006). Water–solids interactions, matrix structural properties and the rate of non-enzymatic browning. *Journal of Food Engineering*, **77**(**4**), 1108–1115.

Aguilar, C.A. & Ziegler, G.R. (1994). Physical and microscopic characterization of dry whole milk with altered lactose content. 1. Effect of lactose concentration. *Journal of Dairy Science*, **77**(**5**), 1189–1197.

Anandharamakrishnan, C., Rielly, C.D. & Stapley, A.G.F. (2008). Loss of solubility of α-lactalbumin and β-lactoglobulin during the spray drying of whey proteins. *LWT – Food Science and Technology*, **41**(**2**), 270–277.

Anandharamakrishnan, C., Rielly, C.D. & Stapley, A.G. (2010). Spray-freeze-drying of whey proteins at sub-atmospheric pressures. *Dairy Science & Technology*, **90**(**2–3**), 321–334.

Augustin, M.A. & Clarke, P.T. (2008). Dry milk products. In: *Dairy Processing & Quality Assurance* (eds Chandan, R.C., Kilara, A. & Shah, N.P), pp. 319–336, John Wiley & Sons.

Baldwin, A.J. & Ackland, J.D. (1991). Effect of preheat treatment and storage on the properties of whole milk powder. Changes in physical and chemical properties. *Netherlands Milk and Dairy Journal*, **45**, 169–181.

Baldwin, A. & Pearce, D. (2005) Milk powder. In: *Encapsulated and Powdered Foods* (ed. Onwulata, C.), pp. 387–434, Taylor & Francis, Boca Raton.

Baldwin, A.J., Baucke, A.G. & Sanderson, W.B. (1980). The effect of concentrate viscosity on the properties of spray dried skim milk powder. *New Zealand Journal of Dairy Science and Technology*, **15**, 289–297.

Barbosa-Cánovas, G.V., Ortega-Rivas, E., Juliano, P. & Yan, H. (2005). *Food powders: physical properties, processing, and functionality*, pp. 83–88, **341**, Kluwer Academic/Plenum Publishers, New York.

Berlin, E. & Pallansch, M.J. (1963). Influence of drying methods on density and porosity of milk powder granules. *Journal of Dairy Science*, **46**(**8**), 780–784.

Bielecka, M. & Majkowska, A. (2000). Effect of spray drying temperature of yoghurt on the survival of starter cultures, moisture content and sensoric properties of yoghurt powder. *Food/Nahrung*, **44**(**4**), 257–260.

Birchal, V.S., Passos, M.L., Wildhagen, G.R. & Mujumdar, A.S. (2005). Effect of spray-dryer operating variables on the whole milk powder quality. *Drying Technology*, **23**(**3**), 611–636.

Bloore, C.G. & O'Callaghan, D.J. (2009). Process control in evaporation and drying. In: *Dairy Powders and Concentrated Products* (ed. Tamime, A.Y.), pp. 332–350, Wiley-Blackwell.

Bockian, A. H., Stewart, G. F., & Tappel, A. L. (1957). Factors affecting the dispersibility of "instantly dissolving" dry milks. *Journal of Food Science*, **22**(**1**), 69–75.

Bodyfelt, F.W., Tobias, J. & Trout, G.M. (1988). Sensory defects of dairy products: An overview. In: *The Sensory Evaluation of Dairy Products*, pp. 67–89, Van Nostrand Reinhold, London.

Boersen, A.C. (1990) Spray drying technology – atomisation and straight through agglomeration. *Journal of the Society of Dairy Technology*, **43**, 5–7.

Caric, M. (1993) Concentrated and dried dairy products. In: *Dairy Science and Technology Handbook 2: Product Manufacturing* (ed. Hui, Y.H.), pp. 257–300, VCH Publishers, New York.

Caric, M. (2002) Milk powders: types and manufacture. In: *Encyclopedia of Dairy Sciences* (eds Roginski, H., Fuquay, J.W. & Fox, P.F.), pp. 1869–1874, Academic Press, London.

Caric, M. (2004) Dairy – milk powders. In: *Food Processing – Principles and Applications* (eds Smith, J.S. & Hui, Y.H.), pp. 319–328, Blackwell Publishing.

Caric, M. & Kalab, M. (1987). Effects of drying techniques on milk powders quality and microstructure: a review. *Food Structure*, **6**(**2**),171–180.

Caric, M. & Milanovic, S. (2002). Milk powders: Physical and functional properties of milk powders. In: *Encyclopedia of Dairy Sciences* (eds Alm, L., Roginsky, H., Fuquay, J. & Fox, P.), pp. 1874–1880, Academic Press, London.

Caric, M., Milanovic, S., Akkerman, C., Kentish, S.E. & Tamime, A.Y. (2009). Technology of evaporators, membrane processing and dryers. In: *Dairy Powders and Concentrated Products* (ed. Tamime, A.Y.), pp. 99–148, Wiley-Blackwell.

Chandan, R.C. (2008). Dairy processing and quality assurance: An overview. In: *Dairy Processing & Quality Assurance* (eds Chandan, R.C., Kilara, A. & Shah, N.P), pp. 1–40, John Wiley & Sons.

Chauhan, B.S. & Kumar, D. (2004). Effect of spray dryer operating parameters on the quality of milk powder. *New Agriculturist*, **15(1/2)**, 107–111.

Chuy, L.E. & Labuza, T.P. (1994). Caking and stickiness of dairy-based food powders as related to glass transition. *Journal of Food Science*, **59(1)**, 43–46.

Codex Alimentarius (1999). *Codex standard for milk powders and cream powder*.Codex Standard, 207.

Conrad, P.B., Miller, D.P., Cielenski, P.R. & de Pablo, J.J. (2000). Stabilization and preservation of *Lactobacillus acidophilus* in saccharide matrices. *Cryobiology*, **41(1)**, 17–24.

Crittenden, R., Laitila, A., Forssell, P., Mättö, J., Saarela, M., Mattila-Sandholm, T. & Myllärinen, P. (2001). Adhesion of bifidobacteria to granular starch and its implications in probiotic technologies. *Applied and Environmental Microbiology*, **67(8)**, 3469–3475.

De Vilder, J., Martens, R. & Naudts, M. (1976). Influence of process variables on some whole milk powder characteristics. *Milchwissenschaft – Milk Science International*, **31**, 396–401.

Deeth, H.C. & Hartanto, J. (2009). *Chemistry of milk – role of constituents in evaporation and drying*. In: *Dairy Powders and Concentrated Products* (ed. Tamime, A.Y.), pp. 1–27, Wiley-Blackwell.

Duffie, J. A., & Marshall, W. R. (1953). Factors influencing the properties of spray-dried materials. *Chemical Engineering Progress*, **49(8)**, 417–423.

Early, R. (1992). Milk powders. In: *The Technology of Dairy Products* (ed. Early, R.), pp. 167–196, Blackie Academic and Professional, Glasgow.

Early, R. (1998) Milk concentrates and milk powders. In: *The Technology of Dairy Products*, 2nd edn (ed. Early, R.), pp. 228–300, Blackie Academic and Professional, Glasgow.

Elversson, J., Millqvist-Fureby, A., Alderborn, G. & Elofsson, U. (2003). Droplet and particle size relationship and shell thickness of inhalable lactose particles during spray drying. *Journal of Pharmaceutical Sciences*, **92(4)**, 900-910.

Enríquez-Fernández, B.E. Camarillo-Rojas, C.R. & Vélez-Ruiz, J.F. (2013). Physical properties of concentrated milk and its influence on powder milk characteristics and spray dryer design parameters. *Journal of Food Process Engineering*, **36(1)**, 87–94.

Erbay, Z., Koca, N., Kaymak-Ertekin, F. & Ucuncu, M. (2015). Optimization of spray drying process in cheese powder production. *Food and Bioproducts Processing*, **93**, 156–165.

Farkye, N.Y. (2006). Significance of milk fat in milk powder. In: *Advanced Dairy Chemistry, Vol. 2, Lipids* (eds Fox, P.F. & McSweeney, P.L.H.), pp. 451–465, Springer.

Fergusson, P.H. (1989). Developments in the evaporation and drying of dairy products. *International Journal of Dairy Technology*, **42(4)**, 94–101.

Fyfe, K., Kravchuk, O., Nguyen, A.V., Deeth, H. & Bhandari, B. (2011). Influence of dryer type on surface characteristics of milk powders. *Drying Technology*, **29(7)**, 758–769.

Gavrielidou, M.A., Valous, N.A., Karapantsios, T.D. & Raphaelides, S.N. (2002). Heat transport to a starch slurry gelatinizing between the drums of a double drum dryer. *Journal of Food Engineering*, **54(1)**, 45–58.

Greenwald, C. & King, C.J. (1981). The effects of design and operating conditions on particle morphology for spray-dried foods. *Journal of Food Process Engineering*, **4**(**3**), 171–187.

Guyomarc'h, F., Warin, F., Muir, D.D. & Leaver, J. (2000). Lactosylation of milk proteins during the manufacture and storage of skim milk powders. *International Dairy Journal*, **10**(**12**), 863–872.

Hahne, L.C.C. & Passos, M.L. (2002). Using the Hausner index for determining the cohesive particle flow regimes in vibro-fluidized beds. In: *Proceedings of 29th Brazilian Congress on Particulate Systems*.

Hall, C.W. and Hedrick, T.I. (1961). The manufacture of instant milk powder. *Dairy Engineering*, **78**(**1**), 7.

Hardy, E.E., Sweetsur, A.W., West, I.G. & Muir, D.D. (1985). Heat stability of concentrated milk: enhancement of initial heat stability by incorporation of food grade lecithin. *International Journal of Food Science & Technology*, **20**(**1**), 97–105.

Hayashi, H., Heldman, D.R. & Hedrick, T.I. (1970). Physical properties of nonfat dry milk as influenced by spray drying conditions. *Transactions of the ASAE*, **13**(**6**), 723–725.

Heino, A.T., Uusi-Rauva, J.O., Rantamäki, P.R. & Tossavainen, O. (2007). Functional properties of native and cheese whey protein concentrate powders. *International Journal of Dairy Technology*, **60**(**4**), 277–285.

Jouppila, K. & Roos, Y.H. (1994). Glass transitions and crystallization in milk powders. *Journal of Dairy Science*, **77**(**10**), 2907–2915.

Karagül-Yüceer, Y., Drake, M. & Cadwallader, K.R. (2001). Aroma-active components of nonfat dry milk. *Journal of Agricultural and Food Chemistry*, **49**(**6**), 2948–2953.

Kelly, J., Kelly, P.M. & Harrington, D. (2002). Influence of processing variables on the physicochemical properties of spray dried fat-based milk powders. *Le Lait*, **82**(**4**), 401–412.

Kelly, A.L., O'Connell, J.E. & Fox, P.F. (2003). Manufacture and properties of milk powders. In: *Advanced Dairy Chemistry – 1 Proteins* (eds Fox, P.F. & McSweeney, P.L.H.), pp. 1027–1061, Springer.

Keogh, M.K., Murray, C.A. & O'Kennedy, B.T. (2003). Effects of ultrafiltration of whole milk on some properties of spray-dried milk powders. *International Dairy Journal*, **13**(**12**), 995–1002.

Kilic, M., Muthukumarappan, K. & Gunasekaran, S. (1997). Kinetics of nonenzymatic browning in cheddar cheese powder during storage. *Journal of Food Processing and Preservation*, **21**(**5**), 379–393.

Kim, E.H.J., Chen, X.D. & Pearce, D. (2005). Effect of surface composition on the flowability of industrial spray-dried dairy powders. *Colloids and Surfaces B: Biointerfaces*, **46**(**3**), 182–187.

Kirn, S.S. & Bhowmik, S.R. (1994). Moisture sorption isotherms of concentrated yogurt and microwave vacuum dried yogurt powder. *Journal of Food Engineering*, **21**(**2**), 157–175.

Koç, B., Sakin-Yılmazer, M., Kaymak-Ertekin, F. & Balkır, P. (2014). Physical properties of yoghurt powder produced by spray drying. *Journal of Food Science and Technology*, **51**(7), 1377–1383.

Kostoglou, M. & Karapantsios, T.D. (2003). On the thermal inertia of the wall of a drum dryer under a cyclic steady state operation. *Journal of Food Engineering*, **60**(**4**), 453–462.

Kyle, W.S.A. (1993) Powdered milk. In: *Encyclopedia of Food Science, Food Technology and Nutrition* (eds McCrae, R., Robinson, R.K. & Sadler, M.J.), pp. 3700–3713, Academic Press, New York.

Labuza, T. P., & Saltmarch, M. (1982). Kinetics of browning and protein quality loss in whey powders during steady state and nonsteady state storage conditions. *Journal of Food Science*, **47**(**1**), 92–96.

Litman, I.I. & Ashworth, U.S. (1957). Insoluble scum-like materials on reconstituted whole milk powders. *Journal of Dairy Science*, **40**(**4**), 403–409.

Lloyd, M.A., Drake, M.A. & Gerard, P.D. (2009). Flavor Variability and Flavor Stability of US-Produced Whole Milk Powder. *Journal of Food Science*, **74**(**7**), S334–S343.

Masters, K. (1991). Spray drying fundamentals: process stages and layouts. In: *Spray Drying Handbook*, 5th edn, pp. 23–64, Longman Scientific and Technical, New York.

Masters, K. (2002). Perfecting the powder. *Dairy Industries International*, **67**(**12**), 27–32.

Mavropoulou, I.P. & Kosikowski, F.V. (1973). Composition, solubility, and stability of whey powders. *Journal of Dairy Science*, **56**(**9**), 1128–1134.

McCrae, C.H. & Muir, D.D. (1992). Heat stability of recombined milk: influence of lecithins on the heat coagulation time-pH profile. *Journal of Dairy Research*, **59**(**02**), 177–185.

Mettler, A.E. (1980). Chemical and physical aspects of powder quality. In: *Milk and Whey Powders*, pp. 92–116, The Society of Dairy Technology, Middlesex.

Neff, E. & Morris, H.A.L. (1968). Agglomeration of milk powder and its influence on reconstitution properties. *Journal of Dairy Science*, **51**(**3**), 330–338.

Nickerson, T.A., Coulter, S.T. & Jenness, R. (1952). Some properties of freeze-dried milk. *Journal of Dairy Science*, **35**(**1**), 77–85.

Oldfield, D. & Singh, H. (2005). Functional properties of milk powders. In: *Encapsulated and Powdered Foods* (ed. Onwulata, C.), pp. 366–386, Taylor & Francis.

Oldfield, D.J., Singh, H. & Taylor, M.W. (1998). Association of β-lactoglobulin and β-lactalbumin with the casein micelles in skim milk heated in an ultra-high temperature plant. *International Dairy Journal*, **8**(**9**), 765–770.

Oldfield, D.J., Teehan, C.M. & Kelly, P.M. (2000). The effect of preheat treatment and other process parameters on the coffee stability of instant whole milk powder. *International Dairy Journal*, **10**(**9**), 659–667.

Palmer, L.S. & Dahle, C.D. (1922). Structure of powdered milk and its possible relation to the keeping quality of whole milk powders. *Journal of Dairy Science*, **5**(**2**), 240–245.

Parris, N., White, A.E. & Farrell Jr,, H.M. (1990). Identification of altered proteins in nonfat dry milk powder prepared from heat-treated skim milk. *Journal of Agricultural and Food Chemistry*, **38**(**3**), 824–829.

Patel, H.A., Singh, H., Anema, S.G. & Creamer, L.K. (2006). Effects of heat and high hydrostatic pressure treatments on disulfide bonding interchanges among the proteins in skim milk. *Journal of Agricultural and Food Chemistry*, **54**(**9**), 3409–3420.

Pisecky, J. (1997). *Handbook of Milk Powder Manufacture*, Niro A/S, Copenhagen, **3**(**1**).

Písecký, J. (2005). Spray drying in the cheese industry. *International Dairy Journal*, **15**(**6**), 531–536.

Rennie, P.R., Chen, X.D., Hargreaves, C. & Mackereth, A.R. (1999). A study of the cohesion of dairy powders. *Journal of Food Engineering*, **39**(**3**), 277–284.

Rogers, S., Wu, W.D., Saunders, J. & Chen, X.D. (2008). Characteristics of milk powders produced by spray freeze drying. *Drying Technology*, **26**(**4**), 404–412.

Rollema, H.S. & Muir, D.D. (2009). Casein and related products. In: *Dairy Powders and Concentrated Products* (ed. Tamime, A.Y.), pp. 235–252, Wiley-Blackwell.

Sanderson, W.B. (1978). Instant milk powders: manufacture and keeping quality. *New Zealand Journal of Dairy Science and Technology*, **13**, 137–143.

Schmitz-Schug, I., Foerst, P. & Kulozik, U. (2013). Impact of the spray drying conditions and residence time distribution on lysine loss in spray dried infant formula. *Dairy Science & Technology*, **93(4–5)**, 443–462.

Selvamuthukumaran, M. & Shukla, S.S. (2006). Optimization of spray drying conditions for production of bifidus milk powder from cow milk. *Journal of Food Quality*, **29(4)**, 305–318.

Shiratsuchi, H., Yoshimura, Y., Shimoda, M., Noda, K. & Osajima, Y. (1995). Contributors to sweet and milky odor attributes of spray-dried skim milk powder. *Journal of Agricultural and Food Chemistry*, **43(9)**, 2453–2457.

Singh, H. (2007). Interactions of milk proteins during the manufacture of milk powders. *Le Lait*, **87(4–5)**, 413–423.

Singh, H. & Creamer, L.K. (1991). Denaturation, aggregation and heat stability of milk protein during the manufacture of skim milk powder. *Journal of Dairy Research*, **58(03)**, 269–283.

Singh, H., Sharma, R. & Tokley, R.P. (1992). Influence of incorporation of soya lecithin into skim milk powder on the heat stability of recombined evaporated milk. *Australian Journal of Dairy Technology*, **47(1)**, 33.

Skanderby, M., Westergaard, V., Partridge, A. & Muir, D.D. (2009). Dried milk products. In: *Dairy Powders and Concentrated Products* (ed. Tamime, A.Y.), pp. 180–234, Wiley-Blackwell.

Sweetsur, A.M. & Muir, D.D. (1981). Role of cyanate ions in the urea-induced stabilization of the caseinate complex in skim-milk. *Journal of Dairy Research*, **48(01)**, 163–166.

Taylor, M.J. & Richardson, T. (1980). Antioxidant activity of skim milk: effect of heat and resultant sulfhydryl groups. *Journal of Dairy Science*, **63(11)**, 1783–1795.

Tuohy, J.J. (1989). Some physical properties of milk powders. *Irish Journal of Food Science and Technology*, **13(2)**, 141–152.

Twomey, M. & Keough, M.K. (1998) Milk powder in chocolate. *Farm Food Spring*, **8**, 9–11.

Valous, N.A., Gavrielidou, M.A., Karapantsios, T.D. & Kostoglou, M. (2002). Performance of a double drum dryer for producing pregelatinized maize starches. *Journal of Food Engineering*, **51(3)**, 171–183.

Vega-Mercado, H., Góngora-Nieto, M.M. & Barbosa-Cánovas, G.V. (2001). Advances in dehydration of foods. *Journal of Food Engineering*, **49(4)**, 271–289.

Verhey, J.G.P. (1973). Vacuole formation in spray powder particles. 3. Atomization and droplet drying. *Netherlands Milk and Dairy Journal*, **27**, 3–18.

Westergaard, V. (1994). *Milk Powder Technology – Evaporation and Spray Drying*, Niro A/S, Copenhagen.

Woodhams, D.J. & Murray, M.J. (1974). Properties of spray dried milk powders. *New Zealand Journal of Dairy Science and Technology*, **13**, 172–178.

11

Packaging of Dried Dairy Products

R. Gopirajah[1] *and C. Anandharamakrishnan*[1,2]

[1] *Department of Food Engineering, CSIR – Central Food Technological Research Institute, Mysore, Karnataka, India*
[2] *Indian Institute of Crop Processing Technology, Ministry of Food Processing Industries, Government of India, Thanjavur, Tamil Nadu, India*

11.1 Introduction

Packaging is an essential vehicle for delivering products to consumers. The increasing consumer demands for product safety, convenience, cost efficacy, environmental awareness and shelf-life extension mean packaging is an essential component of the food processing and distribution system (Marsh & Bugusu 2007). Food packaging acts as a barrier between food and the outside environment, protecting the contained product during storage and transport. The packaging ensures the availability of products irrespective of season, over long distances in various forms, fresh as well as processed (Paine & Paine 1992). Thus, packaging imparts a unique value to food products and help consumers to choose from a variety of similar products. In recent years a variety of packaging techniques with the ability to protect and preserve products have evolved (Simon & Hansen 2001). Such techniques are mainly focused on protecting the contained food from oxygen, water vapour, light, and microbial and chemical contamination. In this context, the development of active packaging, aseptic packaging, gas/vacuum packaging, bio-based packaging, have revolutionized the food industry (Narayanan & Dordi 1998; Ozdemir & Floros 2004).

The packaging of dried food products requires special attention as the moisture content lies in the range of 2–8% and the equilibrium relative humidity (ERH) is 10–30% (Krokida *et al.* 2003), for example dried milk powder contains moisture content in the range of 1–3% with relative humidity (RH) below 20% (Thomas *et al.* 2004). This makes dried food products hygroscopic, adding to their porous nature and high surface-to-weight ratio. Moreover, low-moisture foods are prone to oxidative deterioration (Havemose *et al.* 2006). Hence, a specially designed packaging material impermeable to water vapour becomes necessary. The exposure of milk powder to oxygen and light induces lipid oxidation, producing the undesirable flavour of oxidized fat (Tian *et al.* 2013). Thus, the packaging of dried milk powder is crucial in preventing the formation of various compounds that impart more or less undesirable flavour to the reconstituted milk. The various forms of dried dairy products include skim milk powder (SMP), whole milk powder (WMP), buttermilk powder (BMP) and milk protein concentrates (Hall & Hedrick 1971). The selection of suitable packaging machine and packaging material for these requires a lot of attention, information and research. The keeping quality of dried milk products during storage in a particular environment is also of great importance for consumer acceptance. This chapter presents an overview of developments in the packaging of dried milk products, paying special attention to recent technologies that combat lipid oxidation through active packaging solutions. Also highlighted is the development of new biodegradable polymer nanocomposites for extending shelf life and reducing package waste.

11.2 Dairy packaging trends

Packaging is the most effective means of protecting contained products from the point of manufacture to consumption (Marsh & Bugusu 2007). The food industry is the major user of packaging materials in terms of both variety and quantity (Paine & Paine 1992). About 25–30 years ago, milk was home delivered by milkmen. Later, bottled milk came into existence as the organized sector started milk distribution to maintain quality and hygiene. The retail sale of pasteurized milk by organized dairies was either to consumers from cans to individual containers or bottled (Roy 2002). The different types of pasteurized milk, such as standardized, toned, double toned, cow's milk, buffalo's milk and skim milk, were distinguished by the adoption of aluminium caps of different colours/designs on the bottles. The practice of packing liquid milk in glass bottles remained unchallenged until 1933, when the first carton made of waxed paper was introduced. This was mainly developed to overcome the disadvantages of bottling milk, such as high initial and operational costs, breakage of bottles during handling and transportation, need for cleaning and sterilization of bottles, heavier loads during transportation etc. Later, the development of the packaging industry with the introduction of plastic material (polyethylene in 1940) either alone or in combination with paper resulted in a wide range of containers (termed cartons) and plastic sachets (Dhar 2002). Plastics are used in the food industry in the form of containers/bottles as well as films, co-extruded materials and laminates (Brown 1992). Furthemorer, a variety of packaging materials, including tinplate, aluminum, plastics, paper board and jute, were developed for use in the food industry (Narayanan & Dordi 1998). Tinplate offers excellent resistance to gas, water vapour and light, but the growth of this industry was threatened by the development of alternative materials like plastic containers, films and laminates for their cost-effectiveness. Currently, there are many different forms and types of packaging materials are available, as listed in Table 11.1 (Marsh & Bugusu 2007).

From the different available packages, the selection of a particular package for a product during storage is crucial to achieve the expected shelf life. This mainly depends on factors affecting the transport of low molecular weight compounds from the internal or external environment through the package wall. In case of milk powders, these factors include initial moisture, final acceptable moisture, light protection and the required shelf life. More headspace oxygen promotes oxidation. Milk powder packaging must be oxygen and light impermeable for the best keeping quality (Lloyd *et al.* 2004). On comparing a plastic bag with a glass jar, the oxidation level was higher in the plastic bag (Chan *et al.* 1993). This was because oxidation of the plastic bag occurred through

Table 11.1 Overview of packaging materials used in dairy industries

Product	*Packaging material*
Liquid milk	• Glass bottles are used only for flavoured milk • Plastic sachets made of LDPE film • Paper laminate for tetra packs
Powdered milk	• Tinplate containers, nitrogen-packed and lacquered from the outside • Flexible laminates such as metalized PET/aluminium foil/BOPP • Bag-in-box: laminate-filled powder and packed in cartons
Butter	• Duplex board with vegetable parchment paper • Tinplate containers • Aluminium foil
Cheese/cheese spread	• Tinplate containers lacquered from inside • Packed in aluminium foil and then in duplex board carton • Injection-moulded HDPE container
Ghee	• Tinplate container • Glass bottles • HDPE film pouches
Ice-cream	• Thermoformed/injection-moulded plastic containers • Duplex board cartons that are poly laminated • Laminates of BOPP or PET
Indigenous products such as shrikhand, gulab jamun etc.	• Injection moulded/thermoformed containers • Stand-up laminated pouches

LDPE, low-density polyethylene; PET, polyethylene terephthalate; HDPE, high-density polyethylene; BOPP, biaxially oriented polypropylene.

atmospheric oxygen attacking on it, whereas in the glass jar oxidation occurred due to the incorporated oxygen in the milk powder.

11.3 Forms of packaging materials

Milk and dairy products are packaged in various types and forms of materials depending on the nature of the product, processing conditions, storage parameters and end use format (Karaman *et al.* 2015). Most food production is in rural pockets of the country while the major markets are in urban areas so transportation over long distances has become a necessity. The packaging is often grouped into three categories: primary, secondary and tertiary (Mathlouthi 2013). The primary package is one that is in direct contact with the product for use in retail markets, whereas the secondary packaging contains the primary package, and the tertiary package is designed to contain the secondary package during transportation and warehousing. As dairy products are extremely perishable products the utmost care is required in their preservation during storage, handling and transportation. Dairy products spoil quickly at high temperature and in the presence of oxygen and other contaminating agents present in the atmosphere (Simon & Hansen 2001). The choice of an appropriate packaging material is therefore necessary to enhance the shelf life and influence the purchaser against other competing merchandise. The packaging forms desc ribed in the following sections are important.

11.3.1 Metal cans

These can be round, square, oval or pail-shaped, flat, etc. Cans are often classified into two-piece or three-piece cans. The latter uses tinplate as its basic material, and the can is joined by soldering or welding (Blunden & Wallace 2003). In the case of tin-free-steel (TFS) cans, the body is made using an organic adhesive agent. Two-piece cans include cans punched out by a press machine, while aluminum cans can be made by impact extraction. The prime purpose of packaging in a metal container is physical and chemical protection of the product to be marketed. Metal cans also offer impermeability to moisture, gases and light (Blunden & Wallace 2003). Metal cans are compatible with many products and offer high stacking strength, thermal stability and a good surface for decoration and coating. They have the potential for high-speed manufacturing and filling. Many designs offer easy-opening ends that do not require tools to access the contents. The disadvantage of cans is that the cost of setting up a can production line is high.

11.3.2 Glass bottles

Glass bottles and containers are available in many different shapes, such as large 'free-size' bottles, small 'one-shot' styles, light-weight bottles for soft drinks, heavy-weight hand-crafted-type liquor bottles, returnable bottles etc. Glass bottles have been used to package many liquid foods. Glass offers protection against oxygen and light if it has been treated with anti-actinic compounds. Over the past few years it has replaced cartons because of its recyclability, but the use of glass bottles for dried milk products is very limited. The use of glass packages during the storage of fermented milk products favours the survival of probiotic cultures due to their extremely low oxygen permeability (Kudelka 2005).

11.3.3 Stretch-wrap packaging

In this method food is placed in a tray and film is stretched over the food to cover it. Stretchable polyvinyl chloride (PVC) and polyethylene (PE) films etc. are used as packaging materials (Paulett 1999). Shrink packaging is a form of packaging that shrinks when heated using dry or moist heat. Films that are used for this application include PVC, polypropylene (PP) and PE. Jones (1990) patented thermoplastic stretch-wrap material which offers puncture and tear resistance. It consists of a core layer and a skin layer containing linear low-density copolymer made of ethylene and an olefin (4–10 carbon atoms). The patent claims that this thermoplastic stretch-wrap offers excellent physical properties in terms of tensile and burst strength.

11.3.4 Flexible pouches

Flexible packages come mainly in the form of pouches. A pouch is a container made of a flexible packaging material, such as plastic film, aluminum foil, paper etc., that is used either singly or in continuation. The flexible pouches offer impermeability to water vapour and gases (Alavi *et al.* 2014). The concept of retort pouch was created in the 1940s, arising from a military need since conventional rigid cans posed problems such as difficulty in opening, injury and misuse as an explosives by the enemy. Retortable flexible packages are characterized by their structural components of heat-resistant plastic

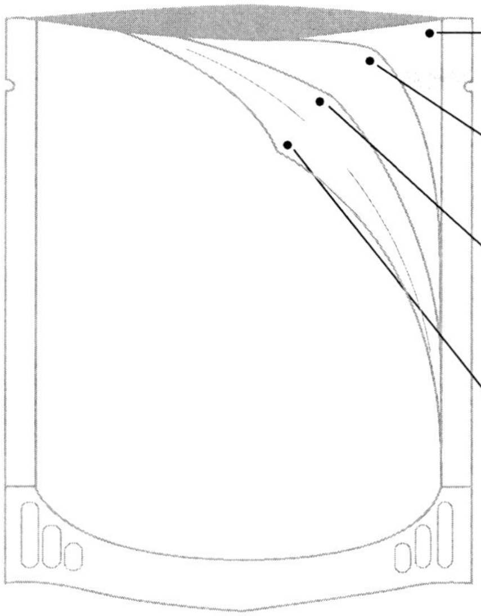

Figure 11.1 The various layers in a retort pouch.

layers with or without aluminium foil and their ability to be thermo-processed to give a shelf-stable food product (Brody & Sacharow 1970). Their cost is also less compared to cans. Retort pouches provide good barrier properties and heat sealing over processing temperatures ranging from 110 to 140 °C. The typical structures of retort pouches and their functional properties are shown in Figure 11.1.

11.3.5 Bag-in-box packages

Bag-in-carton or bag-in-box containers have double construction, with both an inner and an outer package. The former type is used for several food items, including liquids, while the latter is for institutional use and bulk shipment (Avneet 2013). The outside package provides mechanical strength, while the inside bag protects the contents against gases, water vapour and volatiles. It can be made of a single substance or a be multilayer structure.

11.3.6 Cups

The types of cups used as containers include air-pressure formed, thermo-formed and expanded plastic sheets. Recently, a cup with a barrier layer manufactured using pressurized air with a laminated sheet and a composite cup with an inner layer of aluminium foil has been introduced (Avneet 2013). Paper cups, with PE, PP or PET thermoformed inserts, are also used.

11.3.7 Paper-board containers

'Pure-pak'-type containers, with their distinctive gable top, and 'brick-type' smaller containers are the predominant packages used for milk products. Paper laminated to PE is used for ordinary milk while laminated aluminium foils are used for long-shelf life milk

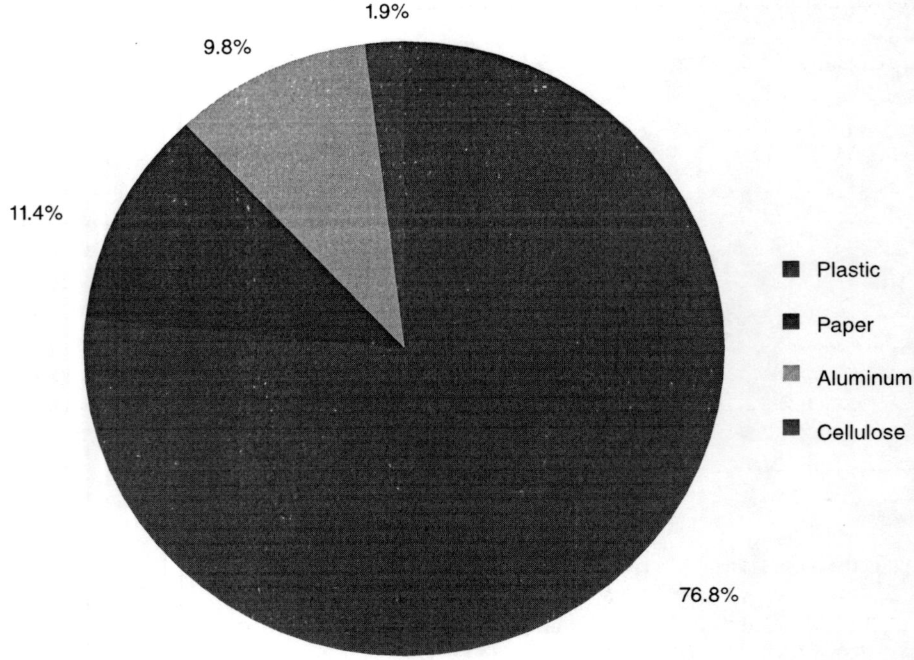

Figure 11.2 Global flexible packaging market voulme share by product, 2012.

products, fruit drinks etc. The market share of flexible packaging materials is shown in Figure 11.2, which indicates that plastics have the major share because of theirs ease of use in various forms. However, plastics have serious economical problems with respect to recycling waste. Environmental pressure and the tendency for plastics not to be used as a food contact material, has led to the development of paper produced from recycled fibers (Triantafyllou *et al.* 2007).

11.4 Packaging of dried milk products

The keeping quality of dried milk products during storage is of great importance to consumer acceptance. Dried milk powder is generally produced by roller or spray drying (Kim *et al.* 2009). Roller-dried milk powder has a creamy colour, is granular and has an oxidized fat odour, whereas, in case of spray-dried powder, an instant heat treatment is used to produce fine white powder with a smaller particle size that reconstitutes better than roller-dried powder. Roller-dried powder has better bacteriological quality due to the more rigorous heat treatment. To maintain the high quality of the product during storage and transportation, it is necessary to prevent chemical reactions causing the formation of various compounds that impart a more or less undesirable flavour to the reconstituted milk (Thomas *et al.* 2004). If the moisture content is below 3%, bacterial growth may be constrained and the only serious deterioration may be due to slow oxidation of the fat, which may be prevented/delayed by holding the product in an oxygen-free atmosphere, coating it with an impermeable material, gas packing or the addition of antioxidants (Rooney 1995).

Milk powder should be packed in a moisture-free atmosphere due to its hygroscopic nature. Lactose is main component of milk and it is in a non-crystalline/amorphous form (glassy state), which makes the milk powder very hygroscopic. If milk powder is exposed to a humid atmosphere, it will readily absorb moisture, become sticky and cake during storage (Hennigs *et al.* 2001), so the packaging of dried milk products is therefore significant. Dried powders tend to absorb water vapour freely from the ambient air surrounding them as the humidity of ambient air is rarely as low (<20%) as that of the dried product. Moreover, these foods often have a porous nature and a high surface-to-weight ratio, thereby increasing the hygroscopicity. Hence, packaging is done in an air-conditioned room with low RH to avoid any moisture uptake. Milk powders are very sensitive to oxidative deterioration so are packaged in containers that are impermeable to water vapour and oxygen. The packaging of dried milk must take account of its sensitivity to moisture, oxidation and microorganisms. Dried milk or powdered milk products include dry whole milk, non-fat dry milk (NFDM), dry buttermilk, dry whey products and dry dairy blends.

11.4.1 Packaging of whole milk powder

Dry whole milk has a shelf life of less than 6 months due to fat oxidation (Havemose *et al.* 2006). Dried milk must be considered a sensitive product because it is often consumed after reconstitution without additional heating. Reducing the oxygen content in the package also reduces oxidative rancidity. Dry WMP packaged in cans or drums for long-term storage is commonly commercially packaged in a modified atmosphere, for example mixed carbon dioxide and nitrogen, 100% nitrogen or reduced oxygen. Driscoll *et al.* (1985) investigated the sensory quality of instant and regular NFDM powders after 4 years of storage in cans or polybags at 10, 21 and 32 °C and modified atmospheres (air, 100% carbon dioxide, 100% nitrogen). At 21 °C, milk powder stored in cans or polybags under air was less desirable in sensory qualities (off-flavours) than milk powder stored under either nitrogen or carbon dioxide at the same temperature. Packaging techniques have also been developed for dry milk powders to eliminate or reduce oxygen and, hence, reduce auto-oxidation. These include gas flushing and use of oxygen absorbers. De (2000) observed that fat decomposition resulting in a tallowy flavour is a major storage defect in WMP and suggested inhibitory measures by removing oxygen from the headspace of the package followed by flushing with either nitrogen or a mixture of nitrogen and carbon dioxide, the latter being restricted to 5–20%.

Oxygen desorption of WMP is slow, principally due to entrapped air in the lactose. The amount of entrapped air is influenced by content in the concentrate, spraying pressure, orifice size, and temperature and treatment of concentrate. Less than 2% final oxygen in the headspace gas of the package is considered satisfactory for most storage conditions (Abbot *et al.* 1961). Good quality WMP with low oxygen content can be expected to withstand room temperature storage for 6 months or more without an oxidized off-flavour. To obtain a low level of headspace oxygen in WMP, a double gassing technique is applied. This procedure involves the collection of filled cans on trays to be conveyed into the vacuum chamber. The air is removed rapidly (60 s), with the gauge indicator decreasing to 29 "Hg of vacuum. After a 2–5 min hold, the pressure is restored with nitrogen to 0.5–1.0 psi above atmospheric pressure. Nitrogen may be replaced with a mixture of nitrogen and carnob dioxide, the latter being restricted to 5–20%. After removal from the chamber, the containers are sealed by soldering a 1–2 mm hole in the lid or crimping the lid (Lea *et al.* 1943). The containers are held for the oxygen desorption. When an oxygen equilibrium has been attained in the head-space, which usually takes a week but may be up to 10 days, the cans are punctured and the vacuum treatment, pressure restored with nitrogen, and sealing steps are repeated. Gas packaging of WMP should not be delayed after drying otherwise quality the deteriorates during the holding period. Warm powder directly from the dryer tends to have a more rapid rate of oxygen desorption under vacuum.

Lloyd *et al.* (2009) studied the effects of various spray-drying processes and packaging systems (polyethylene pouches and crimp-sealed glass vials, with and without oxygen absorbers) on the oxidative stability of lipids in WMP during storage. It was found that oxygen-impermeable packaging containing oxygen absorbers effectively stored the spray-dried WMP for about 6 months. The migration of organic pollutants from recycled paperboard packaging material to the WMP with fat content 27.7% and varying the temperature, paper thickness and pulp composition was studied (Triantafyllou *et al.* 2007). It was observed that short exposure times were enough to achieve equilibrium of migrants between paper and fatty dry foods. The proportion of substances migrating to food was strongly dependent on the nature of the paper samples, fat content of the food, and the chemical nature and volatility of the migrant (Triantafyllou *et al.* 2007). The water sorption properties of dehydrated milk products and the effect of lactose crystallization on milk fat were reported by Jouppila & Roos (1994). Crystallization of amorphous lactose resulted in loss of adsorbed water and the milk fat decreased the rate of crystallization. Thus, as WMP is high in fat content and hygroscopic in nature, an active packaging system using moisture and oxygen scavengers could be an effective way of extending the product shelf life.

11.4.2 Packaging of non-fat dried milk powder

Skimmed milk powder or NFDM powder is packed in containers that retain product quality during storage as well as giving better handling. The shelf life of NFDM ranges from 3 months to 5 years, and is mainly affected by temperature during storage. The successful packaging of skimmed milk powder involves the removal of moisture and oxygen. The addition of inert gas removes the oxygen entrapped in the air cells of the milk powder and thus reduces the oxygen concentration to a lower level. The general procedure is to immediately remove oxygen by subjecting the product to 28 "Hg of vacuum within 24 h of drying, with final packaging within a few days (Lea *et al.* 1943). Less than 2% final oxygen in the headspace of the package is considered satisfactory for most storage conditions. This is accomplished by vacuumizing and gas packing with an inert gas. As per the Prevention of Food Adulteration (PFA) Act, the spray-dried milk should be packed in nitrogen or a mixture of nitrogen and carbon dioxide in hermetically sealed containers (Shrestha *et al.* 2007).

A suitable container for dry milk should be impervious to moisture, light, gases and insects, durable for handling, resistant to corrosion, low cost and relatively easy to fill, seal, handle and empty. The retail package should have a re-closable opening. NFDM for industrial use and storage may be packaged in barrels, drums or bags, and for retail purposes in metal cans, glass jars or cartons. NFDM in domestic commercial trade is commonly packaged in polyethylene bags inside a 4-ply kraft paper bag (Stapelfeldt *et al.* 1997). The outside layer is usually plain, but a crinkled type is available. Freezing, high temperature or low humidity during storage of the bags causes them to become brittle and thus be damaged more easily in handling (Driscoll *et al.* 1985). Manual filling of the bags is most commonly completed by means of a simple device attached to the sifter. Automatic filling machines are available that give an accurate powder weight in one bag before filling the next bag. Later, the sewing machine stitches the bag with 3 to 3.5 stitches per inch. After closing the bags are stacked on pallets.

11.5 Developments in packaging techniques

The packaging type and material are determined by the demands of both the consumer and the product. New technological developments, environmental awareness and changes in the consumer market have forced packaging technologists to consider an increasing number of factors when designing a package (Gante 1997). Unlike traditional packaging, which is totally inert, a modern package should enhance the package quality by interacting with the product and the surrounding environment. This leads to the development of active and intelligent packaging systems to increase the shelf life of processed foods and provide high-quality fresh and safe products.

11.5.1 Intelligent packaging

The need to find ways to control food quality and safety has led to the development of new packaging systems, including intelligent or smart packaging. Intelligent packaging uses different indicators to monitor conditions during production, storage and transport. The different indicators used for intelligent packaging and their methods of detection are given in Table 11.2.

The indicators are classified into two types based on their attachment to the package: internal indicators and external indicators. Internal indicators are attached inside the package either in the head space or in the lid, for example an oxygen indicator to detect package leak, microbial growth, freshness etc. External indicators are attached outside

Table 11.2 Different types of indicators for detecting changes inside packaging components

Type of indicator	Detection method	Use
Oxygen indicator	Colorimetric redox dye (such as methylene blue)	Vacuum package, reduced oxygen package
Time–temperature indicator	Irreversible change in colour/shape in response to preset temperatures	Frozen and chilled foods
Microbial growth indicator	Change in colour of the dye, 2,6-dichloroindophenol	Perishable products such as meat and fish
Freshness indicator	Printing ink consisting of methyl red detects any change in pH and volatile metabolites	Fermented vegetable foods

the package, for example time–temperature indicators. Intelligent packaging systems can be attached as labels printed onto the food packaging material to provide product information, monitor the product quality and trace the critical points (Abreu *et al.* 2012).

Temperature is an important factor influencing the quality and safety of food products during storage, transport and distribution. Time–temperature indicators show an irreversible change in colour or shape in response to temperature history (Riva *et al.* 2001). Ferreira & Van (1999) patented a method to detect spoilage of milk inside a sealed container. The milk, packed in sachets, is monitored against becoming sour using a colour indicator, for example litmus. The indicator is attached to the wall of the sachet and made visible to the colour change. The semipermeable membrane covering the indicator allows the transfer of H^+ and OH^- between the milk and the indicator. The colour change in the indicator shows the consumer whether or not the product is fit for consumption. Furthermore, a new enzymatic time–temperature indicator based on enzyme reaction, diffusion and enzyme demobilization technology has been developed for detecting fresh milk spoilage (Lu *et al.* 2013). Thus, the development of indicators and their applications in milk products will benefit customers, manufacturers and distributors by spot checking and determining product quality.

11.5.2 Active packaging

Recent developments in packaging material have focused on the interaction between the package and the contained food. This leads to active packaging technology, in which the package is actively involved with the food product or interacts with the internal atmosphere to extend the shelf life of a food product while maintaining quality and safety (Ozdemir & Floros 2004). Active packaging technology is designed to enhance the property of the packaging material and therefore increase product shelf life. The active component may be part of the packaging material or an inert attachment to the inside of the pack (Dobrucka & Cierpiszewski 2014). The various functions of active packaging include oxygen scavenging antimicrobial activity, moisture control, ethylene removal, antioxidative reactions etc.

Recently, active packaging films incorporating antimicrobial compounds have been developed for butter preservation. It has been reported that the antimicrobial films had greater resilience and elongation than control films and inhibited the growth of fungi on the surface of the butter (Moraes *et al.* 2007). Similarly, aromatized active films incorporated with herbs, bacon and/or ham flavours have been developed. These have been used for butter packaging, resulting in an altered product with high consumer acceptability and no change in oxidation rates (Moraes *et al.* 2007). In addition, antioxidant active packaging have been shown to be a promising technology for WMP. Since WMP contains 26% fat, it is prone to deterioration due to lipid oxidation. The migration of α-tocopherol from multilayer active packaging to WMP delayed the lipid oxidation of WMP, particularly at higher temperatures (Granda Restrepo *et al.* 2009). In addition, the sensory quality of WMP during storage is improved with the presence of α-tocopherol in multilayer active packaging. Moreover, these films provided an adequate light barrier for WMP and α-tocopherol prevented vitamin A degradation when compared with other films (Granda Restrepo *et al.* 2009). Natural preservatives such as α-tocopherol may be an effective alternative to synthetic ones for incorporation in current packaging to maintain the sensory standards of WMP during storage (Da Cruz *et al.* 2007).

11.5.2.1 Migration mechanism in active packaging

Food packaging systems come in two types: the food/ package/head space system and the food/package system. The former system includes food that comes into contact with the package, such as milk powder and low viscous liquids without head space. Here, the major migration mechanism occurs by diffusion between the food/packaging material and partitioning at the interface. In such cases, inhibitory materials can be incorporated in the packaging material and migrate through diffusion and partitioning. The moisture transfer resulting in gain or loss of moisture in a packaged product is an important quality control and legal issue in foods. Generally, the shelf life of a food inside a package depends on the rate of exchange of moisture through the package to the storage atmosphere plus the rate of change in water activity of food towards critical upper or lower limits with respect to bacteria, fungi, texture, flavour, appearance, eating qualities, aroma, nutritional qualities, cooking qualities and regulatory issues. Thus, the control of release rates and the migration mechanism of an active ingredient are very important considerations in an active packaging system.

11.5.2.2 The use of scavengers (absorbers) to prevent lipid oxidation

Many foods are sensitive to oxygen and the presence of aerobic microorganisms along with oxygen leads to deterioration of the products by oxidation reactions. The presence of oxygen scavengers helps to remove or decrease the level of oxygen inside the package. Oxygen absorbers are a very common method for removing oxygen in a sealed package. They generally lower headspace oxygen to less than 0.01%, which is significant when compared to the 0.3–3% found in packages that have merely been vacuum-sealed or gas-flushed (Robertson 2012). Many systems of oxygen scavengers based on metallic and non-metallic compounds are available, such as iron powder oxidation, ascorbic acid oxidation, enzymatic oxidation (e.g. glucose oxidase) etc. The iron powder system is based on the oxidation of iron and ferrous salts to form stable oxide (Abreu *et al.* 2012). A study of UHT milk packaged with an oxygen scavenger showed a reduction in dissolved oxygen content of about 23–28% and stale flavour volatiles during storage (Perkins *et al.* 2007). Andersson & Lingnert (1997) studied the influence of oxygen concentration at various levels on lipid oxidation in cream powder. The oxygen levels were monitored by observing the formation of volatile oxidation products and oxygen consumptions. The cream powder showed significant difference from the initial oxygen levels, emphasizing the need for scavengers to prevent lipid oxidation.

Similarly, carbon dioxide scavengers made of calcium hydroxide, sodium hydroxide, potassium hydroxide, silica gel, etc. are available. Carbon dioxide is formed due to respiration and deterioration reactions which have to be removed to avoid food deterioration and/or package destruction. The active compound, $Ca(OH)_2$, which is present in commercial carbon dioxide scavenger (FreshLock®), reacts at sufficiently high humidity with the carbon dioxide to produce $CaCO_3$:

$$CaO + H_2O \rightarrow Ca(OH)_2$$

$$Ca(OH)_2 + CO_2 \rightarrow CaCO_3 + H_2O$$

The presence of a high moisture level causes caking of the milk powder. On the other hand, excessive water evaporation through the package may lead to desiccation of the product or may induce lipid oxidation (Vermeiren *et al.* 1999). To prevent this, moisture absorbers, such as desiccating film in the form of sachet or pad, can be used inside the package. These moisture regulators usually consist of a super-absorbent polymer (polyacrylate salts and graft copolymers of starch) in between two packaging layers (Rooney 1995).

Research has shown that light protection is an important factor in protecting milk from lipid oxidation. The initial and final hexanal milk content was calculated as an indicative measure of oxidation. Under fluorescent lighting, a significant difference was found in the hexanal content of light-protected and light-exposed milk (Van Aardt *et al.* 2001). Milk powder stored in daylight showed 100 times greater hexanal content than that stored at room temperature at 30 °C in darkness (Ulberth & Roubicek 1995). Hence, it is very clear that milk powder samples must be protected from light exposure. In addition, as cholesterol is an unsaturated fat, it is susceptible to oxidation, leading to cholesterol oxidation products. Such products are often found in animal foods, including dairy products. Oxidation products of cholesterol are mostly found in dairy products subjected to harsh storage conditions, such as those associated with oxygen and light or oxygen and low water activity (Sieber 2005). β-galactosidase, a lactase enzyme, can be covalently attached to surface-functionalized low-density polyethylene films which act as active packaging materials and reduce the lactose level in packaged dairy products (Goddard *et al.* 2007). The incorporation of enzymes facilitates in-package processing.

11.5.3 Nanotechnology in dairy packaging

Nanotechnology is defined as the manipulation of matter at the atomic or molecular level. The concept of nanotechnology was introduced by Richard Feynman in 1959 at a meeting of the American Physical Society (Luo & Stutzenberger 2008). Since then, nanotechnology has developed into a multidisciplinary field of applied science and technology (Azeredo 2009). The application of nanotechnology in the food packaging sector is constantly seeing new possibilities and great benefits (Anandharamakrishnan & Usha Kiran 2014). Nanoparticles have a large surface area which favours filler–matrix interactions and thus increases the performance of the resulting material. Besides adding strength to the packaging material, nanoparticles, when added to a polymer, can provide antimicrobial activity, enzyme immobilization, biosensing, etc. (Azeredo 2009). Nowadays most packaging materials used for food packaging applications are almost non-degradable, representing a serious global environmental problem. New biopolymer-based packaging materials can solve the waste disposal problem and also enhance the quality of food. Nanotechnology in food packaging helps to extend food shelf life by releasing nano antimicrobes and a nano wrapper envelops foods, preventing moisture and gas transfer. Nanomaterials are used to improve the barrier properties and strength of packaging materials. Nano sensors can be used to detect chemicals, toxins and pathogens in food. In the dairy industry, the application of nanotechnology includes the development of packaging using nanolaminates, nanocomposite bottles and bins with silver nanoparticles. Nano coating, which is an aqueous-based nanocomposite barrier coating, provides an oxygen barrier with a 1–2 μm thick coating for food packaging use (Brody *et al.* 2007). In this context, the development of bionanocomposite has shown potential application in the field of food packaging (Rhim *et al.* 2013). The bionanocomposite allows the use of new, high-performance, lightweight green nanocomposite materials to replace conventional non-biodegradable petroleum-based plastic packaging materials.

11.5.3.1 Bionanocomposites and their applications

The development of bio-based polymers with good physical properties is an interesting area of research, and literature is available on the preparation, characterization and application of bionanocomposites in food packaging (Sorrentino *et al.* 2007; Sozer & Kokini 2009; Rhim *et al.* 2013). So far, the use of biodegradable packaging material has been limited by its high cost and poor performance. In this context, bionanocomposites have proven to be effective in improving the mechanical and barrier properties of biopolymers. Polymer composites are a mixture of polymers with inorganic or organic fillers in the form of flakes, fibers, particulates etc. Polymer nanocomposites are produced by using fillers in the nanoscale range that are randomly and homogeneously dispersed to improve the mechanical and physical properties of the polymer film (Rhim *et al.* 2013). The presence of nano fillers creates a complicated path for water and gas molecules to pass through, as shown in Figure 11.3. Thus, the incorporation of polymer nanocomposites significantly increases mechanical and barrier properties, and produce films superior to those of pure polymers or conventional composites. Also, when these nanocomposites used in food packaging, they are able to withstand the stress of thermal food processing, transportation and storage (Sinha *et al.* 2003).

Bionanocomposites consist of a biopolymer matrix reinforced with nanoscale inclusions, with at least one dimension in the nanometre range (1–100 nm). They exhibit much improved properties due to a high aspect ratio and high surface area of nanoparticles (Zhao *et al.* 2008). Various biopolymers used in the production of bionanocomposites include plant-derived materials such as plant proteins, starch, cellulose, other polysaccharides and animal products such as proteins and polymers synthesized chemically from naturally derived monomers such as poly(lactic acid) (PLA) (Arora & Padua 2010). Overall, the functional properties of bionanocomposites are governed by the extent of dispersion of nanoparticles in the biopolymer matrix and the interaction between nanoparticles and the biopolymer (Kumar *et al.* 2011). Mathematical modelling plays a critical role in determining the properties of bionanocomposites and the effect of different parameters, such as filler aspect ratio and loading levels.

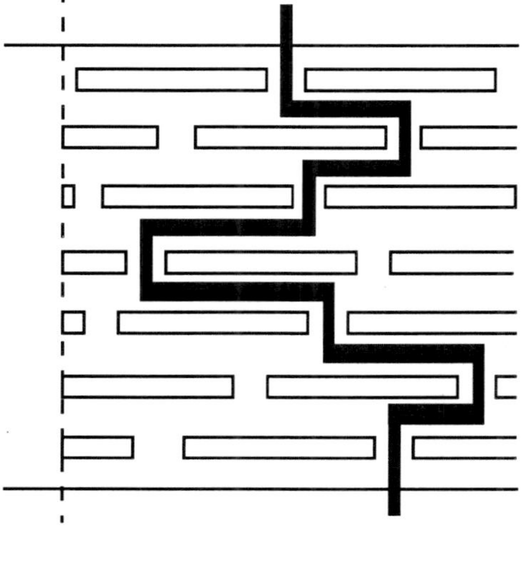

Figure 11.3 Tortuous pathway created by nanocomposites on polymer (Duncan 2011).

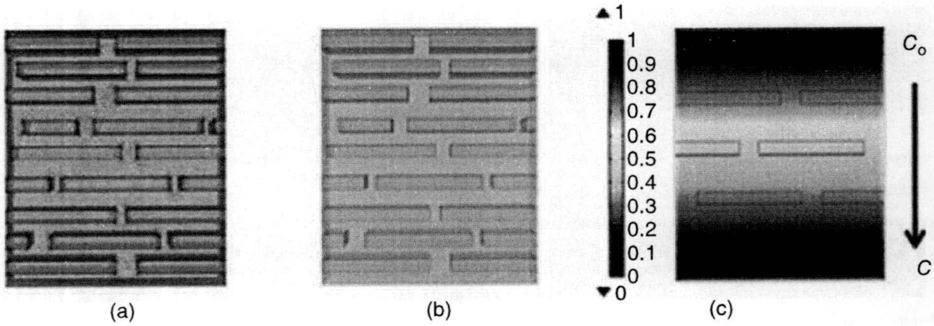

Figure 11.4 Modelling the barrier properties of nanocomposite film using (a) geometry, (b) meshed geometry and (c) predicted concentration profiles.

11.5.3.2 Modelling the barrier properties of polymer-clay nanocomposites

Computational modelling is widely accepted as an effective tool for the design of nanomaterials (Xiao *et al.* 2010). The study of the barrier properties of nanopackaging materials by understanding gas molecule diffusion through the heterogeneous nanomembrane is gaining research interest. The simulation process is purely a diffusion phenomenon involving the penetration of water vapour and gas molecules with a change in initial and final concentration. Figure 11.4 shows the three-dimensional geometry of nanocomposite arrangements, where the mass transfer is governed by Fick's law. The geometry created with structural arrangements of nanomaterials is meshed into discrete elements to solve the equations of mass transfer.

Finite element modelling has been developed to analyse the influence of the morphology and load-on barrier properties of a polymer–clay nanocomposite (Statler *et al.* 2007). The geometry consisted of a polymer–clay packaging film containing uniformly distributed nanometre thick platelets. The basic governing equations for modelling the migration mechanism of the packaging material is based on Fick's law of diffusion, as given in Eqs (11.1) and (11.2). Figure 11.5 shows the arrangement of nano fillers, that

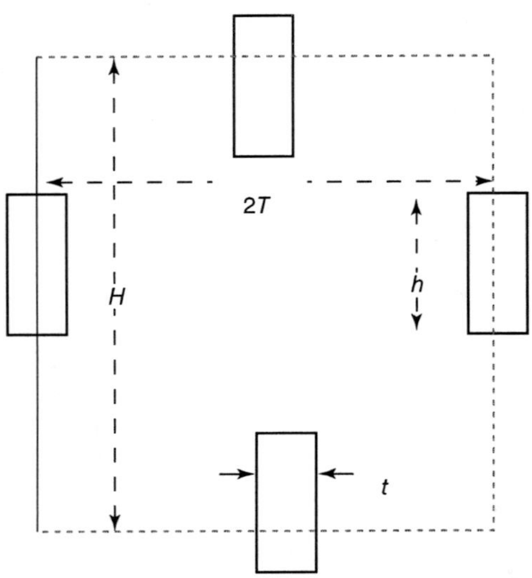

Figure 11.5 Morphology of polymer nanocomposite containing dispersed clay platelets (Statler *et al.* 2007).

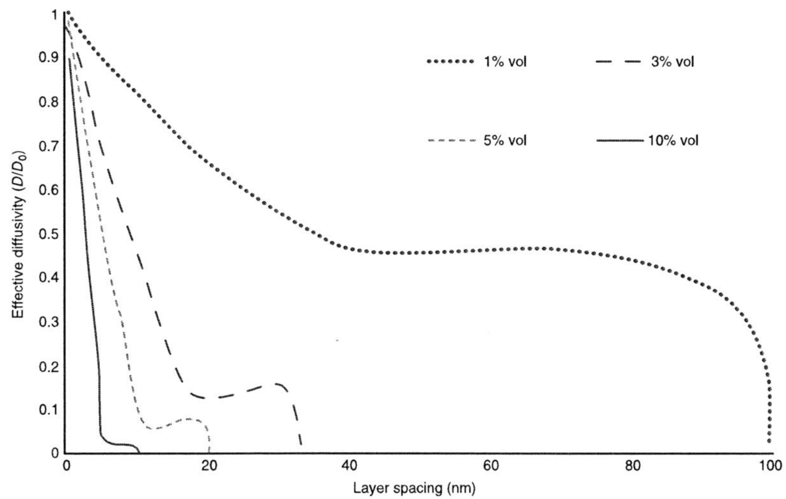

Figure 11.6 Comparison of effective diffusivity with layer spacing at different clay loading levels for different models (Statler *et al.* 2007).

is, clay platelets with layer spacing T, height h and thickness t. The diffusivity of the nanocomposite film is compared with pure polymer for its barrier strength. The barrier property of the nanocomposite varies with the size, shape and volume fraction of the nano fillers or clay platelets. The diffusivity (D) of a polymer can be calculated based on the rate of change in concentration ($C_0 - C$), as given in Equation (11.1):

$$\frac{\partial c}{\partial t} = \nabla \cdot (D \nabla c) \qquad (11.1)$$

$$\left[\frac{D}{D_0}\right] = 1 - \frac{2h}{H} \qquad (11.2)$$

Figure 11.6 shows the computed finite element results of the concentration gradient (D/D_0) for different nanofiller layer spacing (T) and loading levels (volume %). At each loading level the layer spacing is changed from about 1 nm to the value at which the platelets in a given layer touch each other. The results show that D/D_0 decreases with increased layer spacing. Thus, the computational modelling technique can be used effectively for designing biopolymers with improved barrier properties.

11.6 Conclusion

Packaging protects dairy products during transport and storage before they reach the consumer. Chemical changes in dehydrated milk products cause staleness, oxidized flavour, hydrolytic rancidity and insolubility to an extent influenced by the processing treatment, moisture content, free oxygen levels within the container and storage temperature. The exact causes and chemical reactions involved in powder deterioration are complex and partly obscure. However, the keeping quality of dried milk products can be enhanced by proper packaging. The type of packaging material used will depend on cost, barrier properties and convenience, besides functionality. Novel dairy packaging systems involve new technologies such as nanotechnology and intelligent packaging, and will probably dominate the future trends in food packaging. Active packaging and intelligent packaging for food products have evolved to provide enhanced communication and information at optimized cost. Research efforts in this area have been very active and already some packages based on this principle are available in the market.

References

Abbot, J., Waite, R. & Hearne, J. F. (1961). Gas packing milk powder with a mixture of nitrogen and hydrogen in the presence of palladium catalyst. *Journal of Dairy Research.* **28(03)**, 285–292.

Abreu, D.A.P., Cruz, J.M. & Losada, P.P. (2012). Active and Intelligent packaging for the food industry. *Food Reviews International*, **28(2)**, 146–187.

Alavi, S., Thomas, S., Sandeep, K.P. & Kalarikkal, N. (2014). *Polymers for Packaging Applications*, CRC Press.

Anandharamakrishnan, C. & Usha Kiran, K. (2014). Bionanocomposites and their potential applications in food packaging. In: *Polymers for Packaging Applications*, pp. 229, Apple Academic Press Inc.

Andersson, K. & Lingnert, H. (1997). Influence of oxygen concentration on the storage stability of cream powder. *LWT – Food Science and Technology*, **30(2)**, 147–154.
Arora, A. & Padua, G.W. (2010). Review: nanocomposites in food packaging. *Journal of Food Science*, **75**, R43–R49.
Avneet, R. (2013). Materials, technologies for shelf-stable dairy products. *Food and Beverages Magazine*, November. http://www.fnbnews.com/FB-Specials/Materials-technologies-for-shelf-stable-dairy-products.
Azeredo, H.M.C. (2009). Nanocomposites for food packaging applications. *Food Research International*, **42(9)**, 1240–1253.
Blunden, S. & Wallace, T. (2003). Tin in canned food: a review and understanding of occurrence and effect. *Food and Chemical Toxicology*, **41(12)**, 1651–1662.
Brody, A.L. & Sacharow, S. (1970). Flexible packaging of foods. *CRC Critical Reviews in Food Technology*, **1(1)**, 71–155.
Brown, N. (1992). Primary food packaging materials. In: *Plastics in Food Packaging: Properties: Design and Fabrication* (eds Brown, W.E.., Finch, C.R., Alan, S. & Heckman, J.H.), pp. 41, CRC Press.
Chan, S.H., Gray, J.I., Gomaa, E.A. & Harte, B.R (1993). Cholesterol oxidation in whole milk powders as influenced by processing and packaging. *Food Chemistry*, **47(4)**, 321–328.
Da Cruz, A.G., Faria, J.A.F. & Van Dender, A.G.F. (2007). Packaging system and probiotic dairy foods. *Food Research International*, **40(8)**, 951–956.
Dhar, R. (2002). New packaging materials and packaging technology for dairy products. *Indian Dairyman*, **54(2)**, 72–75.
Dobrucka, R. & Cierpiszewski, R. (2014). Active and intelligent packaging food – research and development – A review. *Polish Journal of Food and Nutrition Sciences*, **64(1)**, 7–15.
Driscoll, N.R., Brennand, C.P. & Hendricks, D.G. (1985). Sensory quality of nonfat dry milk after long-term storage. *Journal of Dairy Science*, **68(8)**, 1931–1935.
Duncan, T.V. (2011). Applications of nanotechnology in food packaging and food safety: Barrier materials, antimicrobials and sensors. *Journal of Colloid and Interface Science*, **363(1)**, 1–24.
Ferreira, S.J. & Van, E.J.H. (1999). Indicator and method for identifying that milk as sealably contained has become sour, available at http://www.google.com/patents/EP0890835A1.
Gante, C.R. (1997). Trends in food packaging. *Food Engineering* **2000**, 347–364.
Goddard, J.M., Talbert, J.N. & Hotchkiss, J.H. (2007). Covalent attachment of lactase to low-density polyethylene films. *Journal of Food Science*, **72(1)**, E036–E041.
Granda Restrepo, D.M., Soto Valdez, H., Peralta, E. & Troncoso Rojas, R. (2009). Migration of α-tocopherol from an active multilayer film into whole milk powder. *Food Research International*, **42(10)**, 1396–1402.
Hall, C.W. & Hedrick, T.I. (1971). *Drying of Milk and Milk Products*, pp. 162–164, AVI Publishing Company.
Havemose, M.S., Weisbjerg, M.R., Bredie, W.L.P. & Poulsen, H.D. (2006). Oxidative stability of milk influenced by fatty acids, antioxidants, and copper derived from feed. *Journal of Dairy Science*, **89(6)**, 1970–1980.
Hennigs, C., Kockel, T.K. & Langrish, T.A.G. (2001). New measurements of the sticky behavior of skim milk powder. *Drying Technology*, **19(3–4)**, 471–484.
Jones, J.F. (1990). *Thermoplastic stretch-wrap material* [online patent], available at http://www.google.com/patents/US4923750.

Jouppila, K. & Roos, Y.H. (1994). Water sorption and time-dependent phenomena of milk powders. *Journal of Dairy Science*, **77**(7), 1798–1808.

Karaman, A.D., Özer, B., Pascall, M.A. & Alvarez, V. (2015). Recent advances in dairy packaging. *Food Reviews International*, **31**(**4**), 295–318.

Kim, E.H.J., Chen, X.D. & Pearce, D. (2009). Surface composition of industrial spray-dried milk powders. 2. Effects of spray drying conditions on the surface composition. *Journal of Food Engineering*, **94**(**2**), 169–181.

Krokida, M.K., Karathanos, V.T., Maroulis, Z.B. & Marinos-Kouris, D. (2003). Drying kinetics of some vegetables. *Journal of Food Engineering*, **59**(**4**), 391–403.

Kudelka, W. (2005). Changes in the acidity of fermented milk products during their storage as exemplified by natural bio-yoghurts. *Milchwissenschaft*, **60**(**3**), 294–296.

Kumar, P., Sandeep, K.P., Alavi, S. & Truong, V.D. (2011). A review of experimental and modeling techniques to determine properties of biopolymer-based nanocomposites. *Journal of Food Science*, **76**(**1**), E2–E14.

Lea, C.H., Moran, T. & Smith, J.A.B. (1943). The gas-packing and storage of milk powder. *Journal of Dairy Research*, **13**(**02**), 162–215.

Lloyd, M.A., Zou, J., Farnsworth, H. & Ogden, L.V. (2004). Quality at time of purchase of dried milk products commercially packaged in reduced oxygen atmosphere. *Journal of Dairy Science*, **87**(**8**), 2337–2343.

Lloyd, M.A., Hess, S.J. & Drake, M.A. (2009). Effect of nitrogen flushing and storage temperature on flavor and shelf-life of whole milk powder. *Journal of Dairy Science*, **92**(**6**), 2409–2422.

Lu, L., Zheng, W., Zhiye, L.V. & Tang, Y.L. (2013). Development and application of time–temperature indicators used on food during the cold chain logistics. *Packaging Technology and Science*, 2680–2690.

Luo, P.G. & Stutzenberger, F.J. (2008). Nanotechnology in the detection and control of microorganisms. In: *Advances in Applied Microbiology* (eds Sariaslani, S. & Gadd, G.M.), pp. 145–181, Academic Press.

Marsh, K. & Bugusu, B. (2007). Food packaging – Roles, materials, and environmental issues. *Journal of Food Science*, **72**(**3**), R39–R55.

Mathlouthi, M. (2013). *Food Packaging and Preservation*, pp. 290, Springer Science & Business Media.

Moraes, A.R.F., Gouveia, L.E.R., Soares, N.F.F., Santos, M.M.S. & Goncalves, M.P.J.C (2007). Development and evaluation of antimicrobial film on butter conservation. *Food Science and Technology (Campinas)*, **27**, 33–36.

Narayanan, P.V. & Dordi, M. (1998). Indian food sector and packaging – An overview. In: *Modern Food Packaging* (ed. Cunha, J.F.D*)*, Indian Institute of Packaging, Mumbai, pp. 593–597.

Ozdemir, M. & Floros, J.D. (2004). Active food packaging technologies. *Critical Reviews in Food Science and Nutrition.* **44**(**3**), 185–193.

Paine, F.A. & Paine, H.Y. (1992). *A Handbook of Food Packaging*, Springer Science & Business Media.

Paulett, H.K. (1999). Perforated stretched wrap film [online patent], available at http://www.google.com/patents/US5935681.

Perkins, M.L., Zerdin, K., Rooney, M.L. & D'Arcy, B.R. (2007). Active packaging of UHT milk to prevent the development of stale flavour during storage. *Packaging Technology and Science*, **20**(**2**), 137–146.

Rhim, J.W., Park, H.M. & Ha, C.S. (2013). Bio-nanocomposites for food packaging applications. *Progress in Polymer Science*, **38**(**10–11**), 1629–1652.

Riva, M., Piergiovanni, L. & Schiraldi, A. (2001). Performances of time–temperature indicators in the study of temperature exposure of packaged fresh foods. *Packaging Technology and Science*, **14**(**1**), 1–9.

Robertson, G.L. (2012). *Food Packaging: Principles and Practice*, 3rd edn, CRC Press.

Rooney, M.L. (1995). Overview of active food packaging. In: *Active Food Packaging* (ed. Rooney, M.L.), pp. 1–37, Springer.

Roy, A. (2002). Changing profile of the Indian market. *Indian Dairyman*, **54**(**2**), 47–53.

Rukchon, C., Nopwinyuwong, A., Trevanich, S. & Jinkarn, T. (2014). Development of a food spoilage indicator for monitoring freshness of skinless chicken breast. *Talanta*, **130**, 547–554.

Shrestha, A.K., Howes, T. & Adhikari, B.P. (2007). Water sorption and glass transition properties of spray dried lactose hydrolysed skim milk powder. *LWT – Food Science and Technology*, **40**(**9**), 1593–1600.

Sieber, R. (2005). Oxidised cholesterol in milk and dairy products. *International Dairy Journal*, **15**(**3**), 191–206.

Simon, M. & Hansen, A.P. (2001). Effect of various dairy packaging materials on the shelf life and flavor of pasteurized milk. *Journal of Dairy Science*, **84**(**4**), 767–773.

Sinha R.S. & Okamoto, M. (2003). Polymer/layered silicate nanocomposites: a review from preparation to processing. *Progress in Polymer Science*, **28**(**11**), 1539–1641.

Sorrentino, A., Gorrasi, G. & Vittoria, V. (2007). Potential perspectives of bio-nanocomposites for food packaging applications. *Trends in Food Science & Technology*, **18**(**2**), 84–95.

Sozer, N. & Kokini, J.L. (2009). Nanotechnology and its applications in the food sector. *Trends in Biotechnology*, **27**(**2**), 82–89.

Statler, D.L. & Gupta, R.K. (2007). A finite element analysis of the influence of morphology on barrier properties of polymer-clay nanocomposites. In: *Proceedings of the COMSOL Conference, Boston*.

Thomas, M.E.C., Scher, J., Desobry-banon, S. & Desobry, S. (2004). Milk powders ageing: effect on physical and functional properties. *Critical Reviews in Food Science and Nutrition*, **44**(**5**), 297–322.

Tian, F., Decker, E.A. & Goddard, J.M. (2013). Controlling lipid oxidation of food by active packaging technologies. *Food & Function*, **4**(**5**), 669–680.

Triantafyllou, V.I., Akrida-Demertzi, K. & Demertzis, P.G. (2007). A study on the migration of organic pollutants from recycled paperboard packaging materials to solid food matrices. *Food Chemistry*, **101**(**4**), 1759–1768.

Ulberth, F. & Roubicek, D. (1995). Monitoring of oxidative deterioration of milk powder by headspace gas chromatography. *International Dairy Journal*, **5**(**6**), 523–531.

Van Aardt, M., Duncan, S.E., Bourne, D. & Marcy, J.E. (2001). Flavor threshold for acetaldehyde in milk, chocolate milk, and spring water using solid phase microextraction gas chromatography for quantification. *Journal of Agricultural and Food Chemistry*, **49**(**3**), 1377–1381.

Vermeiren, L., Devlieghere, F., van Beest, M. & de Kruijf, N. (1999). Developments in the active packaging of foods. *Trends in Food Science & Technology*, **10**(**3**), 77–86.

Xiao, J., Huang, Y. & Manke, C.W. (2010). Computational design of polymer nanocomposite coatings: A multiscale hierarchical approach for barrier property prediction. *Industrial & Engineering Chemistry Research*, **49(17)**, 7718–7727.

Zhao, R, Torley, P. & Halley, P.J. (2008). Emerging biodegradable materials: starch- and protein-based bio-nanocomposites. *Journal of Materials Science*, **43**, 3058–3071.

12

Recent Advances in the Drying of Dairy Products

M.W. Woo

Department of Chemical Engineering, Faculty of Engineering, Monash University, Clayton Campus, Victoria, Australia

12.1 Introduction

The production of dairy powders such as milk or cheese powders involves dehydrating tiny atomized dairy droplets into particles. In this process, the dairy feed material is atomized into tiny droplets that come into contact with hot air in a drying chamber. Solid powders are formed when the moisture within the droplet is dehydrated. Several phenomena take place simultaneously during different stages of drying. In the initial drying region, the droplet is still wet and liquid-like. As evaporation occurs, heat transfer into the droplet will maintain the droplet at the wet-bulb temperature. At this stage there is no rigidity to the droplet. This initial stage of drying is very rapid. When the outer layer of the droplet starts to solidify, dehydration is slowly impeded and the particles experience a rise in temperature. Because of the progressive impediment of the dehydration process, the drying rate towards the end of the process is relatively slow and if sufficient time is allowed, the particles will approach the equilibrium moisture level. It is also at this stage that the particles reach very high temperature, corresponding to the air temperatures used in the spray-drying process. In the transition period, when the particle is still semi-rigid, it may become sticky and adhere to the dryer wall on collision. The particle only becomes less sticky if it enters the final stages of drying if sufficient solidification (or moisture removal) is allowed or if the particle is not overheated. This second stage of drying, particularly towards the latter stage closer to the equilibrium moisture level, requires a relatively long drying time when compared to the initial stage of drying.

This whole dehydration process for making dairy powders is typically undertaken in a series of multistage dryers (Westergaard 1994; Pisecky 1997). The atomized droplets are initially semi-dried in a spray dryer before subsequently being dried in a fluidized-bed dryer to remove the remaining moisture. Such a layout will be able to provide sufficient drying time without the need to use excessively high drying temperatures, which might lead to product thermal degradation. This chapter focuses on recent development in understanding and most importantly in operating and designing this multistage drying framework. This chapter is not a systematic description of each phenomenon of the spray-drying process or the common operational techniques in the dairy industry. Rather, the different strategies developed over the years by the industry, as well as the academic research community, specific to dairy powders (and to certain extends food powders) are discussed. These strategies were developed mainly to provide a complementary approach to the operators' experience in running the drying system. The first half of the chapter details developments in spray drying while the latter half focuses on recent advances in operating second-stage fluidized-bed dryers.

12.2 Typical layout of a dairy spray-drying process

As this chapter describes advances in the operation and design of the drying process, it is important to first describe briefly the current typical set-up in a spray-drying facility. Our discussion starts with the spray-dryer unit. Although the preconcentration step, typically using a series of falling-film evaporators, forms a major aspect of the entire process, the focus of this chapter is the drying process.

12.2.1 Multistage drying process

Current dairy spray-drying facilities typically consist of multistage drying. The atomized droplets are initially dried within the spray-drying chamber to about 8–15 wt% moisture (Westergaard 1994). This level of moisture is not sufficiently low to prevent microbial development. It might also not be sufficient to prevent physical changes in the particles during storage. The main aim of this first step is to remove sufficient moisture such that the particle is well formed without being too sticky. There are many reasons why not all the moisture (down to equilibrium) is removed in this stage within the spray dryer. First, towards the lower moisture content of the particle the moisture inhibition is relatively high and moisture removal requires a relatively high level of energy. In view of the relatively short residence time in a spray dryer, typically in the order of seconds or tens of seconds, excessively high temperature will be required to completely dry the particles. However, excessively high temperature is not desired as it might cause thermal degradation of the product and in turn affect the overall product quality. One solution currently used by the dairy industry is to use fluidized-bed drying as the secondary stage of drying for the semi-dried particles after the spray dryer (Pisecky 1997). It should be noted that this is only possible if the particles from the spray dryer are partially dried and not too sticky. In a two-stage drying process the secondary fluidized-bed dryer may take the form of a vibrating fluidized bed. The particles produced from the spray dryer pass into the bed by gravity or via a rotating valve. During the external fluidization drying process mild temperatures are typically used (Westergaard 1994). Use of such low temperatures is feasible as the particles will then have a relatively long residence time as the particles move across the horizontal bed. Cooler air is also used to cool the particles, which may reduce the stickiness of the particles. However, the cooling is controlled such that there is no significant absorption of moisture, which in turn may make the particles even stickier. The fine particles (also known as 'fines') separated from the spray-drying outlet air are typically returned to the fluidized bed or back to the spray-drying chamber. Apart from drying the particles, due to the tumbling and contact behaviour of the particles, the second-stage fluidized bed also acts as an agglomeration unit. The second-stage fluidized bed also helps to achieve good agglomeration of the particles to enhance their dissolution behaviour.

To further enhance the dissolution behaviour of the particles, food-grade surfactants can be sprayed onto the particles within the fluidized bed. The surfactant is typically sprayed onto the particles from the top of the bed and is intended to provide a coating for the particles. This approach is used to make instant full-cream milk powder, in which the particle surface is covered with fat, allowing dissolution of the powder even with cold water. In such an application, cooling of the coated surfactant is important as the surfactant is typically initially dissolved in butter fat. Recent advances in this area have examined the feasibility of incorporating the surfactants directly in the initial dairy feed (Lallbeharry *et al.* 2014). Comparison was undertaken between ionic and non-ionic surfactants, showing potential in the direct application of surfactants to the feed material.

In place of the external fluidized bed, an internal fixed fluidized bed may also be used. This type of fluidized bed is installed at the bottom of the spray-drying chamber cone. The bed usually consists of an annulus of specially designed perforated plate. The angle of the perforation is designed to drive the air in a swirling direction. In the middle of the annulus perforated plate there is a vertical outlet that controls the height of the bed as the particles in the bed overflow. For certain products, a combination of internal and external fluidized bed may be installed.

Schwartzbach & Masters (2001) reported a two-stage dryer (i.e. a spray dryer with a fluidized-bed dryer at the base of spray-drying chamber) with an integrated metallic filter within the spray-drying chamber. By installing the integrated filter, fine particles are contained within the spray-drying chamber, thereby eliminating the issues in handling of fines outside the drying chamber. The concern is that the continuously back-cleaned fines from the installed filters might affect the integrated fluidized-bed operation. However, study of different products on a pilot scale unit was shown not to interfere with the operation of the fluidized bed..

12.2.2 Some unique process layouts

Some unique material-specific process layouts have been reviewed and are available in the literature for whey powder processing. Whey was historically a by-product from cheese manufacturing, but it has found a market as a protein supplement. A review by Pisecky (2005) on the spray drying of whey found that a typical process layout includes a single-stage, two-stage or Filtermat^{TM} Dryer (NIRO), depending on the concentration of the whey stream and the required final outlet moisture. In producing very sticky powders, using the Filtermat^{TM} Dryer, the semi-dried particles are introduced by gravity onto a conveying belt instead of a fluidized bed. Permeate from the filtered whey stream is often spray dried for lactose recovery (Pisecky 2005). The process will typically include a concentration and crystallization step prior to spray drying. Niro developed the TIXOTHERM^{TM} process, which dries the permeate using a combination of paddle dryer, screw dryer and a back mixing fluidized bed. It was reported by Pisecky (2005) that the process saves approximately 30% in energy and 70% in building requirements.

12.3 Advances in operating spray dryers

12.3.1 Controlling the drying process

The spray-drying process is conventionally operated based on the operators' experience, which requires trial and error, particularly when new feed material is used. This approach is useful in most cases, especially long operations for a single type of material. However, when there is a change in the feed composition or for contract manufacturers where the feed is changed regularly, such an approach might lead to trial and error wastage before arriving at the required final product quality. Although some systematic trial and error method is reported in the literature (Woo *et al.* 2007), it is more cost-effective if a predictive approach is used. The following are few approaches available for this purpose.

12.3.1.1 Single droplet to dryer-wide prediction

The spray-drying process consists of the heat and mass transfer of individual droplets within the drying chamber. A recent advancement has been to obtain suitable operating conditions by first examining the heat and mass transfer of a single droplet, rather than the global operation of the drying chamber. Once the drying kinetics of a single droplet have been obtained, this information is fed into a large-scale model of the spray dryer. The model is then used to generate suitable operating conditions specific to the dairy material of interest. Many different research groups as well as commercial industrial companies are adopting and developing this approach.

The kinetics of a single droplet drying broadly refers to a set of data which characterize how drying behaviour changes throughout the drying process as moisture is removed. It is important to capture this behaviour partly because of the effect on evaporation when moisture is progressively lost from the droplet. The types of kinetics measured depend on the approach adopted for the subsequent dryer-wide simulation. It is therefore important for this discussion on the measurement of the drying kinetics to first examine (and classify) the dryer-wide simulation approach and then work backwards to the required kinetic measuring techniques.

One-dimensional simulation One straightforward approach is to undertake one-dimensional simulation of the spray dryer. This approach has been used to model large tower-like pilot units, consisting of short spray dryers fitted with a rotating atomizer, to the laboratory-scale Buchi spray dryer (Truong *et al.* 2005; Langrish 2009; Patel *et al.* 2010). In this approach, the air flow is typically assumed to be a one-dimensional plug flow along the axis of the spray tower. An atomized droplet is assumed to move in a one-dimensional manner in the same axis as the air flow. The premise for the one-dimensional approach is to simultaneously solve first-order ordinary differential equations of mass, energy and momentum for the air and the droplet phase. Details of the mathematical equations involved can be found in the literature cited above. From the solution of these equations, the model generates the particle's moisture and temperature profile along its trajectory in the chamber until it leaves the chamber at the outlet. Similarly, the humidity and temperature profile of the air can be obtained. The differential equations for the droplet and air phase are well established in the literature. The momentum coupling between the phases is also well quantified based on Newton's law and well-known drag relations. The key aspect that determines the accuracy of this approach lies in describing the heat and mass coupling, which is mainly determined by the drying of individual droplets. It is imperative to quantitatively describe how the drying rate changes at different stages of evaporation. This is determined by the required drying kinetics, which has to be experimentally measured.

There are a few different classes of kinetics measurement techniques. One approach is to suspend a single droplet while exposing the droplet to hot convective air. A modified glass filament rig can be used which allows very minute measurement of the changes in the droplet mass during drying (Fu *et al.* 2012). The temperature of the droplet can also be measured by insertion of a fine thermocouple, and the shrinkage behaviour during drying can also be monitored via video monitoring (Fu *et al.* 2013b). Such drying kinetics was also previously measured by drying thin liquid films exposed to very hot convective air (Zbicinski & Li 2006) while the mass was continuously monitored. There has been increasing popularity in using the acoustic levitating technique to measure the required drying kinetics from a single acoustically suspended droplet. Such an approach is currently adopted by GEA (GEA Niro, Denmark) under the trademark DRYNETICS™. On the mechanical aspect, this form of droplet suspension represents a promising situation as there is no foreign intrusion into the particles or any surfaces in contact with the droplet. However, in terms of using this technique for kinetics measurement, there are certain limitations and important aspects to be considered. First, in an acoustic levitator the primary and secondary acoustic streaming will affect the evaporation rate in addition to only air-flow convective heat and mass transfer in a typical spray dryer (Schiffter & Lee 2007a). This leads to the need for a minimal relative velocity in the experiments to negate the secondary acoustic streaming effect, which traps evaporated vapour in the vicinity of the droplet. How closely such drying conditions reflect the actual phenomenon in the spray dryer have to be fully evaluated not just mathematically but also in comparison with the physics involved. Second, the drying rate of a droplet in an acoustic levitator is inferred from the changes in the droplet volume (Schiffter & Lee 2007b), therefore although the kinetics for the initial wet-bulb period can be obtained, the technique will not be able to measure the kinetics of dehydration once the solidification becomes more significant. It is noteworthy that the effect of solute on the drying kinetics becomes significant only when the surface solute concentration increases, therefore measurement of the initial period of drying might closely resemble only the evaporation behaviour of the solvent. Nevertheless, the acoustic levitation technique may be a useful tool to provide qualitative observation on the particle's formation process. In addition to obtaining drying kinetics, the single droplet drying technique, in its various forms, can also be used as a fundamental approach in investigating how the particle's product quality develops under different drying conditions (Fu *et al.* 2013a; Lallbeharry *et al.* 2014). Brief reviews on the use of the different single droplet drying technique can be found in the references cited in Fu *et al.* (2012) and Sadek *et al.* (2015).

Once the drying behaviour of the particles or droplet has been obtained experimentally, the next important aspect is to provide a robust mathematical description of the behaviour. This aspect of spray-drying development has received significant attention in recent years. The key is to have a mathematical description that requires minimal experimental data but can be translated to the dynamic drying environment and different operating conditions in a typical spray dryer. Different classes of droplet drying models are available in the literature. It is not the intention of this discussion to describe all the models in great detail. The most complex of these models is the diffusion-based model in which the heat and mass transfer (external and internal) of the droplet is captured in great detail (Mezhericher *et al.* 2007). While this approach can be utilized to provide very detailed information on the crust and morphological development of the dried particle, due to the high complexity and numerical requirement of the approach, such models have not found implementation in the relatively quick one-dimensional dryer-wide simulations. On the other hand, semi-empirical lump models are commonly used. Two such approaches have been tested in the literature: (i) the characteristic drying-rate curve approach (CDRC) and (ii) the reaction engineering approach (REA). The CDRC approach visualizes the secondary drying stage of a material to exhibit unique falling rate behaviour (Langrish & Kockel 2001). This unique behaviour, the characteristic falling rate behaviour of the material, is expected across different drying conditions, culminating in a normalized dimensionless characteristic drying curve. The REA approach visualizes the dehydration process as a reaction-based phenomenon (Chen & Putranto 2013). At the core of this approach the drying behaviour of a material is characterized by a unique material-specific master curve that describes the activation energy which needs to be overcome by convective heating at different stages of dehydration in order to facilitate moisture removal.

Complementing this information, it is also important to obtain information on how the droplet shrinks during drying. This is because in the one-dimensional simulation the particle movement is tracked through the domain of the spray tower and the size or density of the particle will affect its predicted residence time within the spray tower. Measurements are typically made by video observation of the drying process of a single droplet.

These kinetics and shrinkage measurements are typically incorporated into one-dimensional simulations. The advantage of the one dimensional approach is its simplicity as it can be set up using conventional spreadsheet software (Excel™) or very fast computing mathematical tools (MATLAB™). This approach can therefore be used as a training tool or as a guide to plant operators in adjusting the parameters of the drying process. Another advantage is that this approach allows visualization of the drying history within the chamber, although for large towers the particle moisture will approach the equilibrium moisture anyway. The knowledge of the drying history is important as it allows analysis and better controls of product quality development, particularly for qualities that are heat sensitive.

Equilibrium-based simulation The one-dimensional approach predicts the outlet conditions and product quality kinetically. In other words, the properties were predicted by tracking the moisture loss and the temperature gain by the droplet from the point of atomization along the length of the chamber. In a contrasting approach, the equilibrium approach, the spatial effect of the chamber is not discriminated and the spray tower is treated like a black box. The outlet condition of the air is then computed based on the mass and energy balance between the droplet and the air phase, and the product outlet moisture is computed based on the equilibrium moisture by numerical iterations. The equilibrium approach was used by Ozmen & Langrish (2003) for an equilibrium-limited spray dryer. Given fixed inlet conditions, a quick numerical scheme might be required to determine the outlet conditions of the particles since the outlet particle conditions depend on the outlet air conditions and vice versa. Equilibrium-limited spray drying is typically assumed if the size of the drying chamber is relatively large so the particles have sufficient time to reach equilibrium condition during their flight within the chamber. In contrast, smaller chambers such as the Buchi laboratory-scale chamber, which is widely used for product development and laboratory-scale testing, might be kinetically controlled due to its small size and the particles might not reach the equilibrium conditions on exit from the chamber.

In the one-dimensional simulation approach, the outlet moisture content was tracked and derived kinetically in time, but in equilibrium base modelling the outlet condition is determined thermodynamically. The key measurement required is the particle equilibrium moisture, which can be obtained from the absorption or desorption method using spray-dried particle samples. Such absorption and desorption measurements can be easily undertaken using saturated salt experiments and or another controlled humidity and temperature environment. However, for materials which tend to crystallize under high humidity, there is a risk of crystallization during the equilibrium moisture measurement, which might induce a lower (or zero) equilibrium moisture. It is noteworthy that most spray-dried dairy materials are amorphous and a measured isotherm from a crystalline form of the material might not represent the actual spray-dried powders. In order to overcome this, a fast approach to achieve a controlled environmental humidity and temperature is required. A unique approach is provided in the literature for such measurements (Lin *et al.* 2005).

While the approach of Ozmen & Langrish (2003) utilized and incorporated the equilibrium moisture of the particles in the computation, the heat requirement was computed mainly based on the evaporation of water. Schuck *et al.* (2009) devised another equilibrium moisture-based approach incorporating the additional energy required to induce dehydration in the falling rate period of drying. This approach has been trademarked as the SD^{2}P® approach and is commercially available from INRA (France). The premise of this approach is to use a desorption method in drying the dairy product (or any material) of interest and then compare the drying behaviour of the dairy droplet with that of pure water (Schuck *et al.* 2009). Analysis then reveals the additional energy requirement in dehydrating the dairy droplet due to the falling rate period of drying. This additional energy requirement is incorporated in the larger equilibrium moisture computation approach. Given an outlet product condition, the required inlet air conditions can then be determined. This approach has been tested with pilot-scale dryers in INRA for various dairy products.

A near-equilibrium approach was proposed by NIZO food research in the Netherlands under the commercial trademark of DrySPEC2 to model the drying process in spray dryers (Straatsma *et al.* 1991). This framework can also be used to model second-stage fluidized-bed drying. In essence, this framework is similar to the equilibrium moisture approach but the particles at the outlet of the spray dryer are assumed to always exhibit a slightly higher moisture content from equilibrium. This slightly higher equilibrium moisture was incorporated into the framework empirically and requires a number of experimental runs with the spray-drying unit. Successful testing was reported for different spray-drying operations and materials.

Similarly to the one-dimensional approach, the equilibrium-based approach can be set up as a fast computing approach and requires minimal numerical computation. However, the equilibrium approach might not have the inherent numerical uncertainties in solving the differential equations associated with the one-dimensional approach.

Nevertheless, such numerical uncertainty can be eliminated if a sufficiently small time scale or length scale is used, which is easily done in a well set up numerical framework. One key assumption for the equilibrium approach which is noteworthy is that it forces the particles to reach equilibrium at the exit of the chamber. Over-prediction in drying might occur if the droplet residence time is low relative to the size of the droplet. On the other hand, the one-dimensional approach can automatically account for both kinetically and equilibrium-controlled drying conditions. This is because the kinetics or the rate of dehydration is typically mathematically 'bounded' by the equilibrium moisture; dehydration ceases when the particle moisture reaches the equilibrium moisture.

Experimentally, the three slightly different equilibrium approaches discussed here require the measurement of the material isotherm. The INRA approach requires an additional measurement to determine the additional energy requirement of drying. This information obtained from the measurement is a material characteristic property (Schuck *et al.* 2009), therefore it can be applied to different operating conditions provided that the equilibrium assumption is met. From the information available in the literature (Straatsma *et al.* 1991), the measurements required for the NIZO approach might have some degree of dependency on the operation of the spray dryer used in the measurements, that is, it is more empirical. The correlations developed, however, have been claimed to be applicable across different systems.

Computational fluid dynamics simulation From the preceding section, it can be seen that the previous approaches do not account for any effect of chamber geometry, atomization process and recirculation on the drying process. When more detailed analysis is required, computational fluid dynamics (CFD) simulation can be used to provide more information on the operation. The CFD technique is well established from the fluid mechanics community. It resolves the local air-flow behaviour at specific spatial locations within the drying chamber. When this technique is applied to spray-dryer modelling, the Eulerian–Lagrangian approach is typically used. This involves the particle-in-cell approach in which the air flow within the chamber is computed, and sprayed droplets and particles are then incorporated into the flow field, with their trajectories tracked throughout the numerical domain of the air flow. With this approach, the development of the spray and how the spray dehydrates and disperses can be visualized and analysed in great detail without the 'large' assumption of a plug flow. Drying models similar to those described for the one-dimensional simulation approach are then incorporated into the particle tracking (Woo *et al.* 2008a,b). Some of the recent developments in applying this technique to spray-drying modelling are now described.

There has been ongoing interest in evaluating different approaches in modelling the air-flow turbulence in spray dryers. Such developments can also be found in other applications of the CFD technique. Proper treatment of the turbulence will affect the prediction of the air-flow pattern, which affects the prediction of the droplets or particles trajectory and dispersion. This subsequently determines the prediction of particle drying. Earlier CFD work on spray dryers used the Reynolds average Navier–Stokes (RANS) equation framework to numerically capture flow turbulence. A list of the earlier simulations using this approach can be found in Kuriakose & Anandharamakrishnan (2010) and a detailed discussion is given by Langrish (2009). This approach provides an opportunity to capture the development of turbulent eddies without the need to numerically discretise the drying chamber into excessively fine grids, lowering the computational requirements. With the development of high computing power at lower costs, however, there is a possibility of capturing the turbulence effect using the large eddy simulation (LES) approach, which discretises the domain into relatively fine grids. Even with the large jump in computing power such simulations might still require days to compute. Evaluation work was undertaken using this approach for an industrial-scale dryer (Jongsma *et al.* 2013). However, validation work is yet to be reported for such LES simulations. In between these extremes there are also evaluations using a unique scale adaptive simulation approach (Fletcher & Langrish 2009), which is able to numerically capture the flow structure of the smaller eddies. In that comparative work with the conventional RANS approach it was found that capturing the small eddies did not change the overall flow pattern of the flow field, therefore these different evaluations distil down to one very important question: what is the advantage of capturing such fine details on the small eddies in a spray dryer? In general, a finer numerical simulation will definitely provide greater accuracy in the simulation outcome. More work is required in incorporating particles in these finer scale simulations to gauge the extent of improvement in the accuracy of the simulation, by capturing such small turbulent structures, relative to the increase in the associated computing requirement.

Part of the larger scheme in the development of accurately modelling the air flow is in understanding the transient nature of the air flow within the chamber. Experimental work and a series of numerical studies on large- and small-scale dryers have shown that the air flow in spray dryers is generally unsteady and exhibits a self-sustained oscillatory behaviour. Such transient behaviour was numerically found to be affected by the aspect ratio of the dryer relative to the air-flow rate used (Guo *et al.* 2001; Woo *et al.* 2009a). The main implication from this series of investigations is that transient simulation is required into order to accurately capture the flow behaviour. Transient simulations, however, require significantly longer computation time and resources (Woo *et al.* 2012). Although transient simulations are important numerically, how the transient air-flow behaviour in the spray dryer affects the actual spray-drying behaviour, or specifically the product output yield, is unclear. Some experimental work suggests the transient behaviour might affect the deposition behaviour of the particles. However, the full effect of the transient air-flow behaviour is still uncertain and it will be important to investigate this in the near future.

Recent development has allowed the incorporation of particle agglomeration prediction into the CFD framework (Verdumen *et al.* 2004). This development was partly driven by the common practice of fine return in the spray drying of food and dairy powder to control the flowability and dissolution behaviour of the dried powder. The return of fines and its effect on spray drying can now be simulated and captured using the CFD technique. An important initiative was funded by the European Union under the EDECAD project to develop a agglomeration prediction framework. The approach can account for droplet–droplet coalescence, particle–particle agglomeration and particle–droplet interaction. Specifically, the premise of the framework focused on the development of suitable collision agglomeration criteria to denote successful or non-successful agglomeration. A unique approach to prediction particles dispersion and the probability of particles collision was also developed. As agglomeration involves the contact of sticky particles, part of the work also focused on predicting and measuring the stickiness of particles. This will be described in detail in a later section of this chapter.

The prediction of particle deposition on the internal wall surface of the drying chamber has also received significant attention over the last decade. There are two directions of development in this area: (i) improvement in the development of near-particle transport phenomenon prediction and (ii) improvement in defining the criterion of particle sticking. Earlier work on the CFD simulation of spray dryers relied on the overall momentum of the particles traversing the air flow and hitting the wall, that is, the inertia impact of the particles. Such an approach does not account for the diffusion of particles onto the wall. Exploratory simulation work was reported which incorporated the diffusion effect of particles across the air boundary layer near the wall. In addition, the report also attempted to model the removal of particles from the wall due to coherent near-wall turbulent structures (Jin & Chen 2010). The motivation for such higher level modelling is based on the physical observations in spray dryers where in long durations of operation the deposit layer will normally approach an equilibrium thickness, indicating a delicate balance between deposition and removal. The development mentioned pertains to numerically capturing how particles reach the wall. When a particle is computed to touch the wall, the conventional approach does not discriminate how the particle stickiness affects the prediction of deposition in the CFD framework. A step forward in this area is the use of the glass transition temperature concept to predict particle stickiness (Harvie *et al.* 2002). This concept has been used in large-scale simulations and also adopted in practice to control spray-dryer operations. Details of the physical phenomenon of the glass transition concept will be explained later in realtion to controlling particle stickiness. The evaluation of using the glass transition concept as a criterion, however, revealed that a clear cut-off concept might result in a 'sudden clean' prediction of particle-wall deposition in CFD simulations. More details on this intriguing aspect can be found in Woo *et al.* (2008b). There are recent developments in predicting the particle deposition behaviour combining the concept of glass transition with particle impaction dynamics (Woo *et al.* 2010).

These different features developed for the CFD modelling of spray drying describe the different physical phenomenon that can currently be captured by the technique. A key question that arises is how accurate is the CFD prediction of a spray-drying process? There are several reports which provide very detailed air and particle flow validation of the process for laboratory pilot-scale dryers with good agreement. The overall capability of CFD simulations in capturing the overall flow structure in large industrial-scale spray dryers, incorporating the effect of an internal static fluidized bed, has been validated and reported in the literature (Gabites *et al.* 2010). These different simulations may incorporate some or all of these features described above, therefore the need to incorporate all the phenomenon described above or perhaps the significance of each phenomenon described certainly depends on the configuration of the spray chamber and its operation. Some reports on CFD simulation of large industrial-scale dryers only provided validation on the outlet conditions, although limited experimental data is openly available. The challenge for accurate CFD simulation for industrial spray dryers will certainly depend on the availability of data or the simplification which has to be made.

One important contribution towards validating CFD simulations is the use of a monodisperse spray-drying approach (Rogers *et al.* 2012). The premise of this approach is to generate and dry monodisperse droplets in a straight-through falling trajectory in a spray tower. As the droplets are of the same size, a more consistent trajectory and drying history can be achieved, therefore validation of the drying kinetics implemented in a CFD simulation can be more reliably undertaken. In view of the better control in the drying history of the droplets, this approach is also used to experimentally investigate how drying history affects the quality and functionality of the particles formed (You *et al.* 2014; Sadek *et al.* 2015).

Combining the different approaches The different simulation approaches can be used for different purposes in understanding and operating the spray-drying process. Some useful reviews on the usage of the different levels of simulation for spray drying can be found in the literature (Langrish 2009; Mujumdar *et al.* 2010). For designing spray dryers, the equilibrium moisture approach can firstly be used to estimate the general operating conditions of a spray-drying process. This would resemble a process flow diagram type of development. On that operation basis, the one dimensional simulation approach can be used to scope or estimate the required size of the drying chamber. CFD simulations can then be used to provide a more detailed analysis identifying any possible anomaly in the design of the spray-drying process. Combination of the modelling on different scales, the multiscale approach, has been used or recommended by commercial companies working on product development with the spray-drying process.

Given an existing spray dryer, the equilibrium moisture approach has been shown to be a useful tool to determine spray-drying conditions. This approach and that of the one-dimensional simulation approach can be used as a training tool for the operator due to its simplicity and fast computational speed. This can help the operator to evaluate 'what if' situations. In this respect, the one-dimensional approach offers an additional feature as it provides information on the drying history of the particles or droplets within the chamber. On this framework, quality development models can then be incorporated to provide the operator with more information on how the operating parameters affect the quality of the products. CFD analysis has been shown to be useful in troubleshooting spray-dryer operations (Straatsma *et al.* 1999). One case reported was the use of the CFD technique to identify changes in the air-flow pattern within the system which led to non-dry particles. This visualization was used to complement plant-operating experience in the troubleshooting process.

A strong feature of CFD, which is directly applicable to the dairy industry, is the ability of the technique to evaluate suitable agglomeration. To be more specific, the technique can be used to evaluate suitable fine return position and load. Agglomeration of the particles improves the dissolution behaviour of the final particle product. Commercial companies such as Nestle have been exploring this aspect of the simulation technique (Gianfrancesco *et al.* 2010). Recognizing the importance of this aspect of spray drying to the dairy and food powder community, the European Union commissioned a large collaborative project to develop agglomeration models within the CFD framework (Verdumen *et al.* 2004). One of the latest uses of the CFD technique in spray drying is to evaluate the concept of a low-velocity spray dryer and assess the conditions leading to higher rates of *in situ* particle crystallization in the spray-drying chamber (Woo *et al.* 2011a,b).

12.3.2 Controlling powder stickiness and deposition

Controlling the stickiness of dairy particles is important not only to prevent the deposition of particles onto the internal wall of the drying chamber (Woo *et al.* 2008c, 2009b), but to also control the agglomeration process and the product stability and handling after spray drying. The glass transition concept is a widely accepted basis for describing the stickiness of particles due to their amorphous nature. Higher particle moisture will lower the glass transition temperature of the particles and *vice versa*. The higher the particle temperature above the glass transition temperature, the more sticky the particles will become. This is because the mobility of the molecules within the particle will increase exponentially, making it progressively viscous, with an increase in the particle temperature beyond the glass transition temperature.

The threshold for this positive temperature difference in which the particle becomes sticky is material specific. For particles with high sugar content, this is generally taken as 20 °C above the glass transition point. In general, it is very important to maintain the particle's outlet temperature lower than its sticky point. The particle temperature can be assumed to follow the air outlet temperature (assuming the particle reaches equilibrium) while the glass transition temperature will have to be determined experimentally by measuring its moisture content. From the measured moisture content, the glass transition temperature can be determined using the well-known Gordon–Taylor equation.

It can be appreciated that the approach above might be cumbersome in continuous operation of the spray-drying process due to the need to measure the particle outlet moisture. Schuck *et al.* (2005) proposed a semi-empirical approach by measuring the outlet air absolute humidity and comparing it with the ideal calculated air outlet absolute humidity based on mass balance. The difference between the measured and calculated values was found to be a reasonably accurate guide in delineating particle stickiness. A threshold of 2 g water/kg dry air was obtained below which non-sticky particles will be prodiced and *vice versa*. Although semi-empirical, this approach has a physical basis in that the difference denotes the hygroscopicity of the particle in retaining moisture. A larger value indicates more moisture retained in the particle, leading to higher stickiness. This concept works well for whole milk and skim milk powder. For whey powder, the difference in the absolute humidity value can be further used to delineate the degree of crystallization in the whey powder from the spray-drying process.

GEA recently developed a new approach to predict or measure the stickiness of particles (Petersen *et al.* 2014). Most of the early work on predicting the stickiness of particles was done by measuring the stickiness of already dried particles and then measuring their stickiness at further humidification of the particles. GEA introduced a new concept using a stickiness arm in conjunction with the acoustic levitator by measuring the stickiness of the droplet as it dries. This approach measures the stickiness of the droplet from the beginning of the process. The future aim is to incorporate the stickiness properties into CFD simulations to help design and operate spray dryers. At the moment, the technique shows interesting results. Further assessment will be crucial in determining how representative the measurements made from relatively larger droplet are of actual micron-size particles. Nevertheless, in similar a vein, drying kinetics are currently also derived from a relatively 'larger' system.

In contrast to manipulating the spray-drying operation to avoid particle stickiness, an alternative approach is to manipulate the surface energetics of the dryer wall (Woo *et al.* 2008c, 2009b). Use of low surface energy has been common practice in the food industry to reduce the adhesion of food material to processing equipment. This concept has yet to find commercial application in spray dryers. However, some development has been undertaken in the research community to evaluate the feasibility of this approach by comparing a conventional stainless steel surface (high surface energy) and Teflon (low surface energy). It was found that this approach produce approximately 10–20% reduction in deposition if the particle is sticky or has a relatively viscous outer surface. Further trials for fat-covered particles, typical of dairy powders, revealed a similar advantage in deposit reduction (Keshani *et al.* 2013). For deposition of relatively dry particles, such an approach has minimal effect. Research in this area, so far, has been limited to relatively short runs in small pilot-scale spray dryers. More work is required to gauge how this approach will fare in long continuous runs with relatively heavier deposit build up. It is unclear if this approach can affect the dislodging behaviour of deposited particles in a real spray-drying environment, which may or may not be useful depending on the accumulation rate and degradation on the wall. Nevertheless, measurements have shown that the particle deposit adhered more weakly to the surface with lower surface energy (Teflon).

12.4 Advances in operating fluidized-bed dryers

The fluidized-bed technology used as the second stage of drying is also widely applied across the food industry. Some of the developments reported below, although developed for food industry powder processing, are therefore also applicable to the dairy industry, with modifications. In view of the high residence time within the fluidized bed, developmental work on the feasibility and control of the drying process is not as intensive as for spray dryers. Information on the developments in fluidized beds can be found elsewhere as this technology is aslo used in the chemical industry. Two aspects specific to the food and dairy industry are described here.

12.4.1 Controlling crystallization

There is considerable interest in crystallizing or partially crystallizing spray-dried dairy particles. Partially crystallized particles are found to show improved flowability behaviour and higher storage stability. Recent work by Yazdanpanah & Langrish (2011a) revealed that the exposure of spray-dried milk powder to high humidity conditions promotes outer surface crystallization, which improves the flowability of the powder. However, surprisingly, the inside of the particles is still amorphous-like. It is noteworthy that an amorphous structure is important and promotes fast dissolution, which is required in instant dairy powder, therefore control of the crystallization of spray-dried dairy powders can be used to enhance the overall properties of the product.

On top of this, in chocolate processing the degree of crystallization in the milk powder used is critical in controlling the quality of the chocolate. Aguilar *et al.* (1994) noted that the state of the lactose (amorphous or partially crystalline) in the whole milk powder added during the conching process affects the texture and taste of the final chocolate product. The presence of crystallized lactose enhances the chocolate flavour and reduces the hardness. This was attributed to the release of free-fat in the particles arising from the crystallization of lactose within the milk particles. There are many reports which support this fat-releasing process due to the crystallization of lactose in the milk particles. As well as increasing the free-fat in the milk particles, Aguilar & Ziegler (1994) also showed that the crystallization of the contained lactose, when the milk particles were exposed to high humidity, resulted in denser milk particles.

Lactose in dairy particles may be crystallized prior to the spray-drying process. Such precrystallization is typical undertaken for whey powder processing. Recent work suggests that crystallization of the contained lactose can also potentially be undertaken post spray drying with an external fluidized bed (Yazdanpanah & Langrish 2011b). This is because of the long residence time available in the second or third stage of drying. The strategy is to use warm humid air as the fluidization air in the fluidized bed so that the spray-dried particles will absorb the moisture, giving molecular mobility within the particles for solid-phase transition. One main factor which sets the operation window for this is the maximum air humidity level, which can be used in the fluidization process relative to the velocity used. Too high a humidity will lead to excessive stickiness, resulting in non-stable or unsuccessful fluidization. This concept of a maximum humidity level for spray-dried dairy particles has been illustrated in a series of reports (Nijdam *et al.* 2008; Yazdanpanah & Langrish 2011a–c).

Laboratory trials of this concept for whey particles initially showed that even with the longer residence time afforded by the fluidization bed, the time required for solid-phase transition was still too high to make it feasible as a processing option (Nijdam *et al.* 2008). Latter trials using milk powders, however, showed that this concept is promising and the required crystallization time within the fluidized-bed set-up can be reduced by controlling the humidity of the hot air used for fluidization (Yazdanpanah & Langrish 2011b). The precrystallization of lactose in the milk powder within the spray dryer prior to the fluidized bed was also found to be a key element in reducing the required crystallization time within the second (or subsequent) stage of drying. Also in this line, Yazdanpanah & Langrish (2011c) suggested a strategy that involves a closed humid-loop system returning the exhaust air from the multistage fluidized beds back into the spray dryer. The return of the humid air from the second and third stages of drying was then effectively used in the spray dryer to partially precrystallize the lactose during the particle formation process within the spray dryer. Using the humid air approach, the mechanism of lactose crystallization-induced release of fat was also investigated in detail with the aim of producing suitable dairy particles for the chocolate industry (Yazdanpanah & Langrish 2012). Suggestions and equipment layout concepts from the industry that use this approach can be found in a report by Hynd (1980) that aims to achieve partial or nearly complete crystallization of whey powders to improve the storage and flowability properties of the powder.

12.4.2 Controlling agglomeration

As well as drying the dairy powders in the fluidized bed, another positive advantage of this process is the ability to agglomerate the powders due to the mixing and tumbling movement of the particles. Agglomerated powders offer better handling behaviour and the dissolution rate of the powders is improved as the clumping effect is minimized because the pores generated by the agglomeration of the particles allows water to come into contact with the particles. While the technology has long been used by the industry for this purpose, further reasearch in this area is required to further understand and provide better control of the agglomeration process, particularly in the food powder industry. In a pilot-scale study by Nestle, the effect of drying temperature, fluidization velocity and residence time on the drying and secondary agglomeration of food products was systematically investigated. It was found that adjustment to the residence time of the fluidized bed was the most efficient way to control the size of the agglomerates (Palzer 2007).

The glass transition concept has been widely studied in predicting particle stickiness and caking during storage. More fundamentally, extending the previous understanding, this concept was also found to be a good guide to 'designing' the agglomeration process in fluidized beds (Palzer 2005). In particular, this concept was proposed as a predictive tool to prevent bed collapse when fluidizing food particles with humid air. Combining the glass transition approach in describing the particle rigidity and the viscoelastic aspect of amorphous particles, a theoretical framework was developed to predict the 'collapse conditions' accounting for particle contact time. In another report, Palzer (2009) showed that the same concept can be applied to spray-dried dextrose particles as a model for amorphous particles.

Although such a concept was developed with food powders, the link with dairy powders is that both sprayed dried powders are amorphous due to rapid dehydration in the particle formation stage, therefore, even for dairy powders, the glass transition concept will apply. However, one particular aspect unique to the dairy powder system is the possible presence of fat on the particle surface, which is commonly found in whole milk powder. It is unclear at the moment how this factor will augment the role of the glass transition concept in describing the agglomeration process. Further investigations are certainly required in this area.

12.5 Conclusion

Most of the recent advances in dairy powder drying described in this chapter pertain to achieving better understanding and better control of the drying process as well as other associated processes such as particle crystallization and agglomeration in a more mechanistic manner. This approach is contrary to the purely empirical trial and error approach, but this does not mean that the empirical approach or plant experience is not required. The more mechanistic approaches described here will be a useful tool to complement existing plant experience. In fact, a systematic empirical approach can be found in the literature (Woo *et al.* 2007). Such a combined approach will be particularly useful in extrapolating existing dairy drying processes to new materials and products. This will be particularly important in producing engineered dairy powders to provide health and nutritional benefits. This chapter focuses only on recent specific advances in this area. Readers interested in reading more on the fundamentals of spray drying and general equipment considerations can refer to the useful handbooks of Masters (1979), Mujumdar (2014) and Westergaard (1994).

References

Aguilar, C.A. & Ziegler, G.R. (1994). Physical and microscopic characterization of dry whole milk with altered lactose content. 2. Effect of lactose crystallization. *Journal of Dairy Science*, 77(**5**), 1198–1204.

Aguilar, C.A., Hollender, R. & Ziegler, G.R. (1994). Sensory characteristics of milk chocolate with lactose from spray-dried milk powder. *Journal of Food Science*, **59(6)**, 1239–1243.

Chen, X.D. & Putranto, A. (2013). *Modelling Drying Processes: A Reaction Engineering Approach*, 1st edn, Cambridge University Press, Cambridge.

Fletcher, D.F. & Langrish, T.A.G. (2009). Scale-adaptive simulation (SAS) modelling of pilot-scale spray dryer. *Chemical Engineering Research and Design*, **87(10)**, 1371–1378.

Fu, N., Woo, M.W. & Chen, X.D. (2012). Single droplet drying technique for droplet drying kinetics measurement and particle functionality: A Review. *Drying Technology*, **30(15)**, 1771–1785.

Fu, N., Woo, M.W., Selomulya, C. & Chen, X.D. (2013a). Inactivation of *Lactococcus lactis* ssp. *cremoris* cells in a droplet drying convective drying. *Biochemical Engineering Journal*, **79**, 46–56.

Fu, N., Woo, M.W., Selomulya, C. & Chen, X.D. (2013b). Shrinkage behaviour of skim milk droplets during air drying. *Journal of Food Engineering*, **116**, 37–44.

Gabites, J.R., Abrahamson, J. & Winchester, J.A. (2010). Air flow patterns in an indsutrial milk powder spray dryer. *Chemical Engineering Research and Design*, **88(7)**, 899–910.

Gianfrancesco, A., Turchiuli, C., Flick, D. & Dumoulin, E. (2010). CFD modelling and simulation of maltodextrin solutions spray drying to control stickiness. *Food Bioprocess Technology*, **3(6)**, 946–955.

Guo, B., Langrish, T.A.G. & Fletcher, D.F. (2001). Numerical simulation of unsteady turbulent flow in axisymmetric sudden expansions. *Journal of Fluids Engineering*, **123(3)**, 574–587.

Harvie, D.J.E., Langrish, T.A.G. & Fletcher, D.F. (2002). A computational fluid dynamics study of a tall-form spray dryer. *Food and Bioproducts Processing*, **80(3)**, 163–175.

Hynd, J. (1980). Drying of whey. *Journal of the Society of Dairy Technology*, **33(2)**, 52–54.

Jin, Y. & Chen, X.D. (2010). A fundamental model of particle deposition incorporated in CFD simulations of an industrial milk spray dryer. *Drying Technology*, **28(8)**, 960–971.

Jongsma, F.J., Innings, F., Olsson, M. & Carlsson, F. (2013). Large eddy simulation of unsteady turbulent flow in a semi-industrial size spray dryer. *Dairy Science & Technology*, **93(4–5)**, 373–386.

Keshani, S., Daud, W.R.W., Woo, M.W., Nourouzi, M.M., Talib, M.Z.M., Chuah, L. & Russly, A.R. (2013). Reducing the deposition of fat and protein covered particles with low energy surfaces. *Journal of Food Engineering*, **116(3)**, 737–748.

Kuriakose, R. & Anandharamakrishnan, C. (2010). Computational fluid dynamics (CFD) application in spray drying of food products. *Trends in Food Science & Technology*, **21(8)**, 383–398.

Lallbeharry, P., Tian, Y., Fu, N., Woo, M.W., Selomulya, C. & Chen, X.D. (2014). Effects of ionic and non-ionic surfactants on milk shell wettability during co-spray-drying of whole milk particles. *Journal of Dairy Science*, **97(9)**, 5303–5314.

Langrish, T.A.G. (2009). Multi-scale mathematical modeling of spray dryers. *Journal of Food Engineering*, **93(2)**, 218–228.

Langrish, T.A.G. & Kockel, T.K. (2001). The assessment of a characteristic drying curve for milk powder for use in computational fluid dynamics modelling. *Chemical Engineering Journal*, **84(1)**, 69–74.

Lin, S.X.Q., Chen, X.D. & Pearce, D.L. (2005). Desorption isotherm of milk powders at elevated temperatures and over a wider range of relative humidity. *Journal of Food Engineering*, **68**(**2**), 257–264.

Masters, K. (1979) *Spray Drying Handbook*, George Godwin Limited, London.

Mezhericher, M., Levy, A. & Borde, I. (2007). Theroretical drying model of single droplets containing insoluble or dissolved solids. *Drying Technology*, **25**(**6**), 1025–1032.

Mujumdar, A.S. (ed.) (2014). *Handbook of Industrial Drying*, 4th edn, CRC Press.

Mujumdar, A.S., Huang, L.X. & Chen, X.D. (2010.) An overview of the recent advances in spray-drying. *Dairy Science & Technology*, **90**(**2–3**), 211–224.

Nijdam, J., Ibach, A. & Kind, M. (2008). Fluidisation of whey powders above the glass-transition temperature. *Powder Technology*, **187**(**1**), 53–61.

Ozmen, L.T.A.G. & Langrish, T.A.G. (2003). A study of the limitations to spray dryer outlet performance. *Drying Technology*, **21**(**5**), 895–917.

Palzer, S. (2005). The effect of glass transition on the desired and undesired agglomeration of amorphous food powders. *Chemical Engineering Science*, **60**(**14**), 3959–3968.

Palzer, S. (2007). Drying of wet agglomerates in a continuous fluid bed: Influence of residence time, air temperature and air-flow rate on the drying kinetics and the amount of oversize particles. *Chemical Engineering Science*, **62**(**1**), 463–470.

Palzer, S. (2009). Influence of material properties on the agglomeration of water-soluble amorphous particles. *Powder Technology*, **189**(**2**), 318–326.

Patel, K., Chen, X.D., Jeantet, R. & Schuck, P. (2010). One-dimensional simulation of co-current, dairy spray drying systems–pros and cons. *Dairy Science Technology*, **90**(**2–3**), 181–210.

Petersen, T., Sloth, J., Ullum, T. & Hassager, O. (2014). Method for measuring stickiness in spray drying based on particle impact. Proceedings from the 19th International Drying Symposium, France.

Pisecky, J. (1997). *Handbook of Milk Powder Manufacture*, Niro A/S.

Pisecky, J. (2005). Spray drying in the cheese industry. *International Dairy Journal*, **15**(**6**), 531–536.

Rogers, S., Wu, W.D., Lin, S.X.Q. & Chen, X.D. (2012). Particles shrinkage and morphology of milk powder made with a monodisperse spray dryer. *Biochemical Engineering Journal*, **62**, 92–100.

Sadek, C., Schuck, P., Fallourd, Y., Pradeau, N., Floch-Fouere, C.L. & Jeantet, R. (2015). Drying of a single droplet to investigate process-structure-function relationships: a review. *Dairy Science & Technology*, **95**(**6**), 771–794.

Schiffter, H. & Lee, G. (2007a). Single-droplet evaporation kinetics and particles formation in an acoustic levitator. Part 1: Evaporation of water microdroplets assess using boundary-layer and acoustic levitation theories. *Journal of Pharmaceutical Sciences*, **96**(**9**), 2274–2283.

Schiffter, H. & Lee, G. (2007b). Single-droplet evaporation kinetics and particles formation in an acoustic levitator. Part 2: Drying kinetics and particles formation from microdroplets of aqueous mannitol, trehalose, or catalase. *Journal of Pharmaceutical Sciences*, **96**(**9**), 2284–2295.

Schuck, P., Mejean, S., Dolivet, A. & Jeantet, R. (2005). Thermohygrometric sensor: A tool for optimizing the spray drying process. *Innovative Food Science and Emerging Technologies*, **6**(**1**), 45–50.

Schuck, P., Dolivet, A., Mejean, S., Zhu, P., Blanchard, E. & Jeantet, R. (2009). Drying by desorption: a tool to determine spray drying parameters. *Journal of Food Engineering*, **94**(**2**), 199–204.

Schwartzbach, C. & Masters, K. (2001). Performance of spray dryer with integrated filter and fluid bed. *Drying Technology*, **19**(**8**), 1909–1923.

Straatsma, J., Houwelingen, G.V., Meulman, A.P. & Steenbergen, A.E. (1991). Dryspec2: a computer model of a two-stage dryer. *International Journal of Dairy Technology*, **44**(**4**), 107–111.

Straatsma, J., Van Houwelingen, G., Steenbergen, A.E. & De Jong, P. (1999). Spray drying of food products: 1. Simulation model. *Journal of Food Engineering*, **42**(**2**), 67–72.

Truong, V., Bhandari, B. & Howes, T. (2005). Optimization of co-current spray drying process of sugar-rich foods. Part 1 – Moisture and glass transition temperature profile during drying. *Journal of Food Engineering*, **71**(**1**), 55–65.

Verdumen, R.E.M., Menn, P., Ritzert, J., Blei, S., Nhmaio, G.C.S., Sonne Sprensen, T., Gunsing, M., Straatsma, J., Verschueren, M., Sibeijn, M., Schulte, G., Fritsching, U., Bauckhage, K., Tropea, C., Sommerfeld, M., Watkins, A.P., Yule, A.J. & Schonfeldt, H. (2004). Simulation of agglomeration in spray drying installations: The EDECAD project. *Drying Technology*, **22**(**6**), 1403–1461.

Westergaard, V. (1994). *Milk Powder Technology Evaporation and Spray Drying*, Niro A/S.

Woo, M.W., Daud, W.R.W., Tasirin, S.M. & Talib, M.Z.M. (2007). Optimization of the spray drying parameters – A quick trial-and-error method. *Drying Technology*, **25**(**10**), 1741–1747.

Woo, M.W., Daud, W.R.W., Mujumdar, A.S., Talib, M.Z.M., Wu, Z.H. & Tasirin, S.M. (2008a). Comparative study of drying models for CFD simulations of spray dryers. *Chemical Engineering Research and Design*, **86**(**9**), 1038–1048.

Woo, M.W., Daud, W.R.W., Mujumdar, A.S., Wu, Z.H., Talib, M.Z.M. & Tasirin, S.M. (2008b). CFD evaluation of droplet drying models in a spray dryer fitted with a rotary atomizer. *Drying Technology*, **26**(**10**), 1180–1198.

Woo, M.W., Daud, W.R.W., Tasirin, S.M. & Talib, M.Z.M. (2008c). Effect of wall surface properties on the deposition problem at different drying rates in a spray dryer. *Drying Technology*, **26**(**1**), 15–26.

Woo, M.W., Daud, W.R.W., Mujumdar, A.S., Talib, M.Z.M., Wu, Z.H. & Tasirin, S.M. (2009a). Non-swirling steady and transient flow simulations in short-form spray dryers. *Chemical Product and Process Modeling*, **4**(**1**), Article 20.

Woo, M.W., Daud, W.R.W., Talib, M.Z.M. & Tasirin, S.M. (2009b). Controlling food powder deposition in spray dryers at quasi-steady wall thermal condition: wall surface energy manipulation as an alternative. *Journal of Food Engineering*, **94**(**2**), 192–198.

Woo, M.W., Daud, W.R.W., Mujumdar, A.S., Tasirin, S.M. & Talib, M.Z.M. (2010). Role of rheological characteristics in amorphous food particle-wall collisions in spray drying. *Powder Technology*, **198**(**2**), 251–257.

Woo, M.W., Rogers, S., Selomulya, C. & Chen, X.D. (2011a). Particle drying and crystallization characteristics in a low velocity concurrent pilot scale spray drying tower. *Powder Technology*, **223**, 39–45.

Woo, M.W., Rogers, S., Lin, S.X.Q., Selomulya, C. & Chen, X.D. (2011b). Numerical probing of a low velocity concurrent pilot scale spray drying tower for mono-disperse particle production – Unusual characteristics and possible improvements. *Chemical Engineering and Processing: Process Intensification*, **50**(**4**), 417–427.

Woo, M.W., Le, C.M., Daud, W.R.W., Mujumdar, A.S., Chen, X.D., Tasirin, S.M. & Talib, M.Z.M. (2012). High swirling transient flows in spray dryer and consequent effect on modeling of particle deposition. *Chemical Engineering Research and Design*, **90**(**3**), 336–345.

Yazdanpanah, N. & Langrish, T.A.G. (2011a). Egg-shell like structure in dried milk powders. *Food Research International*, **44**(**1**), 39–45.

Yazdanpanah, N. & Langrish, T.A.G. (2011b). Fast crystallization of lactose and milk powder in fluidized bed dryer/crystallizer. *Dairy Science & Technology*, **91**(**3**), 323–340.

Yazdanpanah, N. & Langrish, T.A.G. (2011c). Crystallization and drying of milk powder in a multiple-stage fluidized bed dryer. *Drying Technology*, **29**(**9**), 1046–1057.

Yazdanpanah, N. & Langrish, T.A.G. (2012). Releasing fat in whole milk powder during fluidized bed drying. *Drying Technology*, **30**(**10**), 1081–1087.

You, X., Zhou, Z.H., Liao, Z.K., Che, L.M. Chen, X.D., Wu, D., Woo, M.W. & Selomulya, C. (2014). Dairy milk particles made with a mono-disperse droplet spray dryer (MDDSD) investigated for the effect of fat. *Drying Technology*, **32**(**5**), 528–542.

Zbcinski, I. & Li, X., (2006). Conditions for accurate CFD modeling of spray-drying process. *Drying Technology*, **24**(**9**), 1109–1114.

13

Industrial Scale Drying of Dairy Products

D. Anand Paul

Nestle Research and Development Centre, Manesar, Gurgaon, India

13.1 Introduction

The market needs of dried dairy products are driven by the export requirements of countries that have surplus milk production. The processing and value addition of milk by means of drying are considered important in relation to enhanced product shelf life and as an attractive marketing strategy. Changes in lifestyle are expected to result in a further increase in the demand for dried milk products, especially in milk-deficient markets (Vadlamudi 2015). With this background, the dairy industry plays a major role in bridging the *demand–supply* gap with respect to dried dairy products. The drying plants that form an integral part of the dairy processing sector are central to the value chain of the dairy industry. The value chain is the flow of commodity and value between producers and consumers (Figure 13.1). In addition, the dairy industry has been the driving force behind the commercialization of most established drying techniques today, for example spray drying and drum drying, whose association with the dairy industry dates back to 18th and 19th centuries, respectively.

The portfolio of dried dairy products is continuously evolving and the key commercial product categories include agglomerated milk powders, skim milk powders, flavoured milk powder, infant formula, whey protein isolates and concentrate, lactose, and tea and coffee whiteners. In accordance with the market demand for dried dairy products, the production capacity of dairy dryers has expanded from about 20 tonnes/h in 1997 to an impressive 30 tonnes/hour in 2012 (GEA 2014).

Drying on an industrial scale has various complexities. An underpinning knowledge of the intricacies associated with industrial drying of dairy products is central to the efficient performance management of this large-scale process. Industrial-scale drying has the following features:

- continuous operation and consistent product quality
- automated process controls to achieve optimized process settings and product characteristics (automation also facilitates real-time monitoring of critical process parameters and thereby facilitates efficient traceability)
- identification of critical control parameters for the drying operation, as a pro-active approach for preventing hazards
- management of bottleneck operations to reduce the downtime for achieving increased productivity at low cost

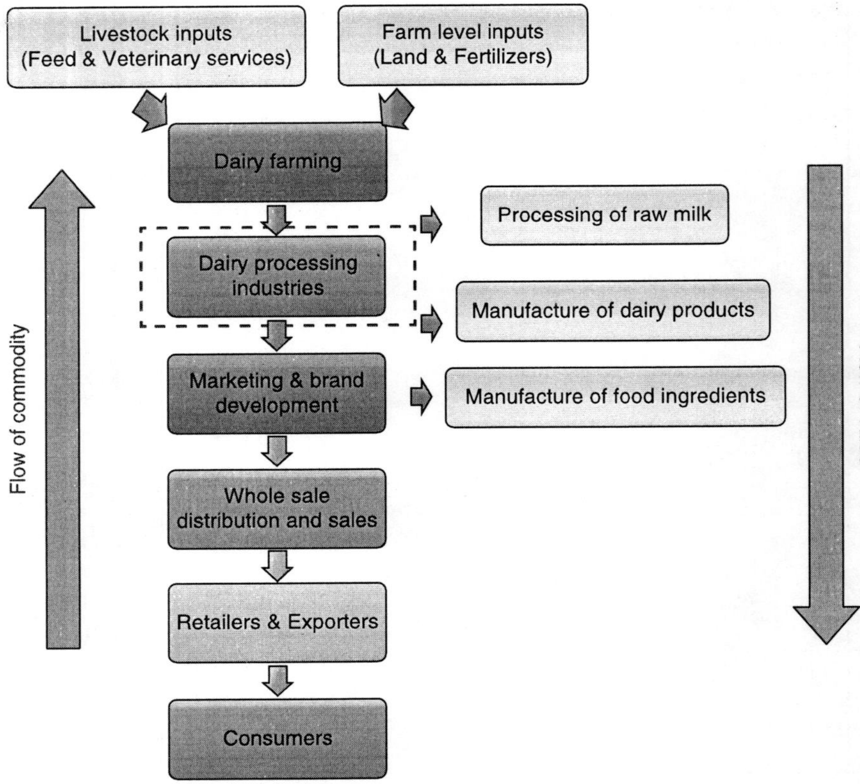

Figure 13.1 Value chain in the dairy industry.

- increased energy efficiency and high productivity at low operation cost
- adherence to good manufacturing practices (GMPs) for ensuring a hygienic drying operation, and establishment of effective cleaning and maintenance procedures for the dryers to increase the mean uptime.

This chapter provides a fundamental insight into all these aspects of industrial-scale drying of dairy products. The dairy industry alternates between spray-drying and roller drum-drying techniques for the production of dried dairy products. The selection of dryer type is straightforward and depends on the rate and volume of production and product characteristics. Spray dryers are preferred when a large volume of powder production is required at a faster rate and drum dryers are used when certain functional characteristics are required in the product. This chapter focuses on the industrial-scale spray drying and drum drying of dairy products, but first the general principles and commonly used terminologies pertaining to the industrial drying of milk products are outlined.

13.2 Process flow in a dairy drying plant

Understanding the overall flow-sheet is essential in terms of achieving the optimum selection of process parameters. It is also important from the perspective of dryer selection, since the dryer type chosen depends on the predrying and postdrying steps (Mujumdar 2006). A typical process flow in a milk powder plant is shown in Figure 13.2. The operation of a milk drying plant consists of the following stages:

(1) Reception of milk, pretreatment and storage.
(2) Pasteurization and standardization of milk before further processing.
(3) Concentration of standardized milk to required solid content before drying by membrane processing or by evaporation with a falling-film evaporator.
(4) Drying of preconcentrated milk by spray-drying or roller-drum drying.
(5) An optional fluidized-bed drying step to achieve the final desired moisture content and agglomeration for better solubility of spray-dried products.
(6) Packaging of the dried dairy ingredients and storage for marketing.

There are other vital upstream and downstream operations in the industrial drying of dairy products, such as the procurement of raw materials, storage and warehousing of products, that are not within the scope of this discussion.

Of the stages listed above, the concentration (3) and post-drying by fluidized-bed drying (5) are considered to be integrated with the drying process (4). The integration of the drying process with the pre- and post-processing steps is considered critical from the view of achieving optimal energy efficiency. The concentration step is significant with respect to energy savings and minimizing dryer dimensions (Mujumdar 2006), as it reduces the amount of water to be evaporated by the dryer. Evaporation under vacuum is the commonly employed concentration technique and results in almost 48–50% solids in the preconcentrated milk, which is considered appropriate to be taken forward for the subsequent drying step. A multiple-effect evaporator, usually of the falling-film or plate type (Pisecky 1995), is employed for the preconcentration of milk in the dairy industry. Although evaporation consumes only 10% of the total energy of the drying process, it is a thermal process that inevitably leads to denaturation of milk proteins. Alternatively, a cold concentration technique such as reverse osmosis is practiced by some dairy plants, which, in addition to reduced thermal damage to the product, also offers a low-energy advantage when compared to the conventional thermal evaporation method. The advantage of using an integrated second of stage drying is described in the following section.

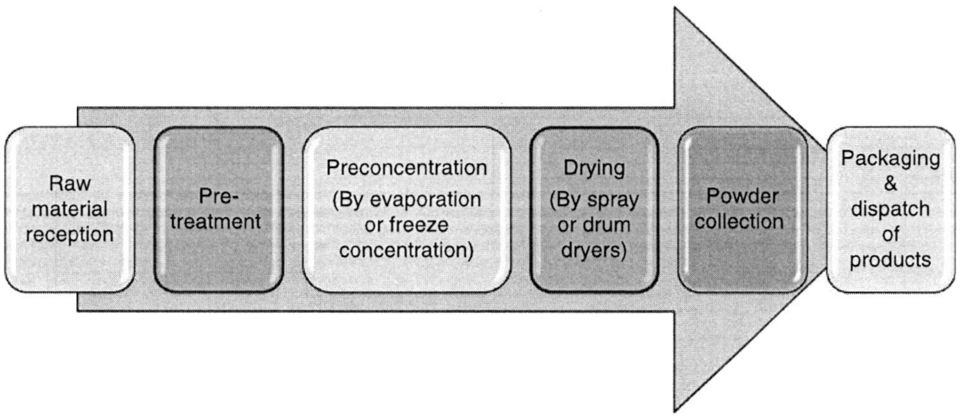

Figure 13.2 Production chain in a plant producing dried dairy products.

13.3 Lexicon of industrial-scale drying

Familiarity with the lexicon of industrial language will be helpful in understanding the following sections.

- **Productivity:** The ratio of output to some or all of the resources used to produce the output (Barners 1980).
- **Cost of production (CoP):** Cost incurred from the production of a product: the sum of the raw and packing material cost, the variable manufacturing cost, the fixed manufacturing cost and the cost of depreciation.
- **Capacity utilization:** The fraction of the equipment's capacity that is used during an operation (Athimulam *et al.* 2006).
- **Uptime:** A measure of the effective and in-time utilization of a piece of equipment. The product of capacity utilization and uptime defines the combined utilization of particular equipment (Koulouris *et al.* 2000; Petrides *et al.* 2002; Athimulam *et al.* 2006).
- **Downtime:** Time lost during activities or events when a bottleneck inhibits production. The downtime can be related to the set-up, start-up, cleaning, change-over or maintenance of the machine/equipment, or to personal time. Downtime can be planned (e.g. for cleaning purposes) or unplanned (e.g. plugging of cyclones, outlet of spray chamber due to moisture build-up and condensation in spray dryers).
- **Bottleneck:** An event that limits the production or nominal speed of the process line. Bottlenecks are process limitations related to either the equipment or the resource (demand for various utilities, labour, raw material). The bottleneck is the process step that has the lowest rate of throughput. The processing step with the highest combined utilization is the most likely to become the process bottleneck (Athimulam *et al.* 2006).
- **Energy efficiency:** The ratio of end-use energy that is generated in proportion to the primary or secondary energy used.
- **Energy consumption (tonnes):** The ratio between the energy consumption and the production volume, expressed as GJ/tonne. It provides a measure of the energy used for production and the environmental impact that results from the industrial operation.
- **Critical control point:** A place, practice, procedure or process that can be controlled to prevent, eliminate or reduce the hazard to acceptable levels, wherein hazard is an unacceptable biological, chemical or physical contamination that renders the food unfit for consumption (Bryan 1993; Silva, 1999).
- **Good manufacturing practice (GMP):** GMP is an aspect of quality assurance that ensures that the products are consistently produced and controlled to the quality standards appropriate to their intended use and as required by the product specification (WHO 2014).
- **Cleaning-in-place (CIP):** A method of cleaning the interior surfaces of the process equipment and associated fittings without dismantling. CIP is carried out with sterile water or cleaning solutions of specific composition, sprayed at high pressure from internal nozzles under controlled conditions.

13.4 Industrial spray drying of dairy products

The production of milk powder is the classic example of industrial scale spray drying. The scale of spray drying in the dairy industry ranges from a few kilograms per hour to more than 100 tonnes per hour per dryer (Bajsic & Kranjcevic 2001). The world's largest spray dryer, in New Zealand, can successfully produce more than 700 tonnes of milk powder per day. This production capacity is translated into the spray dryer's ability to process 4.5 million litres of fresh liquid milk into milk powder (GEA 2014). The need for such high production capacities is justified by the high market demand for spray-dried milk products.

In the dairy industry spray dryers are used for the production of a variety of products, such as instant whole and skim milk powder, infant foods, powdered products of cheese, and whey and dairy whiteners. Other derived products, such as butter milk, lactose, caseinates, cheese, ice-cream and yoghurt, have also been spray dried to meet consumer requirements (Bhandari *et al.* 2008). The principles and different stages of spray drying are well-elucidated in the literature. There is less clarity on industrial-scale spray drying because of the complexities involved in measuring the process parameters within the spray dryer under operation. Nevertheless, the need to improve energy efficiency has driven design developments in industrial spray-drying equipment. While the single-stage spray dryer has high energy requirements, the two-stage version reduces this by the integration of a vibro-fluidized-bed dryer (VFBD) and cooling systems with the spray-drying equipment. During the second stage of drying in the fluidized bed, air at sufficient velocity and temperature (80–120 °C) is supplied for fluidizing and drying the comparatively high moisture content product (~8–9 wt%, depending on the processing conditions and product type) to the desired final moisture content (~3–4 wt%) (Mujumdar *et al.* 2010).

An improvement over two-stage spray drying is the three-stage drying system for milk drying (Tang *et al.* 1997). This comprises a fluidized bed that is attached to the conical bottom of spray drying chamber followed by a VFBD, as in the two-stage spray-drying system. The medium of fluidization for the semi-dried milk powder is hot air at 90–100 °C. The advantage of this system is better control of the drying and cooling operations during the second and third stages. Furthermore, it is possible to recycle the fine powders from the spray-drying chamber and VFBD into the spray-drying chamber. This induces agglomeration of the fine powder particles that eventually results in reduced moisture content and optimal rehydration properties. The three-stage spray-drying system results in 20% better energy efficiency compared to single-stage drying (Westergaard 1994; Huang *et al.* 2001). Another variation that combats the heat-sensitive nature of dairy products is the installation of a conveyor belt drying system at the bottom of the spray-drying chamber (Niro Denmark Ltd). This facilitates continuous drying of partially wet particles on the belt at a lower temperature.

13.4.1 Automation of industrial-scale spray dryers

The automation of industrial spray dryers is essential from many perspectives, such as maintenance of product quality, process stability, maximum throughput at optimal energy efficiency and minimum CoP, safe operation (Sivarajalingam 2009) and the optimization of process parameters such as feed rate, temperature, pressure, residence time, particle size, moisture content and bulk density. Automation also promotes uninterrupted traceability by means of real-time recording of process data (SPX Corporation 2012). Successful automation of spray dryers is indicated by the number of automatically controlled units that one operator can handle, provided they are located in the same complex (Bajsic & Kranjcevic 2001).

With automated controls, a process variable is monitored by a proportional-integral (PI) regulator against a set value in a direct or indirect manner. For instance, the moisture content of the final product exiting the spray dryer is controlled indirectly by maintaining a constant outlet drying air temperature through the regulation of feed rate (i.e. speed of rotation of feed pump) (Sivarajalingam 2009). It is essential that the temperature and pressure sensors are placed in appropriate positions within the spray chamber (e.g. above the point where air inlet pipe is connected to the chamber, at the middle height of the tower and in the upper part of the chamber) (Bajsic & Kranjcevic 2001).

An alternative to PI controllers is to define a control system in simple words. This approach is known as *fuzzy control* and is based on practical application knowledge rather than analytical models. It can be used when there is prior expertise in a process. The syntax for writing the programme code takes the form 'If X is A and Y is B, then C'. For spray dryer operation, this can be designed as shown below.

- Rule1: If the outlet air temperature is low then increase the power of the air heater.
- Rule2: If the air temperature in the drying chamber is low and the inlet air temperature is low then slightly increase the power air heater.

The advantages of fuzzy control are a short development period and greater robustness than a proportional integral derivative (PID) controller due to its dependence on the operating parameters of the system and smaller deviations from the set value over longer duration (Bajsic & Kranjcevic 2001). A schematic representation of the *fuzzy architecture* for a spray-drying process is given in Figure 13.3.

13.4.2 Efficiency of spray-dryer operation

Drying processes are energy intensive and ways of increasing their energy efficiency are often targeted towards obtaining benefits in energy savings (Carić 1994). The efficiency of spray-dryer operation can be evaluated in terms of its resourceful energy utilization and eco-friendliness in addition to obtaining high yield. The efficient operation of industrial-scale spray dryers depends on various factors. During spray drying, the majority of the water content is removed by thermal energy rather than by mechanical dewatering. This renders spray drying an energy intensive operation, consuming about 6000 kJ of thermal energy for evaporating 1 kg of water from the feed material. Moreover, the specific heat required for moisture evaporation is supplied within a short duration during spray drying and this significantly adds to the energy expenditure. In addition to this, the use of an open cycle layout leads to a substantial amount of powder loss along with the exhaust air stream that is discharged to the atmosphere. This is because the open cycle layout draws the drying air from the atmosphere and discharges the exhaust air back into the atmosphere (Filková *et al.* 2006). Thus, the product separation from the air stream is important in promoting spray-dryer efficiency because of its ability not just in reducing powder loss, but also lessening emissions, thus ensuring environmental

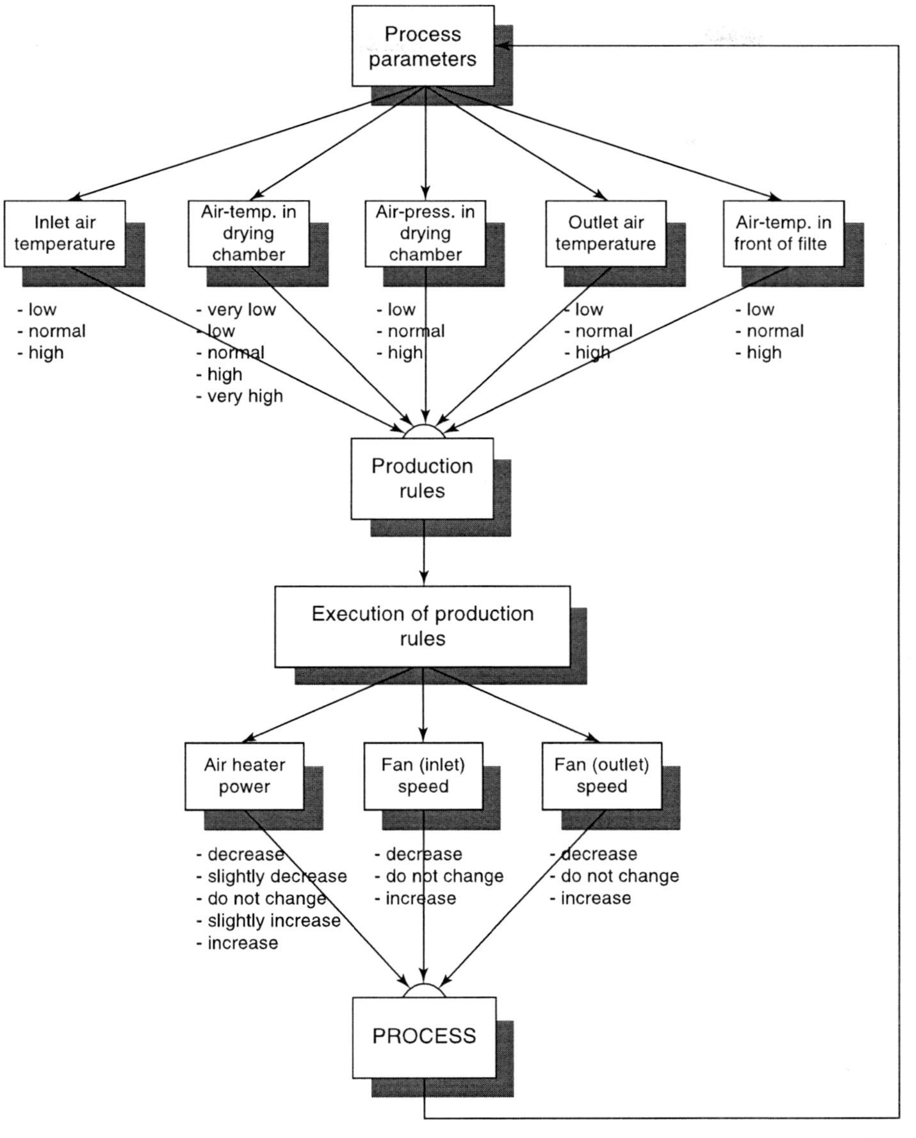

Figure 13.3 The fuzzy-architecture of a spray dryer (Bajsic & Kranjcevic 2001).

protection. Along with the dried particles of the product a large amount of heat energy is also liberated with the exhaust air because of the open cycle operation.

On the other hand, an efficient preconcentration step to remove the maximum possible amount of water from the feed liquid prior to spray drying can potentially lead to energy savings, but due consideration of the viscosity and pumpability of the feed is necessary. An increase in feed solids concentration from 10% to 50% can reduce the heat consumption (kJ/kg powder) by about 88.7% (Filková *et al.* 2006). According to Kessler (1980) an increase in inlet temperature (or, conversely, a decrease in outlet temperature) can influence the thermal efficiency of spray dryers. This is the rationale behind the current usage of multiple-stage spray dryers in the dairy industry. Here, the powder is mostly dried in the spray dryer and then transferred to a fluidized-bed chamber to finish the drying. The fluidized-bed step requires less air to finish the drying compared to the spray dryer, making the drying process more efficient (Brush *et al.* 2011). Multi-stage spray drying not only enhances energy efficiency, but also substantially reduces the size of the spray drying chamber (Mujumdar *et al.* 2010).

A common practice in the food industry to promote the energy efficiency of spray dryers is the use of a bag filter to recover the particulate product from the air stream exiting the cyclone separator. In a recent innovation attempted during the two-stage spray drying of milk fat by a Chinese company, the drying air was exhausted from the middle of the drying chamber. This means usage of cyclones can be omitted, provided an appropriate bag filter is installed in the middle of the chamber (Niro Denmaɪk Ltd; Mujumdar *et al.* 2010).

Yet another approach is the use of heat-recovery equipment (Gibson 1999). The heat integration of a spray dryer with an evaporator demonstrates the potential to enhance heat recovery from the exhaust. However, the limitations in integration include economics, particle loading and fouling, and relatively low exhaust temperatures. Furthermore, the average inlet temperature prior to the heater, which can be as high as 40 °C, and that of exhaust, which can be as low as 65 °C, restricts the amount of heat recovery. The appropriate location for the addition of recovered heat is the point where the incoming air is preheated, either directly or indirectly (Atkins *et al.* 2011).

13.4.3 Bottlenecks in industrial spray-drying

Identifying and managing bottlenecks is vital to increase productivity. In any industrial process, three types of bottlenecks can be identified: size or throughput bottlenecks, scheduling or time utilization bottlenecks and a combination of these two. Throughput bottlenecks occur when equipment or resources limit the amount of product produced per batch or the amount of material processed per batch. Scheduling bottlenecks srise when units or resources limit the number of batches that can be processed per year (Koulouris *et al.* 2000). Being a continuous process, spray drying is predominantly associated with throughput bottlenecks.

Different debottlenecking strategies have been reported in the literature. One approach is the addition or duplication of a spray dryer in the existing line. This reduces the spray-drying cycle time (in hours) by 50% and increases the annual production capacity (in terms of the number of batches produced) by 45%. A second strategy is to include an additional evaporator. This is done on the underlying principle of addressing the bottleneck by integrating the spray-drying process with allied processing steps. The addition of a double-effect forward-feed evaporator to concentrate the feed prior to the spray dryer results in the same effect as that of adding an extra dryer. Because of the reduction in volume in the feed, the process time for the spray-drying operation is assumed to reduce by half. A combination of both these approaches has a synergistic positive effect. Extract water from the storage tank is transferred to the newly added double-effect forward-feed evaporator to be concentrated before it is sent to the two added parallel spray dryers. The process time showed a significant reduction of 63% and annual production capacity increased by 67% (Athimulam *et al.* 2006).

13.4.4 Hygiene in spray-dryer operation

Compliance with GMP is a key mandate in any food industry. Execution of hazard analysis and critical control point (HACCP) plans on the production line and conducting periodic cleaning procedures are central to the maintenance of hygiene in spray-drying plants. Hygiene and cleaning need close attention, especially for milk products because of their hygroscopic nature due to the presence of sugar, protein and fat. The resultant stickiness leads to scale formation on the walls of the spray chamber, known as *fouling*, which can potentially lead to explosions, product quality deterioration and more importantly microbial contamination. Combating microbial contamination is not simple as *Salmonella*, the microorganism of concern in the contamination of dairy products, is ubiquitously present in the surroundings, including farm environments (Van Schothorst 1986) where the milch animals are bred. Preventive measures include good hygienic practices on the shop floor, such as the use of filtered air and stringent demarcation between wet and dry processes. The design of a HACCP plan for the spray-drying process line should consider the possible sources of *Salmonella* contamination and multiplication along with appropriate control measures and monitoring procedures. In this context, the milk residues adhering to the roof of the spray tower are a well-known source of *Salmonella*. Secondary contamination might occur if birds come into contact with the powder or if the secondary air intake is close to the roof. This air may transfer *Salmonella* to the powder if there is improper air filtration. A slight overpressure should therefore be maintained in the spray tower building and packaging areas (International Commission on Microbiological Specifications for Foods 2002).

CIP is an integral function in industrial-scale spray-dryer operation. Accounting for the powder-handling application of the spray-drying equipment, the method of CIP can be either wet or dry. Cleaning by dry mode is predominant because the drying process is in a high hygiene zone. Zoning of industrial processing areas is carried out based on the HACCP procedure, with the objective of protecting products against various sources of contamination from the surrounding milieu. High hygiene zones are those where the contamination of products will lead to serious consequences with regard to the sensitivity of consumers and plausible risks in the distribution chain up to the end point of consumption. The spray-dryer environment is in the high hygiene zone and hence is subjected to dry cleaning or controlled wet cleaning. Dry cleaning is accomplished by manual or automated sweeping of the surface using a high velocity air stream to pass over the concerned surface. The use of compressed air alone may not be effective as this just transfers the residues or dust from one site within the drying equipment to another. An alternative approach includes the use of vacuum cleaners with suitable filters at the air outlet to entrap microorganisms. Scrapers and brushes can also be used, but care should be exercised with respect to the loss of removeable parts.

Wet cleaning is carried out by manual or automated hosing of surfaces using sterile water or cleaning solutions containing defined concentrations of caustic, acid and water. Providing personnel with protective equipment is crucial to prevent contact with the chemicals used for cleaning. However, with recent advances in CIP systems, it is possible to program an automatic cleaning cycle to use the exact concentration of cleaning solution for each equipment area. This can be achieved by integration of cleaning nozzles with the pipework and instrumentation of the spray-drying equipment. The nozzles are actuated either by the pressure of the cleaning solutions or by pneumatic means, with operating pressures up to 10 bar. A normal cleaning cycle comprises a prerinse step with hot water followed by caustic and acid cleaning with subsequent water rinsing steps. Final drying is done by passing warm air through the equipment by means of air heating units and fans (Anonymous 2006) or the use of wet vacuum cleaners. Automated CIP procedures reduce the downtime associated with the cleaning of liquid contact surfaces in the spray-drying plant, such as the feed lines and atomizers, which require frequent cleaning (Bloore & O'Callaghan 2009).

It is recommended that the spray-dryer process area cleaned by dry CIP is devoid of drains. However, when controlled wet cleaning is employed, drains that are appropriately covered may be present in the processing area.

13.4.5 Safety aspects of spray drying

Safety with respect to industrial drying has two perspectives: safe operation of the equipment and compliance with the environmental regulations pertaining to control of emissions and suspended particulate matter. The potential risks associated with spray drying in general are related to the flammability of the solvent being evaporated from the feed material. However, spray drying in the food sector, except the flavour industries, deal with aqueous feed material and hence the risk level is alleviated compared to that prevailing in chemical industries.

It is beneficial to consider the inherent hazards that may occur during a process like spray drying that generates powder that is likely to lead to fire or explosion. Dairy-based powders are flammable and when they exist in the form of a suspension or a cloud of dust, the possibility of explosions is real. In general, milk powder might burn, but without causing an explosion. Nevertheless, accumulations of powder may lead to self-ignition as they adhere to the hot zones of the dryer. Self-ignition of a powder layer occurs within a temperature range. As the layer builds up in thickness, self-ignition can occur at lower temperatures. At the normal spray-drying temperatures, that is, inlet and outlet temperatures of 250 and 80 °C respectively, self ignition is possible when powder deposition occurs. The critical points in a spray dryer with respect to ignition are at the top, where the highest temperatures occur, at the base and on any projections or extensions where thick deposits can build up (Ministry of Labour 1993).

The inflammation temperature and minimum explosion concentration vary for different spray-dried dairy products. While spray-dried milk has an inflammation temperature of 142 °C at a minimum explosion concentration of 54 g/m^3, baby foods have a higher inflammation temperature (205 °C) at a lower explosion concentration (36 g/m^3) (Filková *et al.* 2006). The basis for explosion is that ignition occurs when there is an accumulation or deposition of the dried product whilst the drying air temperature exceeds the inflammation temperature in a layer. Thus, the careful choice of spray-drying temperature plays a role in preventing explosion incidents in dairy plants. The critical thicknesses for the self-ignition of milk products are listed in Table 13.1. In addition, discharges of milk powder from the exhausts of spray drying equipment can be deposited on surrounding surfaces. When wet, these deposits become acidic and can, in extreme cases, cause corrosion (COWI 2000).

Table 13.1 Critical thicknesses for the self-ignition of milk powders (Beever 1985)

Spray-dried product	Minimum thickness for self-ignition of layer at 200 °C (mm)	Minimum thickness for self-ignition of layer at 100 °C (mm)
Skim milk	17	170
Whole milk	10	170
Butter milk	9	130
Whey	13	320

13.5 Industrial drum drying of dairy products

For more than 10 decades drum drying was the preferred choice in the dairy industry for the production of instant cereal-based baby foods. Conventionally, drum dryers are steam-heated continuous dryers that operate on the conduction mode of heat transfer. The predominant mode of heating in drum dryers is by steam, but direct gas and oil heating have also been explored. Circulation of hot water or thermal fluids is also possible (Brown *et al.* 1973; APV 1989). The processing line in drum-drying includes a slurry preparation section, a feeder unit, a mill-sifter section, a section for dry mixing with other ingredients, such as dried milk and fruit powders, and blending additives, such as vitamins and minerals, and a storage section for the dried product in silos. A typical industrial-scale drum dryer of 4' diameter by 10' face length demonstrates a drying rate of 272 kg/h (APV 1989).

A liquid feeder spreads the slurry prepared from cereal- and milk-based ingredients on the surface of a steam-heated drum. The heated drums then rotate together and the dried product is scraped off the drum surface using a scraper. The product at this stage is in the form of a membrane that is placed on a conveyor and transferred to the mill sifter to transform it into powder form. The product is then transported to a cooler and from there it is brought to silos and/or a packaging machine. The drum dryer is suitable for handling low to high viscous products with a moderate to high stickiness (IDM 2013). Different feeding configurations exist for drum dryers and their selection depends on the characteristics of the feed materials to be dried, the dryer size required and the formation of product film of appropriate thickness. The two major configurations are single drum and double drum. The critical control parameters of the drum-drying process, which are described in the next section.

While the single-drum dryer employs a 'dip'-type feeding, the double-drum type uses either a 'nip' (the nip is the tangent line of the two drums) or roller-type feeding device. In the case of single-drum dryers a sufficiently high circulation of feed liquid or slurry in the trough is mandatory to maintain the concentration that is obtained after evaporation. Depending on the nature of feed, top-roller and spray-type feeders can also be used in conjunction with single-drum dryers. The double-drum dryer uses the space between the drums to adjust and control the film thickness. Here, the gelatinization of the product occurs within the pool of material formed between the two drums using two spring-loaded end plates attached to the flat ends of the drums. Drying commences once the gelatinized material exits the pool and forms a thin film on the surface of the drums. Film thickness is controlled by adjusting the gap between the two drums, but with measures to prevent failure of feed material at the gap (Vallous *et al.* 2002). Sub-atmospheric drum dryers are employed for premium-quality products, but the ratio of capital cost to output limits their use on an industrial scale.

13.5.1 Critical control points in industrial drum drying

The critical operational parameters in the drum-drying process include the following (Kalogianni *et al.* 2002):

- **Steam pressure inside the drums:** Steam pressure has a direct influence on drum temperature (Figure 13.4) and the final moisture content of the product. The higher the steam pressure, the greater the temperature of the inner and outer drums. This is because the temperature differential between the slurry and the drum surface decides the boiling rate of the slurry. The boiling point of the slurry is a constant parameter so an increase in steam pressure will increase the drum temperature. This in turn increases the temperature differential, which increases the boiling rate. A high steam pressure should be the objective. Furthermore, in double-drum dryers the steam pressure also has a major influence on the gap width (Gardner 1971). At higher steam pressures, the dryer performance is less sensitive to differences in feed concentration.
- **Speed of rotation of the drums:** The range of selection should be such that the sheet of product becomes neither over-dried and lifted off the drums at early stages nor remain too humid and form an amorphous and cohesive mass on the doctor blades. The increase in drum rotation speed decreases the surface temperature (Figure 13.5), which is explained by the reduction in film thickness and drying period. Thinner films have higher external heat flux and shorter drying time and because of this the drums lower their temperature to meet the larger thermal requirements (Vallous *et al.* 2002).

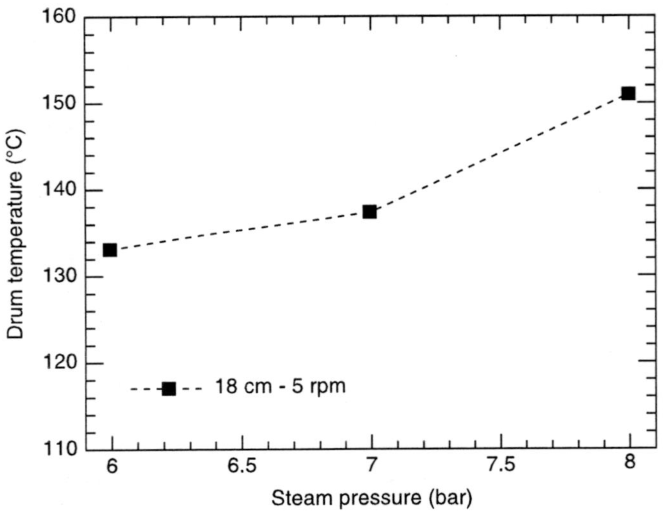

Figure 13.4 Relationship between steam pressure and drum temperature. Modified from Vallous *et al.* (2002).

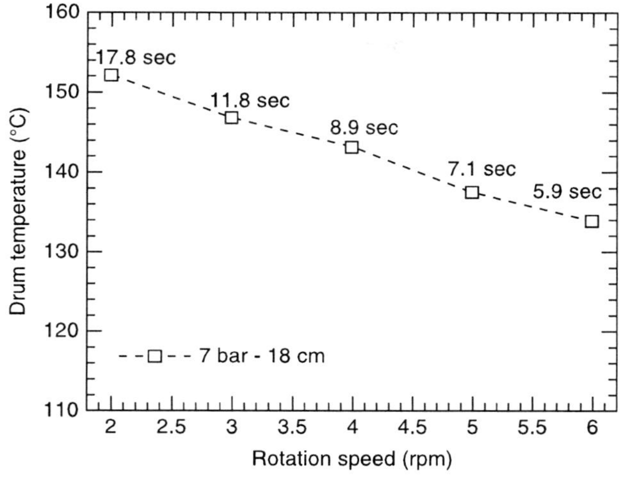

Figure 13.5 Relationship between drum rotation speed and temperature. Modified from Vallous *et al.* (2002).

- **Pool level between the drums:** The level of the pool dictates both the volume of the slurry in the pool and the contact area of the slurry with the drums, so it is directly associated with the heat delivered by the drums. The pool level also determines the residence time of the material in the pool, which dictates the degree of starch gelatinization and therefore the viscosity of the slurry (Anastasiades *et al.* 2002). The drum temperature drops as the pool level goes up. This may be attributed to a higher boiling load with pool level: the temperature of the drums falls because of a higher heat flux from the wall towards the material.
- **Gap between the drums:** This plays a major role in deciding the film thickness of the feed slurry on the surface of the drum. This is significant in double-drum dryers as the drying commences only after the gelatinized material leaves the pool and forms a thin film on the drum surface. However, the gap width is not an independent parameter and decreases with an increase in temperature due to the thermal contraction of the drums (Vallous *et al.* 2002). The gap between the drums and the drum speed of rotation together control the thickness of the film of slurry deposited on the drum surfaces (Fritze 1973; Trystram & Vasseur 1992). This uniform film of slurry will dry uniformly if the variables are constant.
- **Characteristics of feed material:** These are the concentration, flow rate and temperature at which the material reaches the drum surface. A higher percentage of water results in a lower rate of solids recovery. Conversely, a lower percentage of water results in a higher production rate. Suspensions with lower solids content become over-dried and spontaneously detach from the drums before the doctor blades scrap them off, while suspensions with higher solids content are not adequately dried (Kalogianni *et al.* 2002).
- **Flow rate of feed**: This influences the drying rate, with uncontrolled flow rates leading to incomplete moisture removal.
- **Material of construction and thickness of the drum wall:** These influence the heat-transfer rate of the material and hence the moisture removal rate.

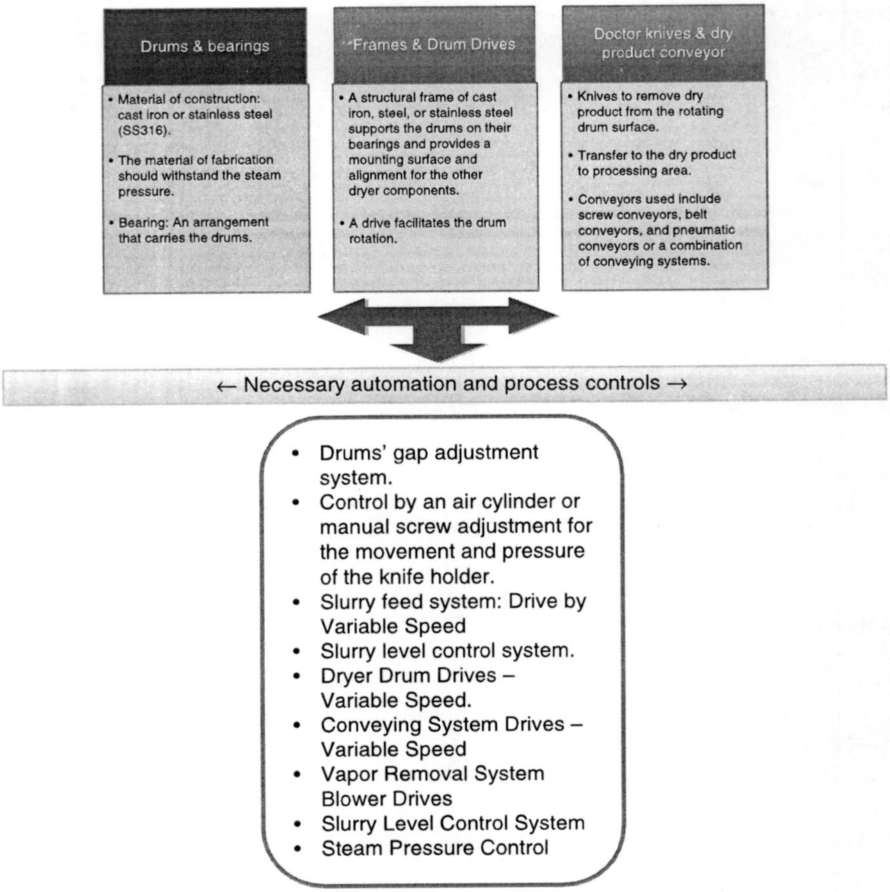

Figure 13.6 Hardware components, and automation and process controls in the drum-drying operation.

- **Dimensions of the drum (length and diameter)**: These decide the area of the heat-transfer surface that establishes contact with the feed slurry. The larger the surface area, the greater the rate of heat transfer and hence the drying rate.
- **Efficiency of condensate removal from the drums:** This influences the heat-transfer rate.
- **Continuous removal of saturated steam vapours:** This facilitates the drying efficiency.

The associated hardware components, and the automation and process control parameters are presented in Figure 13.6.

13.5.2 Energy efficiency of drum drying

Drum-drying technology has various benefits with respect to the ease of process and cost of production. In drum drying, the wet product is dried to the required final moisture content in a single step, thus eliminating the need for any post-processing drying equipment, as there is in spray drying. Furthermore, the thermal efficiency of the drum dryer is high, in the region of 65–75% for small to medium production runs, which corresponds to 1.3–1.5 mass units of steam per mass unit of water evaporated. However, for high productivity requirements in the range of tonnes, the drum dryer is at a disadvantage (APV 1989).

13.5.3 Safe operation of drum dryers

The potential issues that affect the safe operation of drum dryers and lead to fires or accumulations include the following (Febo 2011):

- operation at over-rated capacity: higher capacity increases the operating temperature, which leads to overheating of the system
- lack of a temperature monitoring system
- absence of systems for spark detection/extinguishing
- impractical venting
- explosion risk (which is usually prevalent in the downstream of dryer operation).

13.6 Conclusion

This chapter provides an insight into the industrial-scale drying operation in dairy plants. The complexities involved in the large-scale operation can be handled effectively with an underpinning knowledge of the drying principles and product nature. With the advent of computational simulation tools and automation it is possible to manage drying lines in the dairy industry in a cost- and energy-efficient manner.

References

Anastasiades, A., Thanou, S., Loulis, D., Stapatoris, A. & Karapantsios, T.D. (2002). Rheological and physical characterization of pregelatinized maize starches. *Journal of Food Engineering*, **52**(**1**), 57–66.

Anonymous (2006). *Cleaning-in-Place of Spray Drying Plants*, Technical Bulletin, GEA Niro A/S, Process Engineering Division, Soeborg.

APV (1989). *Dryer Handbook* (DRH-889), APV Crepaco Inc, Rosemont, IL.

Athimulam, A., Kumaresan, S., Foo, D.C.Y., Sarmidi, M.R. & Aziz, R.A. (2006). Modelling and optimization of *Eurycoma longifolia* water extract production. *Food and Bioproducts Processing*, **84**(**2**), 139–149.

Atkins, M.J., Walmsley, M.R.W. & Neale, J.R. (2011). Integrating heat recovery from milk powder spray dryer exhausts in the dairy industry. *Applied Thermal Engineering*, **31**(**13**), 2101–2106.

Bajsic, I. & Kranjcevic, E. (2001). Automation of industrial spray dryer. *Instrumentation Science & Technology*, **29**(**1**), 41–52.

Barners, R.M. (1980). *Motion and time study design and measurement of work*, 7th edn, John Wiley & Sons.

Beever, P.F. (1985). Fire and explosion hazards in the spray drying of milk. *International Journal of Food Science & Technology*, **20**(**5**), 637–645.

Bhandari, B.R., Patel, K.C. & Chen, X.D. (2008). Spray drying of food materials – process and product characteristics. In: *Drying Technologies in Food Processing* (eds Chen, X.D. & Mujumdar, A.S.), pp. 113–159, Blackwell Publishing.

Bloore, C.G. & O'Callaghan, D.J. (2009). Process control in evaporation and drying. In: *Dairy Powders and Concentrated Products* (ed. Tamime, A.Y.), Blackwell Publishing.

Brown, A.H., Van Arsdel, W.B., Lowe, E. & Morgan Jr., A.I. (1973). Air drying and drum drying. In: *Food Dehydration* (eds Van Arsdel, W.B., Copley, M.J. & Morgan Jr., A.I.), pp. 82–160, AVI Publishing, Westport, CT.

Brush, A., Masanet, E. & Worrell, E. (2011). *Energy efficiency improvement and cost saving opportunities for the dairy processing industry, An ENERGY STAR® Guide for Energy and Plant Managers*, Environmental Energy Technologies Division, U.S. Environmental Protection Agency, Berkeley.

Bryan, F.L. (1993). Aplicação do método de análise de risco por pontos críticos de controle, em cozinhas industriais. *Revista Higiene Alimentar*, 7(**25**), 15–22.

Carić, M. (1994). *Concentrated and dried dairy products*, VCH, New York.

COWI (2000). Cleaner Production Assessment in Dairy Processing. United Nations Environment Programme Division of Technology, Industry and Economics, Danish Environmental Protection Agency, Danish Ministry of Environment and Energy, COWI Consulting Engineers and Planners AS.

Febo, H. (2011). *Dryers – Processes & Protection*, NFPA Dust Explosion Hazards Symposium, Factory Mutual Insurance Company, Detroit.

Filková, I., Huang, L.X. & Mujumdar, A.S. (2006). Industrial spray drying systems. In: *Handbook of Industrial Drying*, 3rd edn (ed. Mujumdar, A.S.), pp. 215–256, Taylor & Francis.

Fritze, H. (1973). Dry gelatinized produced on different types of drum dryers. *Industrial Engineering Chemistry, Process Design and Development*, **12**, 142–148.

Gardner, A.W. (1971). *Industrial Drying*, pp. 220–242, Leonard Hill Books, London.

GEA (2014). *Building the World's largest dairy Spray Dryer*, available at http://www.gea.com/global/en/stories/building-the-worlds-largest-dairy-spray-dryer.jsp, accessed 15 November 2015.

Gibson, S. (1999). Improving spray drying efficiency, APV Americas USA.

Huang L.X., Wang Z. & Tang J. (2001). Recent progress of spray drying in China [in Chinese]. *Chemical Engineering (China)*, **2**, 011.

International Commission on Microbiological Specifications for Foods (2002). Sampling plans. In: *Microbiological Testing in Food Safety Management*, pp. 123–144, Kluwer Academic/Plenum Publishers, New York.

IDM (2013). *Dairy Processing on Drum Dryers*, International Dairy Magazine, pp. 14–15.

Kalogianni, E.P., Xynogalos, V.A., Karapantsios, T.D. & Kostoglou, M. (2002). Effect of feed concentration on the production of pregelatinized starch in a double drum dryer. *Lebensmittel-Wissenschaft & Technologie*, **35**(**8**), 703–714.

Kessler, H.G. (1980). Heat conservation in concentration and spray drying of milk products. In: *Drying '80, Vol. 1* (ed. Mujumdar, A.S.), pp. 339–342, Hemisphere/McGraw-Hill, New York.

Koulouris, A., Calandranis, J. & Petrides, D. (2000). Throughput analysis and debottlenecking of integrated batch chemical processes. *Computers & Chemical Engineering*, **24**(**2**), 1387–1394.

Ministry of Labour (1993). *Approved code of practice for the prevention, detection and control of fire and explosion in New Zealand dairy industry spray drying plant*, Occupational Safety and Health Service, Department of Labour, Wellington.

Mujumdar, A.S. (2006). Principles, classification and selection of dryers. In: *Handbook of Industrial Drying*, 3rd edn (ed. Mujumdar, A.S), pp. 4–31, Taylor & Francis.

Mujumdar, A.S., Huang, L.X. & Chen, X.D. (2010). An overview of the recent advances in spray-drying. *Dairy Science and Technology*, **90(2–3)**, 211–224.

Petrides, D., Kououris, A. & Siletti, C. (2002). Throughput analysis and debottlenecking of biomanufacturing facilities: a job for process simulators. *Chimica oggi*, **20(5)**, 10–17.

Pisecky, J. (1995). Evaporation and spray drying. In: *Handbook of Industrial Drying, Vol. 1*, 2nd edn (ed. Mujumdar, A.S.), Marcel Dekker, New York.

Silva, J.A. (1999). As novas perspectivas para o controle sanitário dos alimentos. *Revista Higiene Alimentar*, **13(65)**, 19–25.

Sivarajalingam, S. (2009). *Modelling and Control of a Spray Drying Process: for Control Purposes*, doctoral dissertation, Technical University of Denmark.

SPX Corporation (2012). *Anhydro Spray Drying for the Dairy Industry*, available at http://www.spxflow.com/cn/assets/pdf/ANH_Spray_drying_for_dairy_industry_115_01_03_2012_GB.pdf.

Tang J.X., Huang L.X. & Wang Z.L. (1997). Three stage drying system and its application in dairy product processing [in Chinese]. *Journal of Nanjing Forestry*, **21**, 56–58.

Trystram, G. & Vasseur, J. (1992). The modeling and simulation of a drum dryer. *International Chemical Engineering*, **32**, 689–705.

Vadlamudi, S. (2015). *Propelling Growth in the Dairy Industry – A Perspective* (White paper), CPG Practice Group, Tata Consultancy Services.

Vallous, N.A., Gavrielidou, M.A., Karapantsios, T.D. & Kostoglou, M. (2002). Performance of a double drum dryer for producing pregelatinized maize starches. *Journal of Food Engineering*, **51(3)**, 171–183.

Van Schothorst, M. (1986). Do we have to live with Salmonella? *Food Technology Australia*, **38**, 64–67.

Westergaard, V. (1994). *Milk powder technology: evaporation and spray drying*, pp. 18–121, Niro A/S, Søborg.

WHO (2014). WHO Expert Committee on Specifications for Pharmaceutical Preparations (2014). *WHO expert committee on specifications for pharmaceutical preparations*. WHO Technical Report Series No. 986, pp. 90.

14

Challenges Involved in the Drying of Dairy Powders

U. Kiran Kolli

ITC Limited, Agri-Business Division, Guntur, Andhra Pradesh, India

14.1 Introduction

The fundamental reason behind drying and concentration is to reduce the available moisture content, which supports the growth of microorganisms, and thereby improve the keeping quality. However, in the dairy industry drying provides the added advantage of handling the dairy surplus and converting it into the most convenient form of either powder or semi-fluid product. Moreover, the nutritional and functional quality of the by-products of the dairy industry (cheese, whey, cream, buttermilk) have forced the industry to use drying as a tool for the safe handling and easy recovery of these waste products.

Unlike other food powders, dairy powders are known for their key functional properties besides being used for reconstitution and recombination. The multifunctional and nutritional properties of dairy powders made them suitable for use as functional ingredients in many other industries, for example the confectionery, bakery and meat industries (Sharma *et al.* 2012). Hence, it is imperative to protect the functionality of dairy powders from harsh processing conditions during drying. Process optimization is always important in achieving a balance between production volumes and quality preservation. In general, the final moisture content, the bulk density, the insolubility index and the instant properties are of primary importance. Nowadays the main challenges in the production of powders are the development of specialties (having a high added value), reduction of processing costs, increase in capacity or run times, and reduction of downtime. To address these challenges the production capacity of the available installations must be maximized and the process conditions directed towards minimal fouling of equipment, minimal product losses and reduction in energy consumption. This chapter discusses the challenges that are involved in drying dairy powders along with the concepts behind the challenges.

14.2 Challenges in the drying of dairy powders

14.2.1 Fouling

Thermal treatment is extensively used in the dairy industry and is the most cost-effective technique used to extend the shelf life of milk. It is either used as an ultimate or pretreatment operation depending on the requirement. A typical dairy powder and ingredients manufacturing process consists of four stages:

i. Membrane concentration: A preliminary concentration process to increase the dry matter content of product by removing the water with a membrane.
ii. Falling-film evaporation: The secondary stage of concentration, used to concentrate the product further by allowing the milk to flow downwards into a thin continuous film.
iii. Drying: The concentrate is dried to evaporate the remaining water and converted to a powder by drum drying, spray drying, freeze drying or spray-freeze drying.
iv. Fluidized-bed drying: The final conditioning process, which uses conditioned air to produce the final product with the desired moisture content.

During these stages liquid product has to flow through number of pipes, ducts and heating surfaces before being converted to a final product, leaving behind deposits on the surfaces and annular spaces. This formation of unwanted solid deposits within the process plant is referred to as fouling. Fouling results in huge losses to the industry, with about 80% of the total production costs in dairy industry arising from fouling and the cleaning of process equipment (Bansal & Chen 2006). An important aspect of fouling which contributes to the production cost is the heat resistance created by the solid deposits. These thick deposits act as an insulator, decreasing the overall heat-transfer rate in the drying plant. In addition, fouling increases the downtime of the evaporator or dryer and increases cleaning costs (Verdurmen *et al.* 2003). Hence, as far as drying of dairy products is concerned understanding and dealing with the mechanisms of fouling is vitally important.

14.2.1.1 Mechanisms

The process of fouling can be classified based on the mechanism of the formation of solid deposits as (a) reaction fouling, (b) biological fouling, (c) crystallization or precipitation fouling, (d) particulate fouling and (e) corrosion fouling. Fouling in dairy processes is categorized as reaction, precipitation or particulate fouling. Protein denaturation in the dryers is a possible mechanism of reaction fouling and the subsequent adhesion on the surface leads to particulate fouling, whereas precipitation of calcium phosphate present in milk is an example of precipitation fouling (Epstein 1983; Changani *et al.* 1997; Simmons *et al.* 2007; Rosmaninho *et al.* 2008). The possible mechanisms that are involved in the formation of fouling deposits are as follows (Epstein 1983; Burton 1988):

a) Formation of solid particles by chemical (calcium precipitation) or physical (heat denaturation of proteins) mediated reactions.
b) Transport of the fouling species to the surface of the dryer or heat exchanger.
c) Formation of an initial foulant layer by an adsorption mechanism.
d) Build up of the fouling material by deposition on the initial layer and removal by mechanical forces.
e) Ageing of the foulant, which typically changes the characteristics of the native solid deposits.

Fouling deposits broadly fall into two categories: type A and type B (Burton 1968; Lund & Bixby 1975; Changani *et al.* 1997; Visser & Jeurnink 1997). Type A (protein) fouling occurs at temperatures above 75 °C, reaches a maximum between 95 and 110 °C, then reduces at temperatures above 110 °C (Burton 1968; Bansal & Chen 2006). The deposits are white, soft and spongy (milk film), and their composition is 50–70% proteins, 30–40% minerals and 4–8% fat (Bansal & Chen 2006). Although whey proteins constitute only about 5% of milk solids, they contributes far more than 50% of the fouling deposits in type A fouling, while casein proteins contribute less due to their resistance to thermal processing (Lalande *et al.* 1984; Fox 1989; Visser & Jeurnink 1997). One whey proteins, β-lactoglobulin (β-Lg), makes a large contribution because of its heat sensitivity (Lyster 1970; Lalande *et al.* 1985; Gotham *et al.* 1992; Delplace *et al.* 1994; Byland 1995). Type B (mineral) fouling takes place at temperatures above 110 °C and increases with increasing temperature up to 140 °C. The deposits are hard, compact, granular in structure and gray in color (milk stone), and their composition is 70–80% minerals (mainly calcium phosphate), 15–20% proteins and 4–8% fat (Lyster 1965; Lalande *et al.* 1984). Calcium is present in milk in the form of calcium phosphate and calcium citrate. The first precipitating form of calcium phosphate is usually dicalcium phosphate dehydrate, which may be converted to the more stable calcium phosphate at later stage (Jeurnink *et al.* 1996a). The calcium phosphate may directly deposit on stainless steel surfaces or may deposit on the casein micellar surface or β-Lg molecules in the serum phase (Visser & Jeurnink 1997).

14.2.1.2 Factors affecting fouling

Fouling depends on five major factors: milk composition, the operating conditions of the heat-transfer surface, the type and characteristics of the heat-transfer surface, the presence of microorganisms and the location of the fouling.

Milk composition Milk composition mainly depends on the source and the season. Change in composition is simply a change in constituents, and a change in protein and mineral content changes the degree of fouling (Burton 1968; Belmar-Beiny & Fryer 1993; De Jong 1997). pH is an important factor that contributes to protein stability, hence the pH of milk is a significant factor that contributes to fouling. This is especially true for the milk protein casein, which is insensitive to heat but precipitates on acidification. A decrease in pH reduces the heat stability of this protein and forms larger deposits (Burton 1968). As stated earlier, β-Lg contributes to major fouling deposits. At high temperatures, β-Lg undergoes irreversible structural changes, starting with dissociation, thermal unfolding and exposure of free SH groups, until they are insoluble and finally aggregation of the proteins takes place. This process of aggregation is enhanced in the presence of calcium ions, as shown in Figure 14.1 (Visser & Jeurnink 1997).

Enrichment of milk with calcium is gaining popularity to increase the calcium intake per serving. This enrichment adds more calcium ions, which influence the denaturation temperature of β-Lg, promote aggregation by attaching to β-Lg and enhance deposition (Xiong 1992; Changani *et al.* 1997; Christian *et al.* 2002). Moreover, addition of calcium ions increases the amount of caseins present in the deposits, suggesting an increased instability of casein micelles (Delsing & Hiddink 1983; Daufin *et al.* 1987;

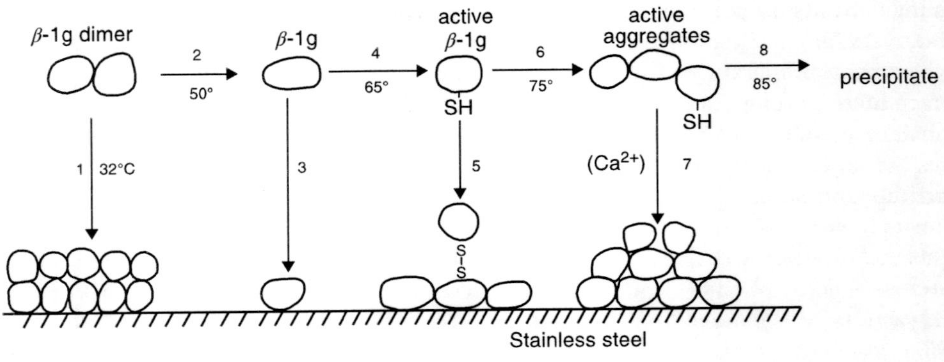

Figure 14.1 Illustration of the structural and conformational changes β-lactoglobulin (β-lg) undergoes during the process of drying in the presence and absence of calcium ions (Ca^{2+}) (Visser & Jeurnink 1997).

Grandison 1988; De Jong 1997; De Jong *et al.* 1999). Both increasing as well as decreasing the calcium content of milk compared to normal milk results in lower heat stability, thereby causing enhanced chances for fouling (Bansal & Chen 2006).

Lactose contributes less to fouling because it is water soluble. It may contribute to fouling at higher temperatures (above 100 °C) on caramelisation or Maillard reaction (Jeurnink *et al.* 1996a; Visser & Jeurnink 1997). Although there are traces of fat found in fouling deposits, milk fat has little effect on fouling (Burton 1968; Lalande *et al.* 1984; Jeurnink *et al.* 1996; Visser & Jeurnink 1997).

Heat-transfer surface operating conditions The temperature of the heat-transfer surface is an important parameter as it affects fouling directly. As discussed earlier, an increase in temperature increases fouling. Fouling starts at temperatures above 68 °C and a shift from type A fouling to type B fouling was observed at temperatures above 110 °C (Burton 1968).

The presence of air in milk increases fouling only when air bubbles form on the heat-transfer surface. These bubbles act as the nuclei for the formation of deposits and therefore greatly enhance the fouling rate (Burton 1968; Thom 1975; Jeurnink & Kruif 1995).

In heat exchangers at lower velocities the thickness of the laminar sub-layer adjacent to the heating surface is greater, so the volume of material subject to higher temperatures and the volume which remains near to the surface for relatively long periods are greater (Burton 1968).

Heat-transfer surface type and characteristics Plate heat exchangers are usually preferred over tubular heat exchangers because of their superior heat-transfer performance, easy maintenance and compactness. In spite of their superiority, plate heat exchangers are more prone to fouling because of their narrow flow channels and contact points between adjacent plates (Bansal & Chen 2006).

Since fouling occurs at the solid–liquid interface, the roughness of the surface plays a vital role until the first deposit layer is established. After the first layer has formed the interaction will shift from surface–deposit to deposit–deposit (Fryer *et al.* 1995). Surface modification by electropolishing is thought to reduce or limit the onset of fouling, but this may not justify the cost of polishing the surface (Burton 1988; Bott 1989; Yoon & Lund 1989). Many studies have examined the effect of modifying surfaces in different ways, such as bombardment of the surface with highly accelerated ions or coating the surface with a film of polymer, glass or polyurethane, but no significant reduction in deposit formation has been found (Yoon & Lund 1989; Muller-Steinhagen & Zhao 1997; Santos *et al.* 2004). Britten *et al.* (1988) found that coatings on heating surfaces did not affect the amount of deposit, but affected the strength of adhesion.

Presence of microorganisms: Deposit formation inside the heat-transfer surfaces of evaporators and plate heat exchangers harbours the growth of microorganisms. This causes another type of fouling, that is, bio-fouling or microbial fouling. Since most of the processing operations before drying are carried out at temperature below 100 °C, thermoduric bacteria, such as *Streptococcus thermophiles*, can grow on the surface walls of heat exchangers and form a film. Microorganisms are subsequently released and contaminate downstream sections. According to Bott (1993), bio-fouling takes place through two different mechanisms: deposition of microorganisms directly on the heat-transfer surface or on the deposit layer formed on the heat-transfer surface. Knight *et al.* (2004) came up with a method of minimizing biofilm build-up using a temperature cycle procedure. This system successfully interrupts the growth cycle of the bacteria and prevents their rapid growth, which is typically logarithmic.

Preheating of milk The basic cause of fouling is aggregation of denatured protein during the thermal processing of milk, therefore it is clear that more denatured protein will be present at the heat exchanger exit. Even the most efficient and advanced heat exchanger technology will not curb fouling as the milk exiting from the heat exchanger will contain more denatured and aggregated protein, which will induce fouling in many downstream locations. One strategy to reduce fouling that could work is preheating milk to 75–95 °C and holding it for 5–600 s before heating (Bell & Sanders 1944; Kessler & Beyer 1991). The reduction in fouling mainly occurs because denatured β-Lg is transformed into aggregated β-Lg during preheating and does not participate in fouling with the other components of milk (De Jong & Van der Linden 1992). Fouling can be further arrested by increasing the volume to surface ratio in the holding section, which improves conversion of the native β-Lg to the aggregated form (De Jong *et al.* 1992; Jeurnink *et al.* 1996b).

14.2.2 Stickiness

The presence of lactose and fat makes dairy powders sticky, given the right conditions of temperature, relative humidity (RH) and water activity. In spray drying, the outlet air conditions cause powder stickiness (Baldwin & Pearce 2005). In liquid dairy products lactose is dissolved in a continuous water phase along with fat and proteins. During spray drying lactose is transformed to a solid-like, amorphous glass directly from the dissolved state (Roos & Karel 1990, 1991). Glass transition is a state transition of amorphous materials that occurs between the solid, glassy and supercooled liquid states. The temperature region where this transition occurs is very important, particularly in relation to stickiness. Many physicochemical changes in dairy powders are influenced by the glass transition of amorphous lactose. In spite of the presence of proteins, which are also amorphous in nature, the glass transitions observed in milk solids correspond closely to those of pure lactose (Jouppila *et al.* 1997; Jouppila & Roos 1994). Hence, very little attention is given to protein. However, if lactose is hydrolysed, the observed the glass transition temperature (T_g) decreases dramatically because of the much lower T_g of the galactose and glucose components. The presence of the water phase acts as a plasticizer and significantly influences T_g (Jouppila *et al.* 1997). The T_g of lactose reduces from 47 to 34 °C when the RH increases from 20% to 30% (Baldwin & Pearce 2005). The lower the $T_{g,}$ the stickier the powder will be. Dairy powders can be prevented from sticking by running the dryer at right RH in the outlet stream. In general, the sticky point of powders with 40–50% lactose (w/w) will be approximately 25 °C above the lactose T_g at the same RH. This point will be influenced by the composition and type of dryer (Baldwin & Pearce 2005). If the powder is too sticky, the outlet RH needs to be reduced by effectively increasing the dryer outlet temperature. A regular sign of a sticky problem is cyclone blockage. Liquid fat in the powder contributes to the stickiness during processing. Fat usually appears on the surface of powders and will become liquid when the operating temperature is above 40 °C (melting point) (Kim *et al.* 2002). This liquid fat causes smearing of the surfaces where there is a high velocity, dust-laden air stream, for example at entry points of exhaust ducts from the main chamber. Introducing cool air to the air stream prior to the cyclones could be a solution (Baldwin & Pearce 2005). Achieving the correct balance between a high outlet air temperature to ensure drying and a low outlet air temperature to avoid smearing is next to impossible. Hence, in practical situations dryer fouling to an extent is accepted and shot runs are standard. In spite of the hurdles associated with drying of high-fat powders, they can be stored well when the storage temperatures are not high (<40 °C) and/or pallets are not stacked too high. High stacking causes caking due to the compressive load.

In contrast, in some products powder sticking to the dryer walls is a desirable phenomenon, for example in the production of precrystallized whey powder. A layer of powder is allowed to build up on the dryer walls, with the correct cone angle. The powder build-up is then restricted and it sloughed off periodically onto the belt where it is further developed into crystalline lactose.

14.2.3 Fires and explosions

Although fire explosions are quite rare in well-run and well-maintained plants, they are a serious threat to plant safety. For an explosion to happen fuel (milk powder, which is combustible) and oxidant (oxygen in the air) have to be present in the right ratio and there needs to be an ignition source. In the dryer, there is a dust-laden atmosphere and there could be spots where there is powder build-up due to fouling. Even a thin powder layer could be dangerous if it is near the inlet air duct. Self-ignition data for some dairy powders are presented in Table 14.1. Some reasons for layer development are:

- drying a cohesive product with the wrong conditions
- wetting the chamber walls (with concentrate) during start up
- poor atomization or spray/air mixing, with the resulting semi-dried particles hitting the walls
- leaking nozzles (a dag can form on the tip of nozzle, which can drop off and pass through the chamber; this is particularly dangerous if the concentrate dries and then self-heats)

Table 14.1 Thickness of layer resulting in self-ignition at two air temperatures

Product	Minimum thickness at 200 °C air temperature (mm)	Minimum thickness at 100 °C air temperature (mm)
Skim milk powder	17	170
Whole milk powder	10	170
Whey protein	13	320
Butter milk powder	8	100

- blockages in cyclones
- poor washing, leaving deposits
- injudicious placement or adjustment of the fines return lines.

In order to prevent fire hazards potential ignition sources should be removed from the vicinity of the drying chamber. These ignition sources include contaminated air supply (e.g. heated dust entering the chamber), friction (e.g. rubbing mechanical parts), static discharge, electrical failure, impact sparks and external heat. Most of these can be avoided by following GMP, regular maintenance and plant checks.

14.2.4 Powder loss

Loss of powder happens in plants for three reasons: start-up and shut down, fouling, and dryer exhaust stack losses.

Fouling is as big a problem in dryers as it is in evaporators and heat exchangers. Dryers need not be cleaned as often as evaporators or heat exchangers and hence can run for many weeks on same product.

In a typical spray-drying process preconcentrated feed from the evaporator is atomized in the drying chamber. Feed comes into contact with hot air and is dried as powder. Some powder is entrapped in the drying air and carried away by the cyclone exhaust gases. These losses are called 'stack losses' (Casey *et al.* 1987) and can be minimized by good equipment design. Normally cyclones are designed to work at a particular gas loading and flow, and any change in these conditions leads to losses. Hence, to maintain the required environment bag filters are kept along with cyclones. Stock losses usually increase with a decrease in the concentration of the feed. It is rare that the concentration of milk protein concentrate and other protein products exceeds 40% as this produces very fine powder which will be prone to losses.

14.2.5 Transport of powder

Transport of the powder from the dryer to the packing system is important when the powder is fragile in nature. If the powder produced is an ordinary skim milk powder, for example, it can be conveyed by conventional methods, including pneumatic transport. If the product is an agglomerated powder, mechanical treatment must be minimized in order to maintain product structure. Pneumatic conveying is harsher on the product than dense phase vacuum conveying, vibrating tubes and bucket elevators. Hence, pneumatic conveyors cannot be used for agglomerated products, instant products or products with low fat content.

14.2.6 Storage of dairy powders

Stickiness and caking are routine problems in storage (Baldwin & Pearce 2005). If the powder is exposed to high RH during storage, there will be loss of sorbed water. The loss of water can be attributed to the difference in water sorption by amorphous and crystalline lactose. Most crystals formed are anhydrous, but at higher storage humidities increasing amounts of α-lactose monohydrate is formed. Crystallization of amorphous lactose in sealed packages and in bulk storage also results in an increase in water activity and acceleration of most deteriorative changes, such as browning reactions and oxidation (Roos 2002).

14.2.7 Plant economics

Dairy powder manufacturing is energy-intensive process. Reduced energy consumption greatly influences the economics of the plant. In spite of strong economic pressures to reduce energy consumption, there is little scope for further major improvement. Large amounts of energy are expended in the process of removing water. Evaporators are much more energy efficient than dryers, using only a fraction of a kilogram of steam (or the energy equivalent) per kilogram of water removed. Dryers, on the other hand, use several kilograms of steam (or steam equivalent) per kilogram of water evaporated. It is not possible to dry whole milk concentrate with a solid content above 48% as it becomes a paste and is not suitable for atomization. In addition, drying of whole milk concentrate puts additional pressure on pumps and affects the solubility of the powder. The use of an evaporator to remove water beyond a certain level is therefore always limited. The solid content of the preconcentrate is not constant but changes from product to product. The higher the protein content, the lower the solids content. For example, in caseinate production the preconcentrate should ideally contain solids of not more than 20% before it goes to spray drying. Since spray drying is energy intensive, fluid beds are used for the final stages of drying to reduce the production cost. The powder remains in the fluid beds for several minutes, allowing time for the last of the water to be removed.

In general the aim of any industry is to generate revenue through more output for a given input. Although it is hard to reduce energy consumption, it is always wise to keep an eye on variables such as inlet out air temperature, viscosity of concentrate, solid content of concentrate, evaporator preheat, concentrate temperature, atomizer pressure, cone angle and concentrate flow rate as they can influence powder properties and economics.

14.2.8 Development of speciality dairy powders

The dairy industry has seen a large shift in demand from nutritional ingredients to functional ingredients. Industries are looking for ingredients for specific applications (e.g. yoghurt, processed cheese, infant formula, clinical nutrition) rather a general purpose. Processing dairy ingredients for functional benefits is always going to be a challenging task and some important challenges are discussed in detail below.

High free-fat-containing whole milk powder plays a crucial role in milk chocolate manufacturing to reduce the amount of cocoa butter required to produce a free-flowing product. The free-fat content of whole milk powder is indirectly proportional to the quantity of cocoa butter, one of the costliest ingredients used in milk chocolate manufacturing. In general roller-dried whole milk powder is preferred over spray-dried whole milk powder for manufacturing milk chocolate because of its high free-fat content of more than 95% (Reimerdes & Mehrens 1994). However, roller-dried milk powder is not stable, but is vulnerable to oxidation when compared to spray-dried whole milk powder. Another alternative to improve free-fat content is blending precrystallized skim milk concentrate with anhydrous milk fat to manufacture less expensive milk chocolates, but this process produces inferior quality chocolate with a persistant oxidation problem. Koc *et al.* (2003) proposed a process to increase the free-fat content and functional properties of spray-dried whole milk powder for milk chocolate manufacture. In this process spray-dried whole milk powder is exposed to high shear and elevated temperatures in a twin-screw continuous mixer, which increased the free-fat content significantly to over 80%.

High-quality non-caking whey powder can be produced by crystallization on a slow-moving belt dryer. A short residence time is the key for lactose crystallization and the production of a quality end product (Henning *et al.* 2006). Low-lactose whey powders are used in ice-cream manufacture as sweeteners. This does not require crystallization but needs special dryers with integrated belts (Filtermat) because of the extremely high hygroscopicity (Henning *et al.* 2006).

A recent development in dairy whiteners is producing the product without whey protein, which is responsible for sedimentation formation during mixing. The instant and soluble properties of dairy whiteners are improved by lecithin treatment. Whey protein concentrate (WPC), an important ingredient in infant foods and other food formulae, can be produced by a combination of ultrafiltration and diafiltration with WPC up to 99% protein (Henning *et al.* 2006).

Cheese powders are produced by spray drying the homogenized cheese slurry, which is primarily obtained by heating and shearing the cheese blend (Henning *et al.* 2006). They are popular over natural cheeses due to their longer shelf life and greater diversity of flavour and functional characteristics.

14.3 Use of modelling as a tool to solve some challenges

The main challenges in the production of dairy powders are the development of value-added products and the reduction of processing costs. For this, the production capacity of available installations needs to be maximized and the process conditions have to be directed towards minimal fouling of equipment, minimal product losses and reduction of energy consumption. Predictive computer models and computational fluid dynamics (CFD) are very helpful in optimizing process conditions to improve yields.

Two predictive models have been proposed for spray drying by NIZO Food Research in the Netherlands. These two models, DrySPEC2 and DrySim, are designed for specific purposes which are described here in detail.

The DrySPEC2 (Dryer System for Property and Energy Control) model is designed to establish a relationship between the process conditions of the spray-drying process, energy consumption and powder properties for the optimization of a two-stage dryer. The model mainly uses mass and thermal balances, psychrometric relations and sorption isotherms, and acts partly as a black box model. Single time calibration is sufficient to tune the model to the specific dryer by considering the heat losses of the installation and the deviation from the equilibrium state between powder and outlet air. The use of this model showed an increase in capacity of up to 20% and an annual energy reduction potential of about 250,000 m^3 of natural gas per installation. It has been successfully been implemented for the production of dairy products such as skim milk, whole milk and whey permeate.

DrySim is a two-dimensional simulation model developed using CFD techniques. It calculates the flow pattern, temperature and moisture content of air, the trajectories of the atomized particles and the drying behaviour of individual particles. Time-averaged Navier–Stokes equations in combination with the standard k-ε turbulence model describes gas flow. A particle tracker, based on the equation of motion, is used to simulate the particle trajectories. Since the particles affect the gas flow (and *vice versa*), an iteration procedure is used to solve the equations. As the drying process involves both external transport phenomena (from particle surface to surrounding air) and internal transport phenomena (diffusion of water within particles), differential equations are used for solving the transport phenomenon separately but simultaneously. This model has contributed to designing new spray dryers and changing the design of existing dryers to solve the problems of fouling and increase existing capacity.

14.4 Conclusion

The dairy industry is facing two main challenges: increasing installation capacity and developing specialty products. The fist can be addressed by minimizing fouling, reducing product losses, determining optimum operation conditions by suitable modelling techniques, and economically running plants with efficient energy consumption systems. Minimizing fouling in the dairy industry is not easy because it is extremely complex and the mechanisms responsible for it are not completely understood. Many theories have beem proposed on fouling behaviour but none of them have led to the elimination of the phenomenon. It is next to impossible to eliminate fouling because denaturation and aggregation reactions are initiated as soon as milk is subjected to heating. The industry is therefore looking to mitigate the problem by using non-thermal techniques such as microwave heating, high-pressure technology and ohmic heating. More effort needs to put into understand fouling completely. The production of specialty dairy powders, such as high-fat milk powder, low-lactose whey powder, whey protein concentrate and dehydrated cheeses, requires special processes, which in turn pose new challenges. Recent scientific and technological advances have addressed many of these challenges but have not outpaced the rising demand.

References

Baldwin, A. & Pearce, D. (2005). Milk powder. In: *Encapsulated and Powdered Foods* (ed. Onwulata, C.), pp, 387–434. Taylor and Francis, New York.

Bansal, B. & Chen, X.D. (2006). A critical review of milk fouling in heat exchangers. *Comprehensive Reviews in Food Science and Food Safety*, **5**(**2**), 27–33.

Bell, R.W. & Sanders, C.F. (1944). Prevention of milestone formation in a high temperature-short time heater by preheating milk, skim milk and whey. *Journal of Dairy Science*, **27**(**6**), 499–504.

Belmar-Beiny, M.T. & Fryer, P.J. (1993). Preliminary stages of fouling from whey protein solutions. *Journal of Dairy Research*, **60**(**4**), 467–483.

Bott, T.R. (1989). Fouling of exchangers: A review. In: *Process Engineering in the Food Industry: Developments and Opportunities* (eds Field, R.W. & Howell, J.A.), Elsevier Applied Science.

Bott, T.R. (1993). Aspects of biofilm formation and destruction. *Corrosion Reviews*, **11**(**1–2**), 1–24.

Britten, M., Green, M.L., Boulet, M. & Paquin, P. (1988). Deposit formation on heated surfaces – effect of interface energetics. *Journal of Dairy Research*, **55**(**4**), 551–562.

Burton, H. (1968). Deposits from whole milk in heat treatment plant – a review and discussion. *Journal of Dairy Research*, **35**, 317–330.

Burton, H. (1988). *Ultra-High-Temperature Processing of Milk and Milk Products*, Blackie Academic & Professional, Glasgow.

Byland, G. (1995). *Dairy Processing Handbook*, Tetra Pak Processing Systems, Lund.

Casey, D., O'Callaghan, D.J., Palmer, J. & Howett, D. (1987). Continuous measurement of stack losses from a skimmed milk drying plant. *International Journal of Dairy Technology*, **40**(**2**), 34–37.

Changani, S.D., Belmar-Beiny, M.T. & Fryer, P.J. (1997). Engineering and chemical factors associated with fouling and cleaning in milk processing. *Experimental Thermal and Fluid Science*, **14**(**4**), 392–406.

Christian, G.K., Changani, S.D. & Fryer, P.J. (2002). The effect of adding minerals on fouling from whey protein concentrate - development of a model fouling fluid for a plate heat exchanger. *Food and Bioproducts Processing*, **80**(**4**), 231–239.

Daufin, G., Labbé, JP., Quemerais, A., Brulé, G., Michel, F., Roignant, M. & Priol, M. (1987). Fouling of a heat exchange surface by whey, milk and model fluids: an analytical study. *Le Lait*, **67**(**3**), 339–364.

De Jong, P. (1997). Impact and control of fouling in milk processing. *Trends in Food Science & Technology*, **8**(**12**), 401–405.

De Jong, P. & Van der Linden, H.J.L.J. (1992). Design and operation of reactors in the dairy industry. *Chemical Engineering Science*, **47**(**13–14**), 3761–3768.

De Jong, P., Bouman, S. & Van der Linden, H.J.L.J. (1992). Fouling of heat treatment equipment in relation to the denaturation of beta-lactoglobulin. *Journal of the Society of Dairy Technology*, **45**(**1**), 3–8.

De Jong, P., Van der Horst, H.C. & Waalewijn, R. (1999). Reduction of protein and mineral fouling. In: *Fouling and Cleaning in Food Processing* (eds Wilson, D.I., Fryer, P.J. & Hasting, A.P.M.), pp. 39–46, '98, EUR 18804, Cambridge.

Delplace, F., Leuliet, J.C. & Tissier, J.P. (1994). Fouling experiments of a plate heat exchanger by whey proteins solutions. *Transactions of the Institution of Chemical Engineers*, **72**, 163–169.

Delsing, B.M.A. & Hiddink, J. (1983). Fouling of heat transfer surfaces by dairy fluids. *Netherlands Milk and Dairy Journal*, **37**, 139–148.

Epstein, N. (1983). Thinking about heat transfer fouling: a 5 × 5 matrix. *Heat Transfer Engineering*, **4**(**1**), 43–56.

Fox, P.F. (1989). The milk protein system. In: *Developments in Dairy Chemistry* (ed. Fox, P.F.), pp. 1–53, Elsevier Applied Science, New York.

Fryer, P.J., Belmar-Beiny, M.T. & Schreier, P.J.R. (1995). Fouling and cleaning in milk processing. In: *Heat-induced Changes in Milk*, 2nd edn (ed. Fox, P.F.), pp. 364–395, Internation Dairy Federation, Belgium.

Gotham, S.M., Fryer, P.J. & Pritchard, A.M. (1992). β-lactoglobulin denaturation and aggregation reactions and fouling deposit formation: a DSC study. *International Journal of Food Science & Technology*, **27**(**3**), 313–327.

Grandison, A.S. (1988). Effects of natural (or seasonal) variation in concentration of components of milk and addition of divalent cations on ultra-high temperature processing characteristics. *International Journal of Dairy Technology*, **41**(**4**),117–119.

Henning, D.R., Baer, R.J., Hassan, A.N. & Dave, R. (2006). Major advances in concentrated and dry milk products, cheese, and milk fat-based spreads. *Journal of Dairy Science*, **89**(**4**), 1179–1188.

Jeurnink, T.J.M. & Kruif, K.G.D. (1995). Calcium concentration in milk in relation to heat stability and fouling. *Netherlands Milk and Dairy Journal*, **49**(**2–3**), 151–165.

Jeurnink, T.J.M., Walstra, P. & Kruif, C.G.D. (1996a). Mechanisms of fouling in dairy processing. *Netherlands Milk and Dairy Journal*, **50**(**3**), 407–426.

Jeurnink, T.J.M., Verheul, M., Stuart, M.A.C. & Kruifm, C.G.D. (1996b). Relation between the denaturation and fouling behavior of beta-lactoglobulin. *International Dairy Federation Special Issue*, 73–82.

Jouppila, K. & Roos, Y.H. (1994). Glass transitions and crystallization in milk powders. *Journal of Dairy Science*, **77**(**10**), 2907–2915.

Jouppila, K., Kansikas J. & Roos, Y.H. (1997). Glass transitions, water plasticization, and lactose crystallization in skim milk powder. *Journal of Dairy Science*, **80**(**12**), 3152–3160.

Kessler, H.G. & Beyer, H.J. (1991). Thermal denaturation of whey proteins and its effect in dairy technology. *International Journal of Biological Macromolecules*, **13**(**3**), 165–173.

Kim, E.H.J., Chen, X.D. & Pearce, D. (2002). Surface characterization of four industrial spray-dried dairy powders in relation to chemical composition, structure and wetting property. *Colloids and Surfaces B: Biointerfaces*, **26**(**3**), 197–212.

Knight, G.C., Nicol, R.S. & McMeekin, T.A. (2004). Temperature step changes: a novel approach to control biofilms of *Streptococcus thermophilus* in a pilot plant-scale cheese-milk pasteurisation plant. *International Journal of Food Microbiology*, **93**(**3**), 305–318.

Koc, A.B., Heinemann, P.H. & Ziegler, G.R. (2003). A process for increasing the free fat content of spray-dried whole milk powder. *Journal of Food Science*, **68**(**1**), 210–216.

Lalande, M., Tisser, J.P. & Corrieu, G. (1984). Fouling of plate heat – exchanger used in ultra-high-temperature sterilization of milk. *Journal of Dairy Research*, **51**(**4**), 557–568.

Lalande, M., Tissier, J.P. & Corrieu, G. (1985). Fouling of heat transfer surfaces related to β-lactoglobulin denaturation during heat processing of milk. *Biotechnology Progress*, **1**(**2**), 131–139.

Lund, D.B. & Bixby, B. (1975). Fouling of heat exchanger surfaces by milk. *Proceedings in Biochemistry*, **10**, 52–55.

Lyster, R.L.J. (1965). The composition of milk deposits in an ultra-high-temperature plant. *Journal of Dairy Research*, **32**(**2**), 203–208.

Lyster, R.L.J. (1970). The denaturation of α-lactalbumin and β-lactoglobulin in heatedmilk. *Journal of Dairy Research*, **37**(**2**), 233–243.

Muller-Steinhagen, H. & Zhao, Q. (1997). Investigation of low fouling surface alloys made by ion implantation technology. *Chemical Engineering Science*, **52**(**19**), 3321–3332.

Reimerdes, E.H. & Mehrens, H.A. (1994). Milk. In: *Industrial Chocolate Manufacture and Use*, pp. 43–54, Springer.
Roos, Y. (2002). Importance of glass transition and water activity to spray drying and stability of dairy powders. *Le Lait*, **82(4)**, 475–484.
Roos, Y. & Karel, M. (1990). Differential scanning calorimetry study of phase transitions affecting the quality of dehydrated materials, *Biotechnology Progress*, **6(2)**, 159–163.
Roos, Y. & Karel, M. (1991). Plasticizing effect of water on thermal behavior and crystallization of amorphous food models, *Journal of Food Science*, **56(1)**, 38–43.
Rosmaninho, R., Rizzo, G., Müller-Steinhagen, H. & Melo, L.F. (2008). Deposition from a milk mineral solution on novel heat transfer surfaces under turbulent flow conditions. *Journal of Food Engineering*, **85(1)**, 29–41.
Santos, O., Nylander, T., Rosmaninhoo, R., Rizzo., Yiantsios, S., Andritsos, N., Karabelas, A., Muller-Steinhagen, H., Melo, L., Boulange-Petermann, L., Gabet, C., Braem, A., Tragardh, C. & Paulsson, M. (2004). Modified stainless steel surfaces targeted to reduce fouling- surface characterixation. *Journal of Food Engineering*, **64(1)**, 63–79.
Sharma, A., Jana, A.H. & Chavan, R.S. (2012). Functionality of milk powders and milk-based powders for end use applications – a review. *Comprehensive Reviews in Food Science and Food Safety*, **11(5)**, 518–528.
Simmons, M.J.H., Jayaraman, P. & Fryer, P.J. (2007). The effect of temperature and shear rate upon the aggregation of whey protein and its implications for milk fouling. *Journal of Food Engineering*, **79(2)**, 517–528.
Thom, R. (1975). Formation of milk deposits in plate heaters. *Milchwissenschaft*, **30(2)**, 84–89.
Verdurmen, R.E.M., Jong, P.D. & Smit, G. (2003). Optimising product quality and process control for powdered dairy products. In: *Dairy Processing: Improving Quality*, pp. 333–365, Woodhead Publishing.
Visser, J. & Jeurnink, T.J. (1997). Fouling of heat exchangers in the dairy industry. *Experimental Thermal and Fluid Science*, **14(4)**, 407–424.
Xiong, Y.L. (1992). Influence of pH and ionic environment on thermal aggregation of whey proteins. *Journal of Agricultural and Food Chemistry*, **40(3)**, 380–384.
Yoon, J. & Lund, D.B. (1989). Effect of operating conditions, surface coatings and pretreatment on milk fouling in a plate heat exchanger. In: *Fouling and Cleaning in Food Processing* (ed. Kessler, H.G.), pp. 59-80, University of Munich, Brussels.

Index

Page numbers in *italic* indicate figures/tables.

r

t